METHODS IN MOLECULAR BIOLOGY™

Series Editor
John M. Walker
School of Life Sciences
University of Hertfordshire
Hatfield, Hertfordshire, AL10 9AB, UK

For further volumes:
http://www.springer.com/series/7651

Plant Metabolic Flux Analysis

Methods and Protocols

Edited by

Martine Dieuaide-Noubhani

INRA Bordeaux, Université Bordeaux, Villenave d'Ornon, France

Ana Paula Alonso

Department of Molecular Genetics, The Ohio State University, Columbus, OH, USA

Humana Press

Editors
Martine Dieuaide-Noubhani
INRA Bordeaux, Université Bordeaux
Villenave d'Ornon, France

Ana Paula Alonso
Department of Molecular Genetics
The Ohio State University
Columbus, OH, USA

ISSN 1064-3745 ISSN 1940-6029 (electronic)
ISBN 978-1-4939-6050-7 ISBN 978-1-62703-688-7 (eBook)
DOI 10.1007/978-1-62703-688-7
Springer New York Heidelberg Dordrecht London

Printed on acid-free paper

Humana Press is a brand of Springer
Springer is part of Springer Science+Business Media (www.springer.com)

Preface

Metabolic flux analysis (MFA) determines the in vivo rates of conversion (fluxes) through active pathways. It concerns primarily flux quantification at the cell or the tissue level but can be extended to the study of the exchange of matter between organs at the whole-plant level. MFA is important for understanding the regulation of metabolism and the response to genetic or environmental perturbation and identifying targets for metabolic engineering.

In *Plant Metabolic Flux Analysis*, expert researchers in the field provide detailed experimental procedures for each step of the flux quantification workflow. Steady-state and dynamic modeling are considered, as well as recent developments for the reconstruction of metabolic networks and for a predictive modeling.

This volume will be of great interest for groups working on metabolism, biomass synthesis, and metabolic engineering. Written in the highly successful *Methods in Molecular Biology*™ series format, chapters include cutting-edge protocols and methodologies that are crucial to getting ahead in MFA.

Villenave d'Ornon, France — Martine Dieuaide-Noubhani
Columbus, OH, USA — Ana Paula Alonso

Contents

Contributors

DOUG K. ALLEN • *Donald Danforth Plant Science Center, St. Louis, MO, USA; United States Department of Agriculture Agricultural Research Service, St. Louis, MO, USA*

ANA PAULA ALONSO • *Department of Molecular Genetics, The Ohio State University, Columbus, OH, USA*

MARIE-HÉLÈNE ANDRIEU • *UMR 1332 de Biologie du Fruit et Pathologie, INRA, Villenave d'Ornon, France*

WAGNER L. ARAÚJO • *Departamento de Biologia Vegetal, Universidade Federal de Viçosa, Viçosa, MG, Brazil*

JEAN-CHRISTOPHE AVICE • *Ecophysiologie végétale, Agronomie et Nutrition N,C,S, Institut de Biologie Fondamentale et Appliquée, UMR EVA INRA, Université de Caen, Caen cedex, France*

VALENTINA BALDAZZI • *UR 1115, Plantes et Systèmes de culture Horticoles, INRA, Domaine St Paul, Avignon Cedex, France*

PATRICIA BALLIAS • *UMR 1332, INRA, Villenave d'Ornon cedex, France*

MARIE-PAULE BATAILLÉ • *Ecophysiologie végétale, Agronomie et Nutrition N,C,S, Institut de Biologie Fondamentale et Appliquée, UMR EVA INRA, Université de Caen, Caen cedex, France*

BERTRAND BEAUVOIT • *UMR 1332 de Biologie du Fruit et Pathologie, INRA, Villenave d'Ornon, France; UMR 1332 de Biologie du Fruit et Pathologie, Université Bordeaux, Villenave d'Ornon, France*

NADIA BERTIN • *UR 1115, Plantes et Systèmes de culture Horticoles, INRA, Domaine St Paul, Avignon Cedex 9, France*

MARIE BEURTON-AIMAR • *LABRI, LaBRI – Université Bordeaux, Talence, France*

BENOIT BIAIS • *UMR 1332, INRA, Villenave d'Ornon cedex, France*

MARÍA L. CÁRDENAS • *CNRS-Aix-Marseille Université, BIP, IMM, Marseille Cedex, France*

JEAN-CHRISTOPHE COCURON • *Center for Applied Plant Sciences, Department of Molecular Genetics, The Ohio State University, Columbus, OH, USA*

SOPHIE COLOMBIÉ • *UMR 1332 Biologie du Fruit et Pathologie, INRA, Villenave d'Ornon, France*

ATHEL CORNISH-BOWDEN • *CNRS-Aix-Marseille Université, BIP, IMM, Marseille Cedex, France*

GILLES CURIEN • *CNRS, UMR 5168, Grenoble, France*

CRISTIANA G.O. DAL'MOLIN • *Australian Institute for Bioengineering and Nanotechnology (AIBN), The University of Queensland, Brisbane, QLD, Australia*

MARTINE DIEUAIDE-NOUBHANI • *INRA Bordeaux, Université Bordeaux, Villenave d'Ornon, France*

ALISDAIR R. FERNIE • *Max-Planck-Institut für Molekulare Pflanzenphysiologie, Golm, Potsdam, Germany*

KARINE GALLARDO • *INRA UMR 1347 Agroécologie INRA/Université de Bourgogne/AgroSup, Dijon Cedex, France*

HÉLÈNE GAUTIER • *UR 1115, Plantes et Systèmes de culture Horticoles, INRA, Domaine St Paul, Avignon Cedex, France*

MICHEL GÉNARD • *UR 1115, Plantes et Systèmes de culture Horticoles, INRA, Domaine St Paul, Avignon Cedex, France*
YVES GIBON • *UMR 1332 Biologie du Fruit et Pathologie, INRA, Villenave d'Ornon, France*
EVA GRAFAHREND-BELAU • *Leibniz-Institute of Plant Genetics and Crop Plant Research Gatersleben (IPK), Gatersleben, Germany*
JORDAN O. HAY • *Biosciences Department, Brookhaven National Laboratory, Upton, NY, USA*
JAN HUEGE • *Department of Physiology and Cell Biology, Leibniz Institute of Plant Genetics and Crop Plant Research (IPK), Gatersleben, Germany*
LARA J. JAZMIN • *Department of Chemical and Biomolecular Engineering, Vanderbilt University, Nashville, TN, USA*
CHRISTIAN JEUDY • *INRA UMR 1347 Agroécologie INRA/Université de Bourgogne/AgroSup, Dijon Cedex, France*
ASTRID JUNKER • *Leibniz-Institute of Plant Genetics and Crop Plant Research Gatersleben (IPK), Gatersleben, Germany*
BJÖRN H. JUNKER • *Department of Physiology and Cell Biology, Leibniz Institute of Plant Genetics and Crop Plant Research (IPK), Gatersleben, Germany*
MOHAMED KOUBAA • *Department of Molecular Genetics, The Ohio State University, Columbus, OH, USA*
CHRISTIAN KRACH • *Department of Physiology and Cell Biology, Leibniz Institute of Plant Genetics and Crop Plant Research (IPK), Gatersleben, Germany*
NICHOLAS J. KRUGER • *Department of Plant Sciences, University of Oxford, Oxford, UK*
FANGFANG MA • *Donald Danforth Plant Science Center, St. Louis, MO, USA*
AMY MARSHALL-COLON • *Center for Genomics and Systems Biology, New York University, New York, NY, USA*
SHYAM K. MASAKAPALLI • *Department of Plant Sciences, University of Oxford, Oxford, UK*
GUILLAUME MÉNARD • *UMR 1332, INRA, Villenave d'Ornon cedex, France*
ANTOINE MONIER • *UMR 1332 de Biologie du Fruit et Pathologie, INRA, Villenave d'Ornon, France*
JOHN A. MORGAN • *School of Chemical Engineering, Purdue University, West Lafayette, IN, USA*
LARS K. NIELSEN • *Australian Institute for Bioengineering and Nanotechnology (AIBN), The University of Queensland, Brisbane, QLD, Australia*
ADRIANO NUNES-NESI • *Departamento de Biologia Vegetal, Universidade Federal de Viçosa, Viçosa, MG, Brazil*
JOHN P. O'GRADY • *School of Chemical Engineering, Purdue University, West Lafayette, IN, USA*
TOSHIHIRO OBATA • *Max-Planck-Institut für Molekulare Pflanzenphysiologie, Potsdam-Golm, Germany*
ROBIN W. PALFREYMAN • *Australian Institute for Bioengineering and Nanotechnology (AIBN), The University of Queensland, Brisbane, QLD, Australia*
KATIA PIANELLI • *UMR 1332 de Biologie du Fruit et Pathologie, INRA, Villenave d'Ornon, France*
C. HART POSKAR • *Department of Physiology and Cell Biology, Leibniz Institute of Plant Genetics and Crop Plant Research (IPK), Gatersleben, Germany*
DUYÊN PRODHOMME • *UMR 1332, INRA, Villenave d'Ornon cedex, France*
LAKE-EE QUEK • *Australian Institute for Bioengineering and Nanotechnology (AIBN), The University of Queensland, Brisbane, QLD, Australia*

R. George Ratcliffe • *Department of Plant Sciences, University of Oxford, Oxford, UK*
David Rhodes • *Department of Horticulture and Landscape Architecture, Purdue University, West Lafayette, IN, USA*
Albrecht Roscher • *Génie Enzymatique et Cellulaire, FRE CNRS 3580, Université de Picardie, Amiens Cedex, France*
Christophe Salon • *INRA UMR 1347 Agroécologie INRA/Université de Bourgogne/AgroSup, Dijon Cedex, France*
Anne-Lise Santoni • *INRA UMR 1347 Agroécologie INRA/Université de Bourgogne/AgroSup, Dijon Cedex, France*
Falk Schreiber • *Leibniz-Institute of Plant Genetics and Crop Plant Research Gatersleben (IPK), Gatersleben, Germany; Institute of Computer Science, Martin Luther University Halle-Wittenberg, Halle, Germany; Clayton School of Information Technology, Monash University, Clayton, VIC, Australia*
Jörg Schwender • *Biosciences Department, Brookhaven National Laboratory, Upton, NY, USA*
Neelanjan Sengupta • *Center for Genomics and Systems Biology, New York University, New York, NY, USA*
Yair Shachar-Hill • *Plant Biology Department, Michigan State University, East Lansing, MI, USA*
Jacqueline V. Shanks • *Department of Chemical and Biological Engineering, Iowa State University, Ames, IA, USA*
Ganesh Sriram • *A. James Clark School of Engineering, University of Maryland, College Park, MD, USA*
Abdelghani Idrissi Taghki • *Génie Enzymatique et Cellulaire, UMR CNRS 6022, Université de Technologie de Compiègne, Compiègne Cedex, France*
Brigitte Thomasset • *Universite de Technologie de Compiegne, CNRS-FRE 3580, Gènie Enzymatique et Cellulaire, Centre de Recherche de Royallieu, Compiègne Cedex, France*
Takayuki Tohge • *Max-Planck-Institut für Molekulare Pflanzenphysiologie, Golm, Potsdam, Germany*
Stéphanie Troufflard • *Département de Physique, Ecole Normale Supérieure, Paris Cedex, France*
Jacques Trouverie • *Ecophysiologie végétale, Agronomie et Nutrition N,C,S, Institut de Biologie Fondamentale et Appliquée, UMR EVA INRA, Université de Caen, Caen cedex, France*
Quyen Truong • *Department of Chemical and Biological Engineering, Iowa State University, Ames, IA, USA*
Gilles Vercambre • *INRA, UR 1115, Plantes et Systèmes de culture Horticoles, Domaine St Paul, Avignon Cedex, France*
Anne-Sophie Voisin • *INRA UMR 1347 Agroécologie INRA/Université de Bourgogne/AgroSup, Dijon Cedex, France*
Tung Vu-Ngoc Nguyen • *LABRI, LaBRI – Univ. Bordeaux, Talence, France*
Thomas C.R. Williams • *Departamento de Biologia Vegetal, Universidade Federal de Viçosa, Viçosa, Minas Gerais, Brazil*
Jamey D. Young • *Department of Chemical and Biomolecular Engineering, Vanderbilt University, Nashville, TN, USA*
Yuting Zheng • *Department of Chemical and Biomolecular Engineering, University of Maryland, College Park, MD, USA*

Chapter 1

Application of Metabolic Flux Analysis to Plants

Martine Dieuaide-Noubhani and Ana Paula Alonso

Abstract

This volume compiles a series of chapters that cover the major aspects of plant metabolic flux analysis, such as but not limited to labeling of plant material, acquisition of labeling data, mathematical modeling of metabolic network at the cell, tissue, and plant level. A short revue, including methodological points and applications of flux analysis to plants, is presented in this introductory chapter.

Key words Metabolic flux analysis, Metabolic network, Subcellular compartmentation, Plant fluxomic, Isotopic labeling, Flux balance analysis, Kinetic modeling, Steady state metabolic flux analysis, Instationary metabolic flux analysis

1 Introduction

Improving the production of compounds of interest, crop yield and/or nutritional value are big challenges in plant research. To reach these goals, a deep understanding of plant biological processes is required, which arouses an increasing interest for the study of plant metabolic networks. However, understanding the regulation of plant cellular metabolism turned to be rather complex (Fig. 1): (1) cell activity results from the operation of a large network of interdependent and duplicated metabolic pathways; (2) these pathways are physically compartmented in different organelles but share metabolic intermediates and cofactors; (3) metabolic steps are subjected to in vivo regulation. Beside the complexity of the metabolic network, its plasticity that allows plants to adapt to environmental conditions, constitutes another brake to the successful engineering of plant metabolism. This is why attempts to engineer plants to increase the production of particular compounds have mixed success, especially the ones targeting primary metabolism. One striking example is the simultaneous overexpression of a yeast invertase and hexokinase into potato tuber with the aim to favor sucrose degradation and starch synthesis. Instead, glycolysis was induced, sucrose was resynthesized

Martine Dieuaide-Noubhani and Ana Paula Alonso (eds.), *Plant Metabolic Flux Analysis: Methods and Protocols*, Methods in Molecular Biology, vol. 1090, DOI 10.1007/978-1-62703-688-7_1, © Springer Science+Business Media New York 2014

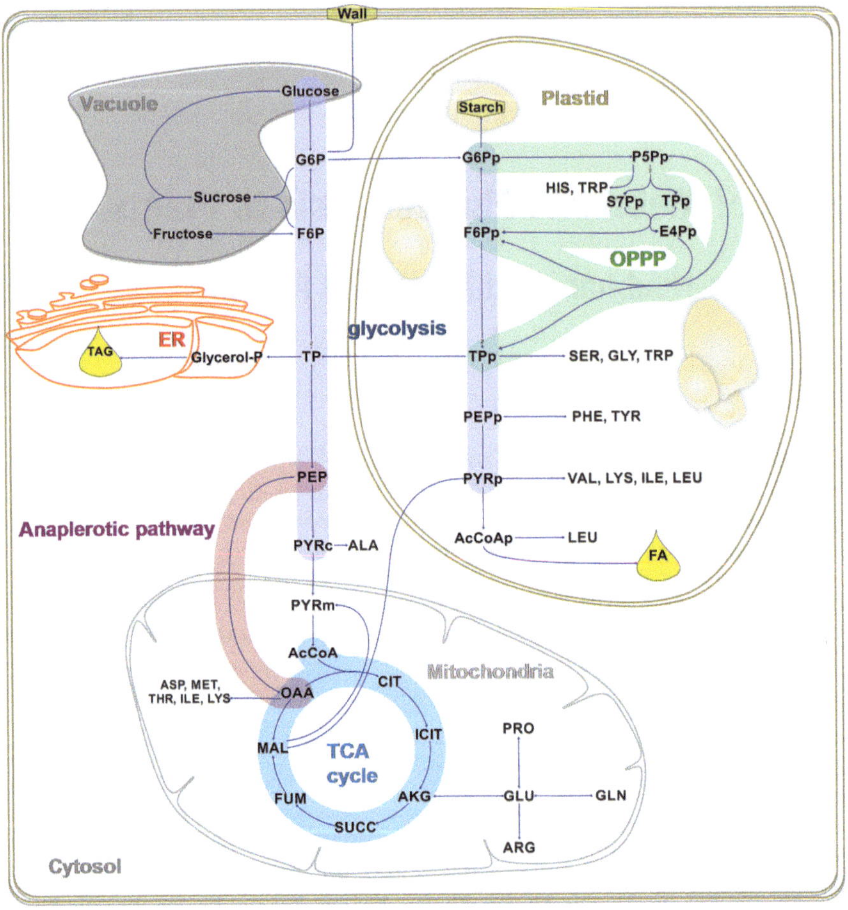

Fig. 1 Main metabolic pathways involved in plant central metabolism. Plant cells are highly compartmented: on the top of regular organelles found in eukaryotes, they contain a vacuole and plastids. The main pathways are glycolysis, pentose-phosphate pathway (PPP), anaplerotic reactions, and tricarboxylic acid (TCA) cycle. *p* plastidic, *c* cytosolic, *m* mitochondrial, *G6P* glucose 6-phosphate, *F6P* fructose 6-phosphate, *TP* triose-phosphate, *P5P* pentose 5-phosphate, *S7P* sedoheptulose 7-phosphate, *E4P* erythrose 4-phosphate, *PEP* phosphoenolpyruvate, *PYR* pyruvate, *AcCoA* acetyl-Coenzyme A, *CIT* citrate, *ICIT* isocitrate, *AKG* α-keto-glutarate, *SUCC* succinate, *FUM* fumarate, *MAL* malate, *OAA* oxaloacetate. The metabolic map was drawn by Dr. Koubaa using the Omix software (Droste et al., Biosystems 2011) (http:/www.13cflux.net/omix)

through the activation of the sucrose phosphate synthase (SPS), thus creating a substrate cycle [1–3], and starch level was decreased. Recent technical advances in transcriptomics, proteomics, and metabolomics have paved the way for the investigation of biological complexity, making it possible to properly examine the overall behavior of living systems. Data thus obtained can be integrated into a model that allows their visualization (Mapman [4]), which helps formulating hypotheses regarding cell function and regulation. However, despite their informative level, the omics datasets fail to integrate the dynamic of metabolic networks. Moreover, as previously underlined [5], most of the metabolites quantified by metabolomic approaches are intermediary metabolites; the observation of their increase or decrease cannot be simply explained without integrating all the reactions involved in their synthesis and degradation. Similarly, no variation in metabolite levels between two physiological conditions does not imply that fluxes are unchanged [5]. On the same way, differences in enzymes activities, often characterized in vitro by the variation of their V_{max}, cannot be easily interpreted as a variation of their in vivo activity because they are subjected to regulations. For instance, the increase of sucrose in the potato tubers overexpressing invertase was shown to be due to the phosphorylation of SPS: its activity increased in vivo despite a constant V_{max} [3]. Another example is the phosphoenolpyruvate carboxykinase (PEPC): in maize root tips, its V_{max} did not change significantly in response to sugar starvation [6], whereas in vivo enzyme activity became undetectable a few hours after the beginning of the treatment [7]. These examples underline the importance of quantifying fluxes in living cells.

2 Metabolic Network Definition and Topological Analysis

The description of a metabolic network is the first and essential step of metabolic flux analysis. Several chapters of the volume "Plant Metabolic Flux Analysis" show that if labeling experiments rigorously do not require network definition, the flux calculation is strictly dependant on its description. Its size depends on the biological question, from a few reactions to the whole genome network. But in all cases, the accuracy of the network is crucial because several pathways can generate similar labeling patterns. For instance, the comparison of the $^{14}CO_2$ production after incubation of a plant organ/tissue with [6-^{14}C]-glucose and [1-^{14}C]-glucose was used in the past to determine the activity of the pentose-phosphate pathway (PPP); the assumption was that the $^{14}CO_2$ produced with [6-^{14}C]-glucose was only due to the respiration, whereas the excess produced with [1-^{14}C]-glucose was due to action of the PPP [8]. However, similar $^{14}CO_2$ production do not exclude the operation of the PPP because additional $^{14}CO_2$ can be produced from [6-^{14}C]-glucose in the synthesis of pentans from UDP-glucuronic acid [9, 10].

Traditionally, the network reconstruction was accomplished using the physiological, biochemical, and molecular genetic literature. More recently, the availability of annotated genomes, plus databases providing information concerning transcription or protein composition, can be used in complement. A recent literature describes genome-scale metabolic reconstructions, mostly concerning unicellular—prokaryotic or eukaryotic—organisms [11–13], and in less extent in pluricellular eukaryotes, such as human [14, 15] or plants [16–18]. For additional literature, consult Dal'Molin et al. (Chapter 19). Methods for metabolic reconstructions are described in ref. [19] and by Beurton-Aimar et al. (Chapter 2).

Once reconstructed, the analysis of the structure of the metabolic network helps defining all the possible routes in the living cell under investigation. Two approaches have been developed: elementary flux modes (EFMs) and extreme pathways (EPs). EFMs can be defined as the smallest (non-decomposable) sub-network that can operate at steady state [20]. Although usually applied to moderately complex networks, the EFM approach has been efficient to explore a complex network, *E. coli* central metabolism, in which 44,000 EFMs were generated by considering 89 metabolites and 110 reactions [21]. EFM analysis was also applied to study the structure of plant metabolic networks [22]. Interestingly, the approach helped pinpointing substrate cycles. For instance [23] identified 14 EFMs in a metabolic network describing the sucrose metabolism in sugar cane, five of them corresponding to substrate cycles consuming ATP or UTP without net synthesis. However, the relationship between structure, function, and regulation is still a matter of debate as well as the applicability of EFM as a predictive tool. A recent study [24] compared the information provided by the modeling of metabolic networks using EFMs (one in plant and one in bacteria) to experimentally calculated fluxes. Interestingly, their results showed that the variations of the coefficients of flux efficiency, calculated from the EFM matrix and observed in response to change in carbon or nitrogen sources, were similar to the variations of the measured fluxes (at least for the internal ones).

3 Metabolic Flux Analysis

The aim of MFA is the detailed quantification of (all) metabolic fluxes in central metabolism. The result is a flux map that shows the distribution of anabolic and catabolic fluxes over the metabolic network. Based on such a map or a comparison of different maps, the effects of genetic or environmental modifications can be assessed. Metabolic flux balancing is one approach that can be used to estimate fluxes through a metabolic network. It is based on a stoichiometric model of the reactions, and internal fluxes are calculated from the quantification of the external ones (i.e., rates of substrate consumption and final products accumulation) [25]. However, this approach is not well suited to plant studies because the metabolic networks are too complex, with redundant and compartmented metabolic pathways but also as a consequence of the regulatory roles of substrates cycles [26, 27]. Isotopic-labeling approaches, based on the analysis of the labeling redistribution in metabolic intermediates, are more suitable to quantify metabolic fluxes in plant cells. Labeling experiments are carried out with radiolabeled (^{14}C) or stable (^{13}C, ^{15}N, . . .) isotopes. The advantage of ^{13}C-labeled substrates resides in the ease with which positional and mass isotopomer information can be obtained through the use of nuclear magnetic resonance (NMR) and mass spectrometric (MS) techniques, respectively. In the case of ^{14}C tracers, positional labeling information can also be obtained, but it is more limited and requires the degradation of the molecules, which is time consuming [28].

3.1 Kinetics of Labeling: Theoretical Aspect

Several reviews described the labeling kinetics [29–31]. Figure 2 illustrates, in the case of a simple linear pathway, the behavior of pool size and labeling when a labeled substrate is introduced, with a fractional enrichment E_S. The concentration of the substrate decreases linearly over time, whereas the final product increases linearly. The concentration of the intermediates first increases before becoming stable. In parallel, the labeling appears in the intermediates according to their order in the pathway, until it becomes constant. Three phases can be defined: the first one in

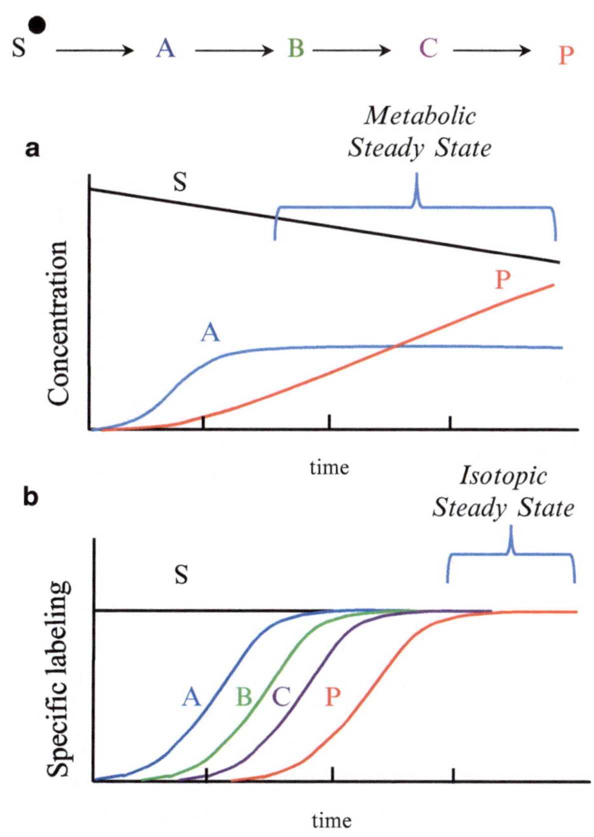

Fig. 2 A scheme of the evolution of the metabolite concentrations (**a**) in a linear pathway and of their labeling (**b**) when a labeled substrate S is fed to the cells

which the size and the labeling of the metabolic pools change, the second one in which the metabolic pools do not vary (metabolic steady state) but labeling varies (dynamic labeling), and the third one in which metabolic and isotopic steady state are achieved. The metabolic network can be described through differential equations involving the reaction rates through each step (these rates do not vary when metabolic steady state is reached), and labeling for each intermediary metabolite. The assumptions of metabolic and isotopic steady state result in algebraic equations instead of differentials equations, which significantly simplifies the mathematical modeling. In the linear pathway presented Fig. 2, all the metabolic intermediates would be equally labeled, and the labeling data would not be informative to quantify the fluxes. However, if one considers a more complex network, the junction of the first pathway to a second one would lead to the dilution of the tracer. Alternatively, labeling could appear in some carbon atoms as a consequence of intramolecular rearrangements. As an important simplification of the network, one can note that between two branch points, reactions are pooled into one same step. For instance in central metabolism, this is the case in glycolysis between fructose 6-phosphate (F6P) and triose-phosphate (TP) (Fig. 1). It is important to note that steady state labeling allows to estimate intermediary fluxes as relative flux values. Input and output fluxes (considered as external fluxes in the model) must be determined independently of the steady state labeling experiment and then included into the model. This can be done by measuring substrate consumption an, final product accumulation. It is important to note that fluxes determined in this way are net fluxes and not unidirectional fluxes. An alternative approach to measure the synthetic rates of a given metabolite involves dynamic labeling (*see* Chapter 3).

3.2 The Labeling Experiment: Experimental Design

The choice of the labeled substrate and experimental approach represent a major difficulty in plants. As underlined in several reviews [29, 31, 32], the selection of the labeled substrate not only depends on the metabolic network under investigation but also on the analytical techniques available. First, this choice affects the isotopomer composition in intermediary metabolites and final products within the network. Second, the detection of a labeled metabolite depends on the labeling amount and the sensitivity of the analytical methods. Therefore, a poor selection of the labeled precursor can lead to undetermined fluxes [33]. Since plant cells/organs mainly receive carbohydrates as the major source of carbon, glucose, or sucrose are usually used as labeled substrates. According to the tissue/organ, carbohydrates have been used alone or in combination with amino acids as a nitrogen source, such as glutamine or alanine. The substrate can be positionally labeled ([1-^{13}C]-glucose, [6-^{13}C]-glucose or [1,2-^{13}C]-glucose) or fully labeled (e.g., [U-^{13}C]-glucose, [U-^{13}C]-sucrose). In the last case, the labeled precursor is introduced simultaneously with the unlabeled form or positionally labeled substrate. Several studies explored the value of optimal substrate designs to investigate plant primary metabolism using different combination of labeled precursors. In Arabidopsis cell cultures, it has been found that [1-^{13}C]-glucose was the best tracer for the majority of fluxes [34], whereas in developing embryos cultures a combination of [U-^{13}C]-glucose and [1,2-^{13}C]-glucose was more informative [35].

The way to feed the substrate is not trivial in plants. Tissues or organs are usually grown in culture medium that mimics, as well as possible, physiological conditions. For instance, maize root tips were grown in a culture medium defined by [36]: glucose was used as the sole source of carbon, with a concentration equal to 300 mM. This high concentration was necessary to keep the respiratory rate constant for a few days. Conditions for oxygenation, salt concentration, nitrogen source also must be considered. For instance, it has been shown that the source of nitrogen supply (organic or inorganic) as well as its quantity affects the metabolism of developing *Brassica napus* and soybean embryos [37, 38]. The incorporation of the labeled precursor can be achieved by the culture of multiple organs at the same time, such as the culture of maize kernels attached to cob tissue to label maize endosperm [39, 40]. Some groups are trying to feed the labeling directly to the plant, using $^{13}CO_2$ that is incorporated through photosynthesis [41–44], or feeding ^{13}C-glucose through the roots and analyzing leaves labeling after 20 days [45]. It is important to note that if one waits long enough $^{13}CO_2$ will label uniformly the metabolic intermediates; hence this approach cannot be used at steady state. Recently, a petiole-feeding bioassay [46] has been shown efficient to introduce labeled sugars and other soluble molecules into young dicotyledonous plants such as soybean and tomato.

The time required to reach steady state depends on the organism, varying from a few minutes in bacteria to several hours (maize root tips [10], potato [47]) and even several days in tomato cells [48], and embryos [49–54]. These long times are due to large vacuolar pools of free sugars, such as glucose and fructose that are only slowly exchanged with their cytosolic pools, but also to biomass products (polysaccharides such as starch, proteins, fatty acids, etc.) that are initially present before the labeling and do not turn over fast enough. To reduce the time required to achieve the isotopic steady state, maize root tips were sugar starved prior labeling [10]: after 4 h, the starch was degraded and glucose reduced to 30 % of its initial value. By this way, labeling steady state was obtained after 12 h of labeling (instead of 20 h). Another approach is to apply diluting/correcting factors to the labeling data in order to account for unlabeled pools [55, 57].

3.3 Analysis of the Labeling

Most of the time, metabolite labeling is analyzed on extracts obtained from labeled samples. Methods for extractions should be adapted to the metabolites studied and are therefore various. In the case of small molecules, such as free sugars, amino acids and organic acids, boiling ethanol [58], or cold methanol [59] are classically used. Phosphorylated compounds are more often extracted in acidic conditions [60]. However, salts remaining in the sample are not always compatible with the analytical techniques (e.g., this is the case for liquid chromatography-mass spectrometry, LC-MS). The analyses of macromolecules such as lipids, proteins, or starch also require specific procedures. Proteins are hydrolyzed in acidic conditions to release amino acids that are further analyzed by NMR ([49]; Chapter 6) or gas chromatography-mass spectrometry (GC-MS) after derivatization [50, 51]. Starch is analyzed as glucose after complete [10, 48] or partial [61] enzymatic hydrolysis. The labeling in triacylglycerol can be determined by NMR [51, 52] or by GC-MS after derivatization to butyl amides [62]. Most of these procedures are detailed in Chapters 5–9.

In the case of ^{14}C-labeling, radioactivity is measured by liquid scintillation counting. Total radioactivity incorporated in a molecule or specific radioactivity (i.e., quantity of radioactivity per mole of molecule or when possible per mole of a particular atom) can be determined. Only a few publications report the analysis of specific carbon, such as the C1 and C5 of glutamate [28] or C1 of glucose [10] obtained after decarboxylation; most of the time, only the labeling of the entire molecule is determined. Thus the information about intermolecular rearrangements that occur during the labeling is rather limited, in contrast to the possibilities offered in the case of ^{13}C-labeling, with the development of NMR, GC-MS, and LC-MS techniques. The use of these methods is well documented in a review from [29]. In brief, three sets of data can be obtained after ^{13}C-labeling (fractional enrichments, positional isotopomers and cumomers, mass isotopomers). Throughout the chapters of Part I in this volume, a panel of methodologies specifically developed for their analysis are described. They rely on 1D or 2D NMR and on GC- or LC-MS techniques.

The sensitivity of the different methods varies with the equipment, but in general the range is mmol–μmol for NMR, nmol–pmol for GC-MS, and pmol–fmol for LC-MS/MS. Therefore MS techniques are much more sensitive than NMR, but sample manipulation is usually easier for NMR. For instance, in contrast to GC-MS, there is no need for derivatization nor analysis of fragments. Another advantage of NMR is that it can directly be used to study metabolic fluxes in vivo, without extraction ([63, 64], Chapter 10). However, using MS techniques that are more sensitive than NMR opens the way to the direct quantification of isotopomers in intermediary (low concentration) metabolites [44, 65, 66].

3.4 Data Interpretation and Flux Calculation

Steady state labeling—The resolution of equations relating fluxes and fractional enrichments, assuming isotopic steady state, can be made using algebraic solving software such as Excel [10] or Mathematica [48]. The main disadvantage of this approach is that it is difficult to account for the redundancy of the information, which happens when experiments produce more data than unknowns. This difficulty is well illustrated in ref. [10]: in their model, the glucose was formed from the hexose moiety of sucrose. In this case, the information given by the analysis of the carbon 1 (C1) was redundant with the information provided by the C6 of these free sugars. To deal with the error on the experimental value, they solved the equations using a graphical approach. The development of dedicated software, such as FiatFlux [67], 13CFLUX [68], or OpenFlux [69] helped with the mathematical interpretation of the redundant labeling data. The modeling is based on the description of the metabolic network, which includes the stoechiometry of the reactions and the exact description of the behavior of each carbon atom between the substrates and the products for each step. Fluxes are then optimized by comparing the predicted labeling obtained from the estimated fluxes to the experimental ones. The main advantage of 13CFLUX versus FiatFlux and OpenFlux is that it allows the interpretation of labeling data from NMR and MS.

Steady state ^{13}C-MFA has been successfully applied to plants, mainly on fast growing plant cells and organs, such as roots [10, 55], cell cultures [34, 48], embryos [49–54], and seeds [40, 70]. By measuring fluxes through active pathways, steady state-MFA revealed unconventional operations of the metabolic network. For example, substrate cycles (aka "futile" cycles) have been shown to operate in several plant organs, such as potato tubers, maize root tips, bananas, tomato cells and fruit, sunflower embryos, and maize kernels [10, 40, 48, 51, 52, 55, 71–73]. These cycles synthesize and degrade a given metabolite, resulting in a net consumption of ATP that can reach 70 % of the total ATP produced by the cell. In developing sunflower embryos, the net flux through the cytosolic glycolysis favors the production of hexose phosphates that are then imported into the plastid to provide the carbon and reductant power necessary for fatty acid synthesis [51]. In developing *B. napus* embryos, ^{13}C-labeling suggested that the ribulose 1,5-bisphosphate carboxylase/oxygenase (Rubisco) would recycle CO_2 to incorporate it into biomass (fatty acids), thus enhancing the carbon conversion efficiency [74]. Furthermore, it has been shown that the TCA cycle do not operate in developing *B. napus* embryos [50]. 13C-MFA was successfully used to study the effect of environment, such as the response to hypoxia [34, 75], and genetic perturbations [54, 56, 70] on primary metabolism. The comparison of metabolic fluxes in tomato cells at three different culture stages [48], in heterotrophic Arabidopsis cells under different oxygenation condition [34], or different genotypes of maize kernels [70] demonstrated the robustness of central metabolism: it was characterized by no significant change of the ratio of metabolic fluxes at the branch point, whereas fluxes in anabolic pathways were more variable.

However, it is important to note that metabolic and isotopic steady state is not always reachable. In plants, ^{13}C-labeling experiments usually require the incubation of organs/cells in conditions that mimic to the ones *in planta*. It is not always possible to define physiological conditions to grow plant organs within a timeframe that is compatible with steady state labeling; the case of the diurnal cycle is one example.

Dynamic modeling—In contrast to steady state modeling, dynamic modeling requires time course measurements of pool sizes and specific labeling for each intermediate and product within the metabolic network. The algebraic equation system described for steady state modeling is here replaced by a differential equation system of equal dimension. Two strategies can be distinguished to estimate fluxes from dynamic modeling: the kinetic modeling and instationary-MFA (INST-MFA).

The kinetic modeling requires the knowledge of enzyme kinetic parameters; the fluxes are estimated by fitting the model parameters to the experimental data (i.e., the time course of labeling in the molecules of interest). It has been successfully applied in plants to study choline metabolism [76, 77] and the phenylpropanoid pathways [78–80]. For more detailed applications of kinetics modeling in plants, *see* ref. 27. Such kinetics models, combined to metabolic control analysis [81, 82] can be used in a predictive manner, providing information on the distribution of flux control in the metabolic network [80].

INST-MFA explores the kinetics of the positional labeling to increase the data available. To reduce the required information and the computational time to simulate isotopic labeling, the isotopomer network is decomposed in elementary metabolite units or EMU [83, 84]. This approach has been successfully applied to study metabolic fluxes in cyanobacterium *Synechocystis*, a photoautotrophic green algae, using $^{13}CO_2$ labeling [85, 86]. This new development would be a solution to studying metabolic fluxes when isotopic steady state cannot be applied. For instance, it is particularly attractive to study the metabolism in autotrophic cells in which CO_2, the natural substrate, cannot be used for steady state modeling because it uniformly labels the metabolic intermediates.

3.5 The Question of the Compartmentation in MFA

For a long time, compartmentation was not taken into account in MFA; the cell was considered to be a unique compartment. However, the role of subcellular compartmentation in the regulation of the metabolism is now well established. It is important to note that plant cells are probably the most highly compartmented cells (Fig. 1) with plastids, vacuoles, and cell wall that are not found in all eukaryotic cells. To that subcellular compartmentation, should be added redundant reactions and pathways (e.g., glycolysis occurs in the cytosol and in the plastids, invertase in the cytosol and in the vacuole, etc.) with, in several cases an uncertainty about the exact localization of the enzymes. For instance, the localization of the different parts of the pentose phosphate pathway (cytosol or plastid) depends on the plant species, organs, and tissue. Consequently there are, for a few metabolites, several routes possible for their biosynthesis, which may result in different labeling patterns according to their subcellular localization. Thus errors in flux estimation can occur when metabolites are considered as a single pool. For instance, the difficulty to interpret the labeling of internal glucose illustrates this problem [87]. However, compartmentation is not easy to take into account because the subcellular extraction of metabolites cannot be achieved with classical fractionation: metabolites are lost or modified consequently to dilution or enzymatic activities. Recent developments of nonaqueous fractionation in the domain of metabolomics [88–90] open the possibility of applying this approach in combination with ^{13}C-labeling to subcellular fluxomics. It is important to note that nonaqueous fractionation does not produce pure organelle fractions and therefore relies on deconvolution techniques; its applicability to labeled subcellular metabolites has not yet been determined.

However, the distinction of fluxes through compartments is permissible when metabolites are localized in a single compartment or produced from one pathway only. For instance, starch is exclusively produced in the plastid (Fig. 1), therefore its labeling pattern represents the labeling of the plastidic glucose 6-phosphate (G6P). On the other hand, the labeling of sucrose, cell wall, and protein glycans provide information on the labeling of cytosolic G6P. This information is crucial not only to define the labeling entering each metabolic pathways present in the plastid and the cytosol but also to estimate the rates of hexose phosphate exchange between these two compartments [10, 49, 51, 52, 62, 75]. Similarly, labeling of C16 and C18 fatty acids are used to estimate the labeling of plastidic acetyl-CoA, whereas the labeling in additional carbons (C20 or longer fatty acids) reveals the labeling of cytosolic acetyl-CoA [50, 53, 91]. Metabolites that provide information on the labeling of metabolic intermediates with respect to their subcellular localization have been listed by Allen et al. [62], and flux maps of compartmented metabolic networks are found in the plant literature [10, 34, 40, 49, 51, 52, 57]. However it is noteworthy that information about compartmented fluxes is often incomplete, specifically when metabolic intermediates are rapidly exchanged between the compartments under consideration. Moreover, the specific localization of the metabolites is not always well established. In a recent publication, Allen and collaborators [92] compared the labeling of the amino acids of the two subunits of the

RuBisCo: the small subunit is formed in the plastid, whereas the large subunit is produced in the cytosol. According to their results, most of the amino acids are labeled identically, thus indicating a similar origin or a rapid exchange between subcellular compartments. However, the labeling of three amino acids, such as alanine, was found to be significantly different between the two RuBisCo subunits. Although this study cannot be easily extended to all plant tissues, such an observation suggests that alanine is not a good marker for the labeling of cytosolic pyruvate.

4 A Predictive Modeling to Estimate Fluxes Without Labeling Experiments

13C-MFA is clearly a powerful method to calculate fluxes in plant cells/organs. However, it is not applicable to all plant systems, for several reasons. The first one, which has been underlined in Subheading 3.2, is the difficulty to label some plant tissues. Then, 13C-MFA suffers from a lack of data when the complexity of the metabolic network increases. For instance, this is the case when subcellular compartmentation is taken into consideration, or when the number of metabolic pathways is increased to account for the whole cellular metabolism.

Kinetic modeling is based on the modeling of enzyme kinetic parameters (*see* Chapter 15). It has been successfully applied to model the branch-point between methionine and threonine biosynthetic pathways in *Arabidopsis thaliana* ([93]; *see* Chapter 16), and sucrose metabolism in sugarcane [94]. Interestingly, robot-based platforms to measure enzymes activities have been developed this last decade. The adaptation of cycling assay to microplate technology and the automation of the data analysis allow to estimate kinetic parameters of a wide range of enzymes involved in primary metabolism in a minimum time [95].

When the complexity of the metabolic network increases, constraint-based modeling proposes to predict rather than to calculate fluxes, using a minimal dataset: flux balance analysis (FBA) or flux variability analysis [96, 97], a recent development of FBA, requires the knowledge of the stoechiometry of the metabolic network and the definition of an objective function (*see* Chapters 17 and 18). Extensively developed and used in bacteria, with an objective function to optimize the biomass yield, FBA has been first applied to plant cells to study the metabolic flux distribution in response to oxygen depletion and enzyme deletion in barley seeds [96]. More recently, FBA has been associated to genome-scale reconstruction to study metabolic network in Arabidopsis cells [16, 17] and in C4 plants ([18], Chapter 19).

5 Flux Analysis at the Whole Plant Level

MFA is dedicated to the study of metabolic fluxes at the cell/tissue scale, with the hypothesis, when tissues or organs are studied, that cells therein have the same activities. However, metabolism at the whole plant level reveals more complexity. Considering C metabolism for example, plants have very specialized organs with the source tissues, essentially the leaves, that produce and export carbon, whereas other organs are heterotrophic and require import carbohydrates for their development. This problematic is not specific to carbon and can be extended to other nutrients (i.e., nitrogen and sulfur) also essential for plant growth [98]. Thus, plant development and quality is the result of a complex interaction between environment, genetic, and plant structure (including organs, tissues, and cells) that could interact or compete for carbon, nitrogen, water, etc. Ecophysiologists are attempting to develop models integrating all these parameters. Chapters 20 and 21 are dedicated to these approaches. Interestingly, studies from the Génard's group are oriented towards the modeling of fruit development, such as peach [99] and tomato [100, 101]: the goal is to estimate the fruit size and composition in response to the environment and genotype. Several parameters are involved [102], such as cell division, endoreduplication, fluxes of water and solutes during the expansion phase, and concentration of the major compounds (carbohydrates, organic acids, oil, proteins).

References

1. Sonnewald U, Hajiraezaei MR, Kossmann J et al (1997) Expression of a yeast invertase in the apoplast of potato tubers increases tuber size. Nat Biotechnol 15:794–797
2. Trethewey RN, Geigenberger P, Hajirezaei MR et al (1998) Combined expression of glucokinase and invertase in potato tubers leads to a dramatic reduction in starch accumulation and a stimulation of glycolysis. Plant J 15: 109–118
3. Trethewey RN, Riesmeier JW, Willmitzer L, Stitt M, Geigenberger P (1999) Tuber-specific expression of a yeast invertase and a bacterial glucokinase in potato leads to an activation of sucrose phosphate synthase and the creation of a sucrose futile cycle. Planta 208(2):227–238
4. Thimm O, Blasing O, Gibon Y (2004) MAPMAN: a user-driven tool to display genomics data sets onto diagrams of metabolic pathways and other biological processes. Plant J 37: 914–939
5. Fernie AR, Geigenberger P, Stitt M (2005) Flux an important, but neglected, component of functional genomics. Curr Opin Plant Biol 8:174–182
6. Brouquisse R, James F, Raymond P et al (1991) Study of glucose starvation in excised maize root tips. Plant Physiol 96:619–626
7. Dieuaide-Noubhani M, Canioni P, Raymond P et al (1997) Sugar-starvation-induced changes of carbon metabolism in excised maize root tips. Plant Physiol 115:1505–1513
8. Gibbs M, Beevers H (1955) Glucose dissimilation in the higher plant. Effect of age of tissue. Plant Physiol 30:343–347
9. ap Rees T (1980) Assessment of the contribution of metabolic pathways to plant respiration. In: Stumpf PK, Conn EE (eds) The biochemistry of plants, vol 2. Academic, London, pp 1–29
10. Dieuaide-Noubhani M, Raffard G, Canioni P, Pradet A, Raymond P (1995) Quantification of compartmented metabolic fluxes in maize

root tips using isotope distribution from 13C- or 14C-labeled glucose. J Biol Chem 270:13147–13159

11. Oh Y-K, Palsson BO, Park SM et al (2007) Genome-scale reconstruction of metabolic network in *Bacillus subtilis* based on high-throughput phenotyping and gene essentiality data. J Biol Chem 282:28791–28799
12. Orth LD, Conrad TM, Na J et al (2011) A comprehensive genome-scale reconstruction of Escherichia coli metabolism. Mol Syst Biol 7:535
13. Nogales J, Gudmundsson S, Knight EM et al (2012) Detailing the optimality of photosynthesis in cyanobacteria through systems biology analysis. Proc Natl Acad Sci USA 109:2678–2683
14. Duarte NC, Becker SA, Jamshidi N et al (2007) Global reconstruction of the human metabolic network based on genomic and bibliomic data. Proc Natl Acad Sci USA 104(6):1777–1782
15. Bordbar A, Palsson BO (2011) Using the reconstructed genome-scale human metabolic network to study physiology and pathology. J Intern Med 271:131–141
16. Poolman MG, Miguet L, Sweetlove LJ, Fell DA (2009) A genome-scale metabolic model of Arabidopsis and some of its properties. Plant Physiol 151(3):1570–1581
17. de Oliveira Dal'Molin CG, Quek LE, Palfreyman RW et al (2010) AraGEM, a genome-scale reconstruction of the primary metabolic network in Arabidopsis. Plant Physiol 152: 579–589
18. de Oliveira Dal'Molin CG, Quek LE, Palfreyman RW et al (2010) C4GEM, a genome-scale metabolic model to study C4 plant metabolism. Plant Physiol 154:1871–1885
19. Thiele I, Palsson BO (2010) A protocol for generating a high-quality genome-scale metabolic reconstruction. Nat Protoc 5:93–121
20. Schuster S, Dandekar T, Fell DA (1999) Detection of elementary flux modes in biochemical networks: a promising tool for pathway analysis and metabolic engineering. Trends Biotechnol 17:53–60
21. Klamt S, Stelling J (2002) Combinatorial complexity of pathway analysis in metabolic networks. Mol Biol Rep 29:233–236
22. Poolman MG, Assmus HE, Fell DA (2004) Applications of metabolic modelling to plant metabolism. J Exp Bot 55:1177–1186
23. Rohwer JM, Botha FC (2001) Analysis of sucrose accumulation in the sugar cane culm on the basis of in vitro kinetic data. Biochem J 358:437–445
24. Beurton-Aimar M, Beauvoit B, Monier A et al (2011) Comparison between elementary flux modes analysis and 13C-metabolic fluxes measured in bacterial and plant cells. BMC Syst Biol 5:95
25. Bonarius HPJ, Schmid G, Tramper J (1997) Flux analysis of underdetermined metabolic networks: the quest for the missing constraints. Trends Biotechnol 15:308–314
26. Roscher A, Kruger NJ, Ratcliffe RG (2000) Strategies for metabolic flux analysis in plants using isotope labelling. J Biotechnol 77:81–102
27. Morgan JA, Rhodes D (2002) Mathematical modeling of plant metabolic pathways. Metab Eng 4:80–89
28. Salon C, Raymond P, Pradet A (1988) Quantification of carbon fluxes through the tricarboxylic acid cycle in early germinating lettuce embryos. J Biol Chem 263:12278–12287
29. Ratcliffe RG, Shachar-Hill Y (2006) Measuring multiple fluxes through plant metabolic networks. Plant J 45:490–511
30. Dieuaide-Noubhani M, Alonso AP, Rolin D et al (2007) Metabolic flux analysis: recent advances in carbon metabolism in plants. In: Baginsky S, Fernie AR (eds) Plant systems biology. Birkhäuser, Basel, pp 213–243
31. Libourel IGL, Shachar-Hill Y (2008) Metabolic flux analysis in plants: from intelligent design to rational engineering. Annu Rev Plant Biol 59:625–650
32. Kruger NJ, Ratcliffe RG (2012) Pathways and fluxes: exploring the plant metabolic network. J Exp Bot 63:2243–2246
33. Chang YJ, Suthers PF, Maranas CD (2008) Identification of optimal measurements sets for complete flux elucidation in metabolic flux analysis experiments. Biotechnol Bioeng 100:1039–1049
34. Williams TCR, Miguet L, Masakapalli SK et al (2008) Metabolic network fluxes in heterotrophic Arabidopsis cells: stability of the flux distribution under different oxygenation conditions. Plant Physiol 148:704–718
35. Libourel IGL, Gehan JP, Shachar-Hill Y (2007) Design of substrate label for steady state flux measurements in plant systems using the metabolic network of Brassica napus embryos. Phytochemistry 68: 2211–2221
36. Saglio PH, Pradet A (1980) Soluble sugar, respiration, and energy charge during aging of excised maize root tips. Plant Physiol 66:516–519
37. Junker BH, Lonien J, Heady LE, Rogers A, Schwender J (2007) Parallel determination of

enzyme activities and in vivo fluxes in Brassica napus embryos grown on organic or inorganic nitrogen source. Phytochemistry 68: 2232–2242

38. Allen DK, Young JD (2013) Carbon and nitrogen provisions alter the metabolic flux in developing soybean embryos. Plant Physiol 161:1458–1475
39. Glawischnig E, Tomas A, Eisenreich W et al (2000) Auxin biosynthesis in maize kernels. Plant Physiol 123:1109–1119
40. Alonso AP, Val DL, Shachar-Hill Y (2011) Central metabolic fluxes in the endosperm of developing maize seeds and their implications for metabolic engineering. Metab Eng 13: 96–107
41. Römisch-Margl W, Schramek N, Radykewicz T et al (2007) 13CO2 as a universal metabolic tracer in isotopologue perturbation experiments. Phytochemistry 68:2273–2289
42. Huege J, Sulpice R, Gibon Y et al (2007) GC-EI-TOF-MS analysis of in vivo carbon-partitioning into soluble metabolite pools of higher plants by monitoring isotope dilution after 13CO2 labelling. Phytochemistry 68: 2258–2272
43. Chen W-P, Yang X-Y, Harms GL et al (2011) An automated growth enclosure for metabolic labeling of Arabidopsis thaliana with C-13-carbon dioxide—an in vivo labeling system for proteomics and metabolomics research. Proteome Sci 9:9–22
44. Szecowka M, Heise R, Tohge T et al (2013) Metabolic fluxes in an illuminated Arabidopsis rosette. Plant cell 25:694–714
45. Ettenhuber C, Radykewicz T, Kofer W et al (2005) Metabolic flux analysis in complex isotopolog space. Recycling of glucose in tobacco plants. Phytochemistry 66:323–335
46. Lin Y-H, Lin M-H, Gresshoff PM, Ferguson BJ (2011) An efficient petiole-feeding bioassay for introducing aqueous solutions into dicotyledonous plants. Nat Protoc 6:36–45
47. Roessner-Tunali U, Liu JL, Leisse A et al (2004) Kinetics of labelling of organic and amino acids in potato tubers by gas chromatography mass spectrometry following incubation in 13C labelled isotopes. Plant J 39:668–679
48. Rontein D, Dieuaide-Noubhani M, Dufourc EJ, Raymond P, Rolin D (2002) The metabolic architecture of plant cells. Stability of central metabolism and flexibility of anabolic pathway during the growth cycle of tomato cells. J Biol Chem 277:43948–43960
49. Sriram G, Fulton DB, Iyer VV et al (2004) Quantification of compartmented metabolic fluxes in developing soybean embryos by employing biosynthetically directed fractional 13C labeling, two dimensional [13C, 1H] nuclear magnetic resonance, and comprehensive isotopomer balancing. Plant Physiol 136:3043–3057
50. Schwender J, Shachar-Hill Y, Ohlrogge JB (2006) Mitochondrial metabolism in developing embryos of Brassica napus. J Biol Chem 281:34040–34047
51. Alonso AP, Goffman FD, Ohlrogge JB, Shachar-Hill Y (2007) Carbon conversion efficiency and central metabolic fluxes in developing sunflower (Helianthus annuus L.) embryos. Plant J 52:296–308
52. Alonso AP, Dale VL, Shachar-Hill Y (2010) Understanding fatty acid synthesis in developing maize embryos using metabolic flux analysis. Metab Eng 12:488–497
53. Allen DK, Ohlrogge JB, Shachar-Hill Y (2009) The role of light in soybean seed filling metabolism. Plant J 58:220–234
54. Lonien J, Schwender J (2009) Analysis of metabolic flux phenotypes for two Arabidopsis mutants with severe impairment in seed storage lipid synthesis. Plant Physiol 151: 1617–1634
55. Alonso AP, Vigeolas H, Raymond P, Rolin D, Dieuaide-Noubhani M (2005) A new substrate cycle in plants. Evidence for a high glucosephosphate-to-glucose turnover from in vivo steady-state and pulse labeling experiments with [13C]glucose and [14C]glucose. Plant Physiol 138:2220–2232
56. Alonso AP, Raymond P, Hernould M et al (2007) A metabolic flux analysis to study the role of sucrose synthase in the regulation of the carbon partitioning in central metabolism in maize root tips. Metab Eng 9:419–432
57. Masakapalli SK, Kruger NJ, Ratcliffe RG (2013) The metabolic flux phenotype of heterotrophic Arabidopsis cells reveals a complex response to changes in nitrogen supply. Plant J. doi:10.1111/tpj.12142
58. Stitt M, ap Rees T (1978) Pathways of carbohydrate oxidation in leaves of *Pisum sativum* and *triticum aestivum*. Phytochemistry 18: 1905–1911
59. Roessner-Tunali U, Hegemann B, Lytovchenko A et al (2003) Metabolite analysis for the comparison of irrigated and non-irrigated field grown tomato of varying genotype. Plant Physiol 133:84–99
60. Brouquisse RM, Evrard A, Rolin D, Raymond P, Roby C (2001) Regulation of protein degradation and protease expression by mannose in maize root tips. Pi sequestration by

mannose may hinder the study of its signaling properties. Plant Physiol 125:1485–1498
61. Keeling PL, Wood JR, Tyson RH, Bridges IG (1988) Starch biosynthesis in developing wheat grain. Evidence against the direct involvement of triose phosphates in the metabolic pathway. Plant Physiol 87:311–319
62. Allen DK, Shachar-Hill Y, Ohlrogge JB (2007) Compartment specific labeling information in 13C metabolic flux analysis of plants. Phytochemistry 68:2197–2210
63. Roscher A, Emsley L, Raymond P, Roby C (1998) Unidirectional steady state rates of central metabolism enzymes measured simultaneously in a living plant tissues. J Biol Chem 273:25053–25061
64. Troufflard S, Roscher A, Thomasset B et al (2007) In vivo 13C NMR determines metabolic fluxes and steady state in linseed embryos. Phytochemistry 68:2341–2350
65. Alonso AP, Piasecki RJ, Wang Y, LaClair RW, Shachar-Hill Y (2010) Quantifying the labeling and the levels of plant cell wall precursors using ion chromatography tandem mass spectrometry. Plant Physiol 153:915–924
66. Koubaa M, Cocuron JC, Thomasset B, Alonso AP (2013) Highlighting the tricarboxylic acid cycle: liquid and gas chromatography-mass spectrometry analyses of ^{13}C-labeled organic acids. Anal Biochem 436:151–159
67. Zamboni N, Fischer E, Sauer U (2005) FiatFlux—a software for metabolic flux analysis from 13C-glucose experiments. BMC Bioinformatics 6:209
68. Wiechert W, Möllney M, Petersen S, de Graaf AA (2001) A universal framework for 13C metabolic flux analysis. Metab Eng 3: 265–283
69. Quek LE, Wittmann C, Nielsen LK, Krömer JO (2009) OpenFLUX: efficient modeling software for 13C-based metabolic flux analysis. Microb Cell Fact 8:25
70. Spielbauer G, Margl L, Hannah LC et al (2006) Robustness of central carbohydrate metabolism in developing maize kernels. Phytochemistry 67:1460–1475
71. Hatzfeld WD, Stitt M (1990) A study of the rate of recycling of triose phosphates in heterotrophic *Chenopodium rubrum* cells, potato tubers and maize endosperm. Planta 180: 198–204
72. Hill SA, ap Rees T (1994) Fluxes of carbohydrate metabolism in ripening bananas. Planta 192:52–60
73. N'tchobo H, Dali N, Nguyen-Quoc B et al (1999) Starch synthesis in tomato remains constant throughout fruit development and is dependent on sucrose supply and sucrose synthase activity. J Exp Bot 50:1457–1463
74. Schwender J, Goffman F, Ohlrogge JB et al (2004) Rubisco without the Calvin cycle improves the carbon efficiency of developing green seeds. Nature 432:779–782
75. Alonso AP, Raymond P, Rolin D, Dieuaide-Noubhani M (2007) Substrate cycles in the central metabolism of maize root tips under hypoxia. Phytochemistry 68:2222–2231
76. McNeil SD, Nuccio ML, Rhodes D, Shachar-Hill Y, Hanson AD (2000) Radiotracer and computer modeling evidence that phospho-base methylation is the main route of choline synthesis in tobacco. Plant Physiol 123: 371–380
77. McNeil SD, Rhodes D, Russell BL et al (2000) Metabolic modeling identifies key constraints on an engineered glycine betaine synthesis pathway in tobacco. Plant Physiol 124:153–162
78. Matsuda F, Morino K, Miyashita M, Miyagawa H (2003) Metabolic flux analysis of the phenylpropanoid pathway in wound-healing potato tuber tissue using stable isotope-labeled tracer and LC–MS spectroscopy. Plant Cell Physiol 44:510–517
79. Heinzle E, Matsuda F, Miyagawa H, Wakasa K, Nishioka T (2007) Estimation of metabolic fluxes, expression levels and metabolite dynamics of a secondary metabolic pathway in potato using label pulse-feeding experiments combined with kinetic network modelling and simulation. Plant J 50:176–187
80. Marshall-Colón AM, Sengupta N, Rhodes D, Dudareva N, Morgan J (2010) A kinetic model describes metabolic response to perturbations and distribution of flux control in the benzenoid network of Petunia hybrida. Plant J 62:64–76
81. Kacser H, Burns J (1973) The control of flux. Symp Soc Exp Biol 27:65–104
82. Fell DA (1997) Understanding the control of metabolism. Portland Press, London
83. Antoniewicz MR, Kelleher JK, Stephanopoulos G (2007) Elementary metabolite units (EMU): a novel framework for modeling isotopic distributions. Metab Eng 9:68–86
84. Young JD, Walther JL, Antoniewicz MR, Yoo H, Stephanopoulos G (2008) An elementary metabolite unit (EMU) based method of isotopically nonstationary flux analysis. Biotechnol Bioeng 99:686–699
85. Shastri AA, Morgan JA (2007) A transient isotopic labeling methodology for 13C metabolic flux analysis of photoautotrophic microorganisms. Phytochemistry 68:2302–2312

86. Young JD, Shastri AA, Stephanopoulos G, Morgan JA (2011) Mapping photoautotrophic metabolism with isotopically nonstationary 13C flux analysis. Metab Eng 13: 656–665
87. Kruger NJ, Le Lay P, Ratcliffe RG (2007) Vacuolar compartmentation complicates the steady-state analysis of glucose metabolism and forces reappraisal of sucrose cycling in plants. Phytochemistry 68:2189–2196
88. Farré EM, Tiessen A, Roessner U et al (2001) Analysis of the compartmentation of glycolytic intermediates, nucleotides, sugars, organic acids, amino acids, and sugar alcohols in potato tubers using a nonaqueous fractionation method. Plant Physiol 127:685–700
89. Benkeblia N, Shinano T, Osaki M (2007) Metabolite profiling and assessment of metabolome compartmentation of soybean leaves using non-aqueous fractionation and GC-MS analysis. Metabolomics 3:297–305
90. Krueger S, Giavalisco P, Krall L et al (2011) Topological map of the compartmentalized Arabidopsis thaliana leaf metabolome. PLoS One 6(3):e17806
91. Schwender J, Ohlrogge JB (2002) Probing in vivo metabolism by stable isotope labeling of storage lipids and proteins in developing *Brassica napus* embryos. Plant Physiol 130: 347–361
92. Allen DK, Laclair RW, Ohlrogge JB, Shachar-Hill Y (2012) Isotope labelling of Rubisco subunits provides *in vivo* information on subcellular biosynthesis and exchange of amino acids between compartments. Plant Cell Environ 35:1232–1244
93. Curien G, Ravanel S, Dumas R (2003) A kinetic model of the branch-point between the methionine and threonine biosynthesis pathways in *Arabidopsis thaliana*. Eur J Biochem 270:4615–4627
94. Uys L, Botha FC, Hofmeyr J-HS et al (2007) Kinetic model of sucrose accumulation in maturing sugarcane culm tissue. Phytochemistry 68:2375–2392
95. Gibon Y, Blaesing OE, Hannemann J et al (2004) A robot-based platform to measure multiple enzyme activities in Arabidopsis using a set of cycling assays: comparison of changes of enzyme activities and transcript levels during diurnal cycles and in prolonged darkness. Plant Cell 16:3304–3325
96. Grafahrend-Belau E, Schreiber F, Koschützki D, Junker BH (2009) Flux balance analysis of barley seeds: a computational approach to study systemic properties of central metabolism. Plant Physiol 149:585–598
97. Hay J, Schwender J (2011) Computational analysis of storage synthesis in developing Brassica napus L. (oilseed rape) embryos: flux variability analysis in relation to 13C metabolic flux analysis. Plant J 67:513–525
98. Salon C, Lepetit M, Gamas P et al (2009) Analysis and modeling of the integrative response of *Medicago truncatula* to nitrogen constraints. C R Biol 332:1022–1033
99. Génard M, Dauzat J, Franck N et al (2008) Carbon allocation in fruit trees: from theory to modelling. Trees 22:269–282
100. Génard M, Bertin N, Borel C et al (2007) Towards a virtual fruit focusing on quality: modelling features and potential uses. J Exp Bot 58:917–928
101. Bertin N, Lecomte A, Brunel B, Fishman S, Génard M (2007) A model describing cell polyploidization in tissues of growing fruit as related to cessation of cell proliferation. J Exp Bot 58:1903–1913
102. Martre P, Bertin N, Salon C, Génard M (2011) Modelling the size and composition of fruit, grain and seed by process-based simulation models. New Phytol 191:601–618

Chapter 2

Metabolic Network Reconstruction and Their Topological Analysis

Marie Beurton-Aimar, Tung Vu-Ngoc Nguyen, and Sophie Colombié

Abstract

This chapter focuses on the way to build a metabolic network and how to analyze its structure. The first part of this chapter describes the methods of the network model reconstruction from biochemical data found in specialized databases and/or literature. The second part deals with metabolic pathway analysis as a useful tool for better understanding the complex architecture of intracellular metabolism. The graph analysis and the stoichiometric network analysis are important approaches for understanding the network topology and consequently the function of metabolic networks. Among the methods presented, the Elementary Flux Modes analysis will be more detailed. Finally, we illustrate in this chapter an example of network reconstruction from heterotrophic plant cells metabolism and its topological analysis leading to a huge number of Elementary Flux Modes.

Key words Metabolic networks, Metabolic pathway analysis, Stoichiometry matrix, Topological analysis, Elementary Flux Modes, Plant cell metabolism

Abbreviations

AccoA	Acetyl coenzymeA
aKG	2-Oxoglutarate
Ala	Alanine
Asp	Aspartate
Cit	Citrate
coA	CoenzymeA
DHAP	Glycerone phosphate
E4P_p	Erythrose 4-phosphate
FA	Fatty acids
Fru	Fructose
F16bP	Fructose-1,6-biphosphate
Fum	Fumarate
F6P	Fructose-6-phosphate

Martine Dieuaide-Noubhani and Ana Paula Alonso (eds.), *Plant Metabolic Flux Analysis: Methods and Protocols*, Methods in Molecular Biology, vol. 1090, DOI 10.1007/978-1-62703-688-7_2, © Springer Science+Business Media New York 2014

GAP	Glyceraldehyde-3-phosphate
Glc	Glucose
Gln	Glutamine
Glu	Glutamate
G1P	Glucose-1-phosphate
G6P	Glucose-6-phosphate
HP	Hexose phosphate
Mal	Malate
OAA	Oxaloacetate
PEP	Phosphoenolpyruvate
PG	Phosphoglycerate
Pyr	Pyruvate
P5P	Pentose-5-phosphate
Ri5P	Ribose-5-phosphate
Ru5P	Ribulose-5-phosphate
Suc	Sucrose
Succ	Succinate
sucP	Sucrose phosphate
S7P	Sedoheptulose-7-phosphate
TP	Triose phosphate
X5P	Xylulose-5-phosphate
"_p"	For metabolites located in plastid
"_vac"	For metabolites located in vacuole

1 Introduction

A challenge in systems biology is to identify the relationship between structure, function and regulation in complex networks that can be reconstructed from genomic or biochemical data. Metabolic network reconstruction and simulation allows for an in depth insight into comprehending the molecular mechanisms. A reconstruction consists in merging metabolic pathways with their respective reactions and enzymes in the perspective of analyzing the entire network assimilated to a metabolic model. Dynamic mathematical modelling of large-scale networks meets difficulties as the mechanistic details and kinetic parameters are rarely available. Conversely, structural (topological) analyses require only reaction stoichiometries and reversibility information, which are often well-known.

This chapter describes two types of methods: network model reconstruction from biochemical data and stoichiometric network analysis as special class of structural (topological) analysis methods. Stoichiometric modelling has emerged in 1990 for microorganisms and has become an important approach in plant science only recently [1]. This approach is used to explore and to understand deeply the function of metabolic networks. The aim of this chapter is to describe advantages and limitations of model reconstruction and stoichiometric analysis with respect to their potential applications for realistic biological network.

2 Reconstruction

A reconstruction involves collecting all of the relevant metabolic information of an organism and then compiling it in a way that makes sense for various types of analyses to be performed. A reconstruction often combines the relevant metabolic and genomic data of an organism. A reconstruction also allows for metabolic comparisons to be performed between various organisms of the same species as well as between different organisms.

The links between the genome and metabolism can be made by searching gene databases, such as KEGG or GeneDB, with a request by enzyme or protein names for particular genes. Nowadays, increasing work tend to generate the entire metabolic network from the available genomic data. These so-called genome-scale models are discussed at the end of the chapter and are developed in detail in Chapter 19 of this volume. Concerning classical reconstruction of one or several pathways, for instance to build the central metabolic network, international databases contain various metabolic pathways such as glycolysis, TCA cycle, pentose phosphate pathway, etc. But at this time, assembling different pathways to design a global network is mainly made by hand and it is a not trivial task. Plant models have been designed using metabolic databases, biochemical textbooks, and the primary literature, and are essentially confined to the well known pathways of central metabolism.

2.1 First Step of a Reconstruction: Finding Resources

Even a lot of genome annotations are available through free international databases, with cross-linked data, plant genomes are still rare. Today, the TAIR (http://www.arabidopsis.org/) Web site has listed about 20 projects in plant genetics (maize, rice, soybean, etc.) At this time, Arabidopsis and maize are fully annotated and recently, the Tomato Genome Consortium (http://solgenomics.net) has published in Nature the last version of their annotation [2].

To give a more detailed description of a few gene/enzyme/reaction/pathway databases that are crucial to a metabolic reconstruction, we can site:

1. *Kyoto Encyclopedia of Genes and Genomes* (KEGG, http://www.kegg.com/) is a bioinformatics database containing information on genes, proteins, reactions, and pathways. The 'KEGG Organisms' section, which is divided into eukaryotes and prokaryotes, encompasses many organisms (but less than 20 plant species) for which gene and DNA information can be searched

by typing in the enzyme of choice. This resource is useful to build associations between metabolism enzymes, reactions, and genes.

2. *BioCyc* (http://biocyc.org/) is a collection of 1,962 pathway/genome databases (version 16.1, 2012), with each database dedicated to one organism. Today about 20 databases of eukaryotes are available. For example, AraCyc is a highly detailed database on the genome and metabolic reconstruction of *Arabidopsis thaliana*, including thorough descriptions of signaling pathways and regulatory networks. The AraCyc database can serve as a paradigm and a model for any reconstruction. Additionally, MetaCyc, an encyclopedia of experimentally defined metabolic pathways and enzymes, contains 1,877 metabolic pathways and 10,247 metabolic reactions (2012).

The BioCyc database can be managed with the help of *Pathway Tools* framework, (http://bioinformatics.ai.sri.com/ptools/) a bioinformatics software package that assists in the construction of pathway/genome databases. Developed by Peter Karp and associates at the SRI International Bioinformatics Group, Pathway Tools comprises several separate units. First, PathoLogic takes an annotated genome for an organism and infers probable metabolic pathways to produce a new pathway/genome database. This can be followed by application of the Pathway Hole Filler, which predicts likely genes to fill "holes" (missing steps) in predicted pathways. Afterward, the Pathway Tools Navigator and Editor functions let users visualize, analyze, access, and update the database. Thus, using PathoLogic and encyclopedias like MetaCyc, an initial fast reconstruction can be obtained automatically, and then using the other units of Pathway Tools, a very detailed manual update, curation, and verification step can be carried out.

Other relevant databases where the request can be led by the protein name or the EC number (representing the catalytic function of the enzyme of interest) have to be mentioned:

3. *ENZYME* (http://enzyme.expasy.org/) is an enzyme nomenclature database (part of the ExPASY proteonomics server of the Swiss Institute of Bioinformatics). After searching for a particular enzyme on the database, this resource gives the reaction that is catalyzed. Additionally, ENZYME has direct links to various other gene/enzyme/medical literature databases such as KEGG, BRENDA, PUBMED, and PUMA2 to name a few.

4. *BRENDA* (http://www.brenda-enzymes.info/) is a comprehensive enzyme database to search an enzyme by name or EC number. Moreover, when an enzyme search is carried out, BRENDA provides a list of all organisms containing the particular enzyme of interest.

One can note that several inconsistencies exist between gene, enzyme, and reaction databases and published literature sources regarding the metabolic information of an organism. A reconstruction is a systematic verification and compilation of data from various sources that takes into accounts all of the discrepancies.

2.2 Next Step of the Reconstruction: Checking Data

After the initial stages of reconstruction, a systematic verification is made in order to track inconsistencies, i.e., that all the entries listed are correct and accurate. Furthermore, previous literature can be researched in order to support any information obtained from one of the many metabolic reaction and genome databases.

To achieve the reconstructed network, the collected pathways and reactions have to be properly connected. For instance pathways have to be connected by transport reactions between subcellular compartments. The presence or absence of certain reactions of the metabolism will affect the amount of reactants/products that are present for other reactions within the particular pathway. Francke et al. [3] provide an excellent example as to why the verification step of the project needs to be performed in significant detail. During a metabolic network reconstruction of *Lactobacillus plantarum*, the model showed that succinyl-CoA was one of the reactants for a reaction that was a part of the biosynthesis of methionine. However, an understanding of the physiology of the organism would have revealed that due to an incomplete tricarboxylic acid pathway, *Lactobacillus plantarum* does not actually produce succinyl-CoA, and the correct reactant for that part of the reaction was acetyl-CoA.

Therefore, systematic verification of the initial reconstruction will bring to light several inconsistencies that can adversely affect the final interpretation of the reconstruction, which is to accurately comprehend the molecular mechanisms of the organism. Then the network analysis suggests finding hypothetical routes between specific inputs and outputs that only emerge in the context of complex network.

2.3 Main Definitions

Before network analysis, we first need to define some terms related to the metabolic networks, composed with biochemical reactions.

The *stoichiometry* specifies the reactants, educts (substrates) and products participating in a reaction as well as the molar ratios in which they are consumed or produced. Stoichiometry is a quantitative relationship between substrates of a chemical reaction. The stoichiometric coefficient of a metabolite by convention is positive if it is produced when the reaction proceeds in the forward reaction, and negative otherwise.

The reaction *directionality* specifies the reversibility or not of the reaction. While a chemical reaction is thermodynamically reversible, in biochemical networks, some reactions are irreversible and proceed only in one direction due to biological constraints.

Knowledge on the reversibility of reactions allows constraining the number of feasible pathways in the network.

Reaction *kinetics* describes the dynamics of the reaction based on the reaction mechanism and the enzyme properties. More often, these characteristics are imprecise, context dependent, or unknown.

3 Analysis

The investigation of the structure of metabolic networks has recently attracted increasing interest. The formal way to deal with networks is to use graph modelling. The next paragraph gives basic knowledge about graph analysis.

3.1 Graph Analysis

Several authors have proposed methods to analyze biological networks as classical graphs in computer science. Graph theory uses the scheme of network connectivities, which is a simplified representation of real reaction networks. The graph theory reaches back to the end of the eighteenth century, to solve the problem of finding the best way through Konigsberg crossing its seven bridges only once (Leonard Euler "Konigsberg bridge problem", 1736, see F. Schreiber for details in [4]). Nowadays, the widespread use of computers to analyze graphs has encouraged developments of algorithms to compute properties of large graphs. Mainly, topological analysis of a graph provides information about relationship between elements belonging to it. It is why topological properties are commonly computing to characterize graph organization. A graph $G = (N, E)$ consists of a set of nodes N and a set of edges E, where each edge is assigned to two nodes (not necessarily disjoint). To manage computation of graph properties, two main coding can be used: adjacency matrix or adjacency list. Figure 1a displays a simple example showing a simplified sucrose pathway, Fig. 1b gives the adjacency matrix built from this drawing. The adjacency list could be extract from this matrix following this rule, each element in the list is a set of two nodes linked by an edge.

3.1.1 Diameter and Degree of Nodes

Topology of graph can be studied with parameters such as the diameter or the degree of nodes. *The diameter* is the maximum distance of any pair of nodes. The average path length is the average distance between all pairs of nodes. The distance between two nodes i and j is the length of the shortest path between these two nodes, i.e., the minimal number of edges that needs to be traversed to travel from i to j. More details about equations to compute path length can be found in Steuer and Lopez [4]. It has been shown that, whatever the organism, large networks such as metabolic networks have a short average path length (around three reactions) between two metabolites in the network [5–7]. It is worth to note that networks like metabolic networks generally exhibit average

a

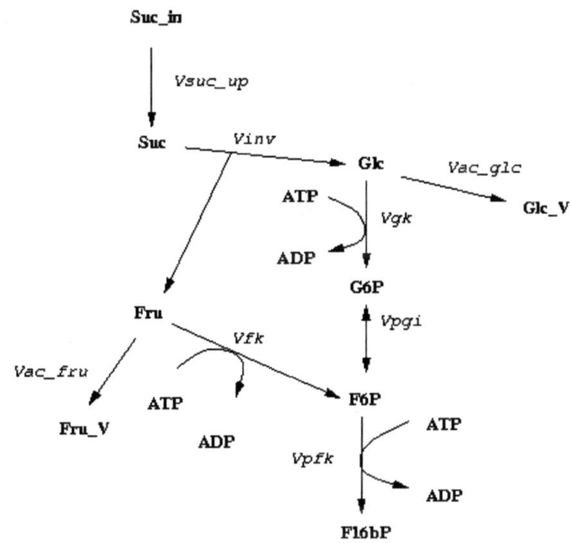

b

M \ R	*Vsuc_up*	*Vinv*	*Vgk*	*Vpgi*	*Vfk*	*Vpfk*	*Vac_gk*	*Vac_fru*
Suc_in	1	0	0	0	0	0	0	0
Suc	0	1	0	0	0	0	0	0
Gk	0	1	1	0	0	0	0	0
G6P	0	0	1	1	0	0	0	0
F6P	0	0	0	1	1	1	0	0
Fru	0	1	0	0	1	0	0	1
ATP	0	0	1	1	1	1	0	0
ADP	0	0	1	1	1	1	0	0
Fru_V	0	0	0	0	0	0	0	1
Gk_V	0	0	0	0	0	0	1	0
F16bP	0	0	0	0	0	1	0	0

Fig. 1 (**a**) Metabolic network of a simplified sucrose pathway. (**b**) The adjacency matrix built from this drawing

path length. Several authors, mainly coming from social sciences, have described properties of specific networks called small-world networks. To resume their description, in a small world network the average path length decreases slower or equal than the logarithm of the network size [4]. Metabolic networks can be considered as belonging to this category [7]. For example Fell et al. [5] have shown that a metabolic network of the TCA cycle posses such characteristics. More recently, Barabasi and Oltvai [8] have also studied the possibility to characterize biological networks as small world networks. They consider the degree distribution of nodes which is more heterogeneous than those of a random network and the clustering coefficient to measure the cliquishness of the local neighborhoods. The average degree of nodes and the clustering coefficient are the two another parameters to analyze characterizing network topology. The degree of a node belonging to the graph G is equal to the number of adjacent edges to this node. The clustering coefficient measures the probability that two vertices with a common neighbor are connected. The clustering coefficient can be computed as following:

$$C_i = \frac{2E_i}{E_{\max}} \tag{1}$$

with

$$E_{\max} = k_i(k_i - 1) \tag{2}$$

where C_i is the clustering coefficient of the node n_i, and E_i is the number of edges; $E_{\max}$ is the maximum expected number of edges between the k_i neighbors of n_i. Then these measures are compared to those of a random network generated with the same number of nodes. Finding differences for both parameters, the average degree of nodes and the clustering coefficient between the random network and the studied network, signifies that a specific organization can be found in the studied network, led for example by functionality constraints.

3.1.2 Digraph Modelling

One complex characteristic of metabolic networks is that they are hypergraphs. Often enzymatic reactions imply more than one substrate and produce more than one metabolite (see the description of the simple sucrose pathway Fig. 1a in the previous section). A way to solve that is to convert the graph in a digraph that is a graph with two types of nodes. Petri net is a well known format of digraph. The graph is defined as a set of nodes called "places" for the molecules and a set of nodes called "transitions" for the reactions. Molecules are considered as discrete token and applying Gillepsie formula [9] the kinetics can be taken into account as probability to fire reactions. Srivastava et al. [10] gives an example of simulation with this tool. But even if the Petri net modelling provides an efficient simulation of the network behaviors, the proof that all behaviors have been explored has to be brought. Moreover, numerical values like reaction kinetics useful to validate the simulation are often missing.

3.2 Metabolic Pathway Analysis

Analysis of metabolic pathway has become increasingly important to assess inherent network properties in reconstructed biochemical reaction networks [11]. From the last 10 years, the amount of available data has increased drastically. With the development of recent systems biology approaches, the number of reactions taking into account for the design of one network has moved from tens to one or several hundreds. In this case, it is not possible to identify by hands all the interactions through this kind of networks. Using matrix formalism, Seressiotis and Bailey [12] have suggested identifying pathways through the network using linear algebra algorithms. Before explaining more deeply different computational methods, the next paragraph describes the specific matrix which is the stoichiometry one, using reaction nodes and metabolite nodes.

3.2.1 The Stoichiometry Matrix

The stoichiometry of a complete system can be summarized in the matrix form, the so-called *stoichiometry matrix* denoted by the symbol N, only refers to the m internal metabolites of the network composed of n reactions. By convention, the stoichiometry matrix will have m rows and n columns. The stoichiometric matrix marks all interactions between metabolites and reactions:

- If a reaction i involves the metabolite j, the number of required molecules to process is given at the jth rows and the ith column.
- Else the value is zero.

Figure 2 gives the stoichiometry matrix for the given example in Fig. 1a. As previously mentioned a stoichiometric matrix takes into account only the internal metabolites. Following the metabolic assumptions, metabolic network can be considered at the steady state that is all the *internal metabolites* have to be balanced in a valid pathway. At the opposite, *external* metabolites are considered as always available and so unnecessary to be mentioned in the model. Designers have to pay attention to the decision to declare a metabolite as internal or external determines the frontiers of the analyzed network. Decision to balance or not a metabolite can drastically change the results of analysis.

To compute valid pathways, the stoichiometry matrix N is combined with the rate vector v to form a compact equation describing the rates of change of the molecular species S:

$$\frac{dS}{dt} = N \cdot v \tag{3}$$

M\R	Vsuc_up	Vinv	Vgk	Vpgi	Vfk	Vpfk	Vac_fru	Vac_glc
Suc	1	−1	0	0	0	0	0	0
Glc	0	1	−1	0	0	0	0	−1
Fru	0	1	0	0	−1	0	−1	0
G6P	0	0	1	−1	0	0	0	0
F6P	0	0	0	1	1	−1	0	0

Fig. 2 The stoichiometry matrix of the network displayed in Fig. 1

3.2.2 Steady-State Assumption

A steady state is a situation in which all state variables are constant in spite of ongoing processes that strive to change them. For an entire system to be at steady state, i.e., for all state variables of a system to be constant, there must be a flow through the system. Several constraints have to be respected, there must be no accumulation of mass over the time period of interest and the same mass flow rate will remain constant in the flow path through each element of the system. Thermodynamic properties may vary from point to point, but will remain unchanged at any given point. Metabolism usually involves fast reactions and high turnover of metabolites compared to regulatory events. Therefore analysis of metabolic networks is often based on the assumption that metabolite concentrations and reactions rates are constant. This assumption leads to the fundamental mass balancing equation [13].

$$\frac{\mathrm{d}S}{\mathrm{d}t} = N \cdot v = 0 \tag{4}$$

By applying Metabolic Flux Analysis (MFA), one tries to shrink the possible solution space of Eq. 4 by measuring some of the reactions rates, such as uptakes or excretion rates, in steady state. Ideally, one unique solution remains for the actual flux distribution. In general MFA is useful to analyze specific flux distributions but it is not able to characterize the complete admissible steady state solution space.

With the metabolite balancing equation, resulting from steady state assumption, the space of all possible flux distributions in a reaction network is studied in a constraint-based modelling approach.

In order to deduce what the metabolic network suggests, recent research has centered on approaches such as Extreme Pathways (EPs), Elementary Flux Mode (EFM) analysis, Flux Balance Analysis (FBA) and a number of other constraint-based modelling methods. Pathway analysis deals with the discovery and analysis of feasible routes in metabolic networks.

3.3 Finding Feasible Pathways Through Network

EFMs and EPs closely match to the same approach; they are the smallest sub-networks that allow a metabolic reconstruction network to function at steady state [14, 15]. An EFM is the minimal biochemical pathway that, at steady state, catalyzes a set of net reactions between input(s) and output(s). EFMs characterize the complete space of admissible steady-state flux distributions by particular flux vectors. The EFMs method provides a rigorous formalism for describing and assessing metabolic pathways. According to Stelling et al. [15] EFMs can be used to understand cellular objectives for the overall metabolic network. Furthermore, EFMs analysis takes into account stoichiometries and thermodynamics when evaluating whether a particular metabolic route or network is feasible and likely for a set of proteins/enzymes [14].

Its ability to assess the functional and structural properties of metabolic networks means that EFM analysis is suitable for both biotechnology and physiology. EFMs can be used to recognize operational modes and cycles. EFM analysis has already been used for microbiological systems in metabolic engineering [16]. In plants, EFM analysis appears to be a useful theoretical tool for structural modelling. It has been successfully applied to biological systems characterized by a high redundancy level without any key (i.e., rate limiting) enzyme, e.g., the Calvin cycle [17], sucrose metabolism [18], and starch metabolism [19]. The main difference between EPs and EFMs is the way to code reversibility of reactions. The EPs method requires to double reversible reaction in two irreversible ones, one backward and one forward. Actually, each EP can be mapped onto a corresponding EFM while the inverse is not true. Moreover, if a reaction is deleted the subset of EFMs not involving this reaction is the complete set of EFMs in the reduced network [14]. In contrast the set of EPs needs to be recalculated whenever a partial reaction is removed or the direction modified. Considering the small network in Fig. 1a, four Elementary Flux Modes are computed and displayed in Fig. 3, with bold arrays.

Sucrose is the only one entry of the network, then the uptake of sucrose and its cleavage in glucose and fructose (corresponding to the reactions *Vsuc_up* and *Vinv*) are present in the four EFMs. Then these sugars can be stored in vacuole (reactions *Vac_fru* and *Vac_glc*, Fig. 3a) or both of them metabolized in fructose-1-6-bisphosphate (Fig. 3d). The EFMs in Fig. 3b, c are a mix of these two situations. Note that the irreversibility of the kinase reactions (*Vfk* and *Vgk*) constraints the network and prevent the reversible reaction *Vpgi* to be active in the forward the direction. But in real biological systems, two other reactions (catalyzed by the sucrose synthase and the sucrose phosphate synthase) are present and lead to a cycle of sucrose in this pathway.

3.4 Other Metabolic Analysis

3.4.1 Minimal Cut Sets

EFMs computation provides the set of feasible pathways of functional fluxes. Conversely, Minimal Cut sets (MCs) method allows computing a dual set, i.e., the set of reactions which disconnects the network and so disable the fluxes. Klamt and Gilles [20] and Ballerstein et al. [21] proposed an algorithm to compute the cut sets respecting the constraints of metabolic networks design: reaction reversibility/irreversibility and steady state assumption. Computing the MCs for the simple example described in Fig. 1a leads to this list of six MCs: 1: VSuc_up, 2: Vinv, 3: Vgk, Vac_glc, 4: Vpfk, Vac_glc, 5: Vfk, Vac_fru and 6: Vpfk, Vac_fru. In this simple network, the number of MCs is higher than that of EFMs. But this method takes advantage in the case of large networks. In this case, the number of MCs is expected to be smaller than the number of EFMs. However, at this time just a few networks have been explored with this approach.

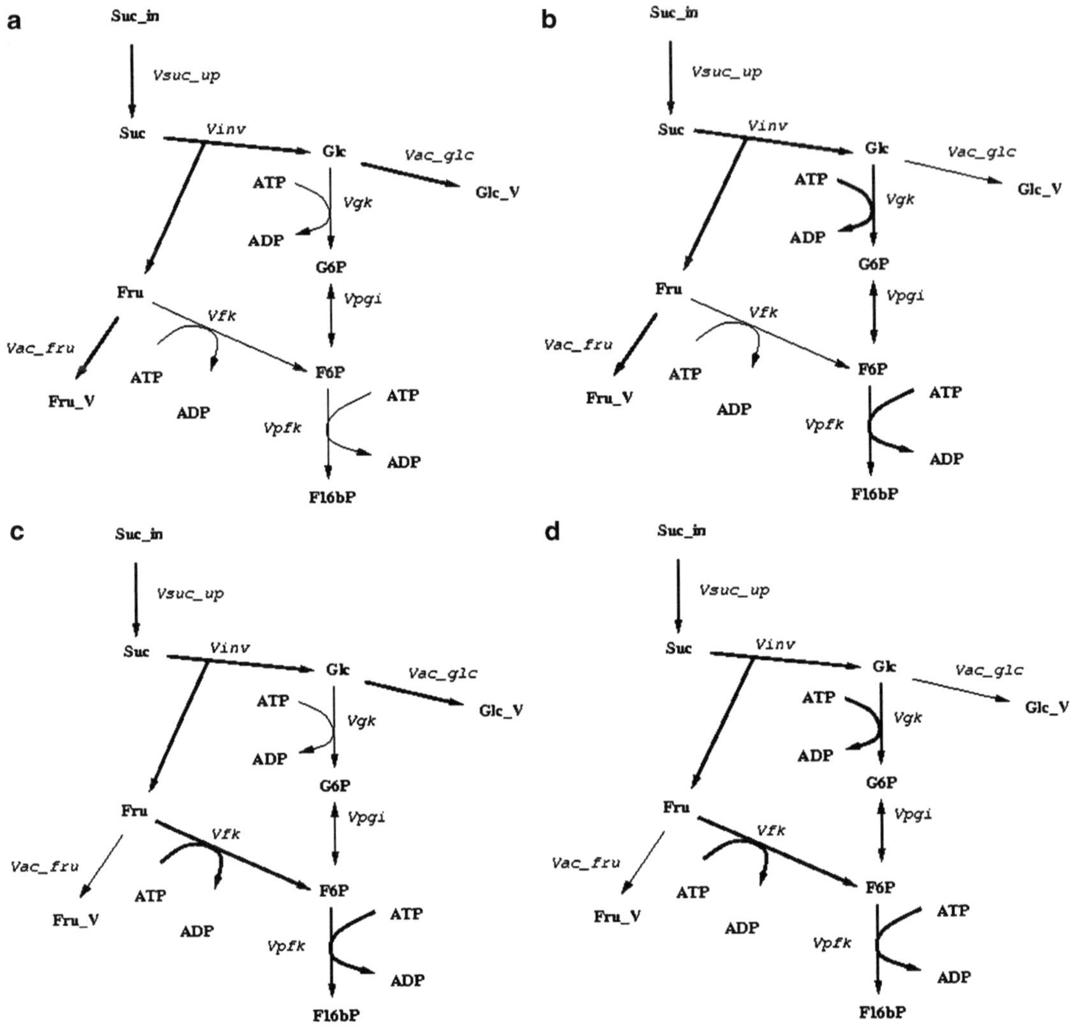

Fig. 3 The four Elementary Flux Modes of the network displayed in Fig. 1

3.4.2 Flux Balance Analysis

Another way to analyze metabolic networks is to simulate fluxes by Flux Balance Analysis (FBA). Linear programming is usually used to obtain the maximum potential of the objective function that you are looking at, and therefore, when using flux balance analysis, a single solution is found to the optimization problem [15]. In a flux balance analysis approach, exchange fluxes are assigned to those metabolites that enter or leave the particular network only. Those metabolites that are consumed within the network are not assigned any exchange flux value. Also, the exchange fluxes along with the enzymes can have constraints ranging from a negative to positive value (e.g., −10 to 10).

FBA can highlight the most effective and efficient pathway through the network in order to achieve a particular objective function. In addition, gene knockout studies can be performed using flux balance analysis. The enzyme that correlates to the gene that needs to be removed is given a constraint value of 0. Then, the reaction that the particular enzyme catalyzes is completely removed from the analysis. Flux balance modelling which the main advantage is that it is relatively easy to scale up to cover very large networks can also be applied to study genome-scale metabolic models. FBA is explored deeply in Chapter 17 of this book.

3.4.3 Kinetic Modelling

In order to perform a dynamic simulation with such a network it is necessary to construct an ordinary differential equation system that describes the rates of change in each metabolite's concentration or amount. To this end, a rate law, i.e., a kinetic equation is required for each reaction. Often these rate laws contain kinetic parameters with uncertain values. In many cases it is desired to estimate these parameter values with respect to given time-series data of metabolite concentrations. The system is then supposed to reproduce the given data. For this purpose the distance between the given data set and the result of the simulation, i.e., the numerically or in few cases analytically obtained solution of the differential equation system is computed. The values of the parameters are then estimated to minimize this distance. One step further, it may be desired to estimate the mathematical structure of the differential equation system because the real rate laws are not known for the reactions within the system under study. This part is detailed in Chapter 16 of this book.

4 Example: Heterotrophic Plant Cells

Finally, we illustrate in this chapter an example of network reconstruction from heterotrophic plant cells metabolism and its topological analysis leading to a huge number of EFMs.

4.1 Description of the Heterotrophic Plant Cell Network

Let us first look at the metabolic network of heterotrophic plant cells described in Fig. 4 and appendix. The reconstruction of this metabolic network is detailed in our previous paper [22]. The network includes all the main pathways of the central carbon metabolism in plants: glycolysis (black), the TCA cycle (blue), the pentose phosphate pathway (pink), the starch and sucrose pathways (green) and the storage reactions towards the vacuole (brown) described in AraCyc and in relevant papers [23–25]. Irreversible reactions (see Appendix) are indicated by unidirectional arrows in Fig. 4. The glycolytic pathways are illustrated with reversible glycolysis in the cytosol and with irreversible glycolysis in plastids because amyloplasts lack fructose-1,6-bisphosphatase [26, 27]. Due to its autotrophic nature, the plant synthesizes its own respiratory substrates (mainly carbohydrates) which then serve as

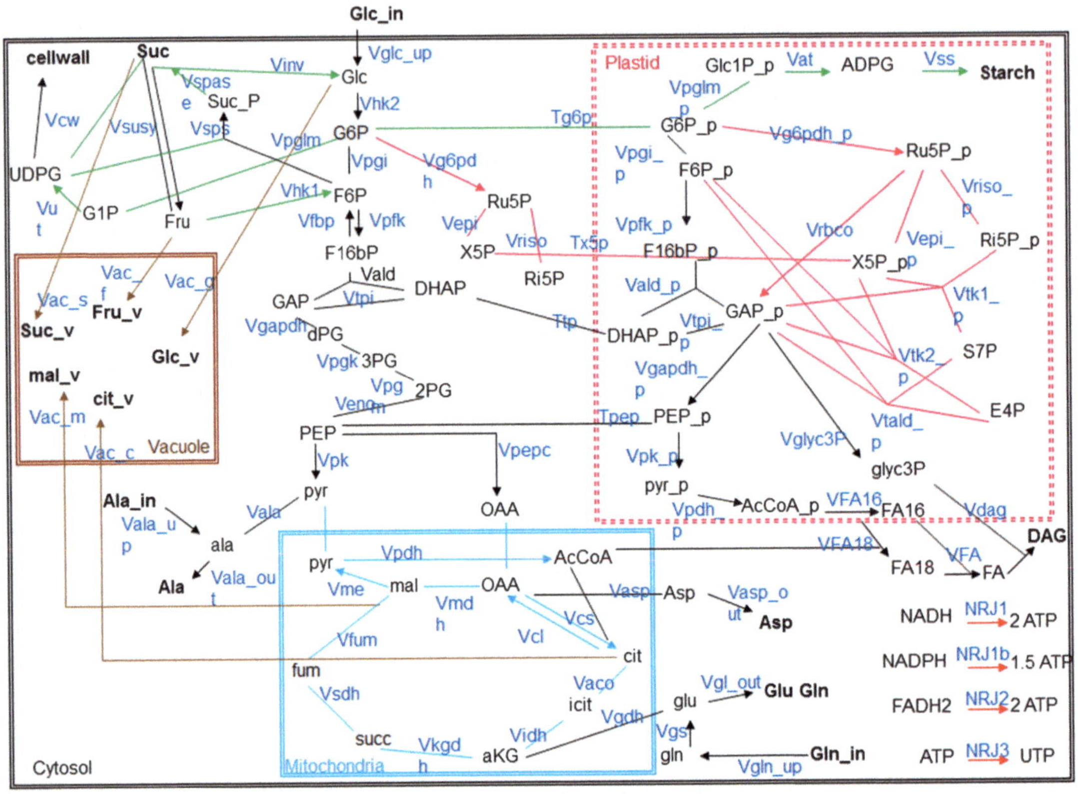

Fig. 4 Metabolic network of heterotrophic plant cells. From Beurton-Aimar et al. (2012) [22]. Freely available for redistribution

substrates for the TCA cycle. The TCA cycle provides precursors for several biosynthetic processes, such as nitrogen fixation and biosynthesis of amino acids [28]. The pentose phosphate pathway includes the irreversible oxidative branch, whereas the non-oxidative branch is reversible (recycling of pentose-phosphates from fructose phosphate and triose-phosphate). The enzymes of the oxidative branch, which also exist in cytosol, could lead to the synthesis of pentose phosphates in both cytosol and plastids [29, 30]. In the starch and sucrose pathways, sucrose is metabolized in cytosol, whereas starch is metabolized in plastids from imported hexose phosphates (G1P or G6P). Several effluxes are illustrated in Fig. 4: protein synthesis from several amino acids (glutamate and glutamine, aspartate and alanine), lipid synthesis (diacyl glycerol) from plastidial pyruvate and trioses, synthesis of cell wall polysaccharides from UDP-glucose, sugars (glucose, fructose and sucrose) and storage of organic acids (malate and citrate) in vacuoles. The energy reactions (NRJ, red arrows in Fig. 4) are essential to balance the cofactors of the system. Subcellular compartments, such as mitochondria and plastids, can lead to potentially reversible transport of metabolites such as G6P, X5P, PEP, and DHAP [30–33].

There are 70 different metabolites in this network at steady state. Among them, 15 metabolites are external, meaning that they are carbon sources or carbon sinks (nutrients, waste products, stored and excreted products, and precursors for further transformation). These are exogenous glucose and amino acids (glutamine and alanine), CO_2, sugars (sucrose, glucose and fructose) and organic acids (citrate and malate) stored in vacuoles, amino acids for protein synthesis (aspartate, alanine, glutamate and glutamine), cell wall polysaccharides, starch and lipids (cf. Appendix). Assuming that their concentration cannot affect the system, the other small molecules (e.g., oxygen, ammonium, phosphate, pyrophosphate and water) are not included in the metabolic network. The other 55 metabolites, including cofactors (ATP, NADH, NADPH, and $FADH_2$) are internal, which means that they are expected to be balanceable at steady state. For the sake of simplicity, oxidized equivalents (NAD, NADP, and FAD) and ADP are omitted in the reaction equations. As mentioned above, the network contains 78 reactions. The direction (reversibility or irreversibility) of the 78 reactions is derived from thermodynamic properties, 33 reactions are reversible.

4.2 Consistency and Properties of the Network

This step of analysis ensures that all the reactions present in the reconstruction are properly balanced. In order to validate the reconstruction, it is important to check the network consistency. That means to detect the blocked reactions or the missing elements which can compromise the use of the network model. The network consistency can be detected by analytical methods. Tools like CellNetAnalyzer (CNA, Matlab, http://www.mpi-magdeburg.mpg.de/projects/cna/cna.html) allow checking the network consistency. Similarly, the important reactions (the model is modified if these reactions are removed) or the correlated reactions (reactions operating together) can be detected first in the network, with analytical methods.

From the tools available for EFM computation, the original METATOOL software (module 5.0.3 beta in CellNetAnalyzer), a program with a graphical user interface for the analysis of metabolic networks based on Matlab (Mathworks, Inc.), was used for computation [34]. The result of the computation on the metabolic network of heterotrophic plant cells described in Fig. 4 is a set of 114,614 EFMs. Because each EFM represents a unique metabolic route, the 114,614 EFMs indicate a large number of metabolic routes that are potentially active at metabolic steady state. Thus, a robust network of high redundancy is obtained. Note that even though the number of EFMs is already quite big, Klamt and Stelling [35] showed that this number is actually smaller than the strict combinatorial computation of the metabolic reactions. Moreover, after MCS computing, 93,009 Minimal Cut Sets have been obtained providing a smaller set of data than the set of EFMs.

EFM analysis is very helpful to distinguish whether the EFM is producing or consuming ATP. Thus, it allows us to visualize futile cycles, as well as catabolic and anabolic EFMs. The analysis revealed that futile cycles exist between sucrose and hexose, as well as between hexose phosphates and triose phosphates. METATOOL software provides a list of external metabolites for each EFM. Although each EFM is unique, several EFMs lead to the same overall metabolite. This means that a specific external metabolite can be obtained from several different EFMs. This multiplicity is an indicator of the robustness of the metabolic network.

We used the results of the EFM analysis of the metabolic network of heterotrophic plant cells (about 80 reactions and 70 metabolites from the main pathways of the central carbon metabolism) in comparison with the experimental fluxes determined in *Brassica napus* embryos using 13C labelling measured with a glucose and organic/inorganic nitrogen source [36], by introducing the calculated coefficients of flux efficiency [22]. The main experimental results were cross-validated our EFM analysis so this work has shown that 13C-MFA experimental results can be qualitatively validated by the EFM analysis.

5 Conclusion

To sum up, a reconstruction that is fully accurate can lead to greater insight about understanding the functioning of the organism of interest. A reconstruction model serves as a first step to deciphering the complicated mechanisms. Metabolic network reconstruction and analysis can be used to understand how a cell functioning. Pathway analysis suggests finding hypothetical routes between specific inputs and outputs that only emerge in the context of complex network, and Elementary Flux Modes allow identifying all feasible routes in the network.

In conjunction with "classical" biochemical knowledge, the huge amount of data produced by the sequencing of complete genomes can be used for the reconstruction of metabolic networks. Metabolic models can be reconstructed at a genome-scale using all the reactions catalyzed by the enzymes encoded in an annotated genome (see Chapter 19). Henceforth, by sequence comparison methods and genome context methods, for greater than half of the newly sequenced genes, the physiological functions can be predicted. With such information available, pathway analysis (mainly FBA) can be performed at the scale of whole-cell models. However this remains a nontrivial task: and only few papers deal with genome_scale models of plant, with genome scale models [1, 37–39]. Several problems arise in the construction of metabolic models from genome annotation databases, including network gaps caused by incomplete or imprecise genome annotation, mass-balance errors caused by reaction stoichiometry errors in the annotation database, or the presence of excess, non-functional reactions. Another difficulty is that genome annotation databases contain no information about reaction directionality.

Appendix

List of reactions corresponding to the network in Fig. 4:

Glc_up : Glc_in => Glc .
ala_up : ala_in => ala .
gln_up : gln_in => gln .
Vpgi : G6P <=> F6P .
Vhk1 : ATP + Fru => F6P .
Vhk2 : ATP + Glc => G6P .
Vpfk : ATP + F6P => F16bP .
Vfbp : F16bP => F6P .
Vald : F16bP <=> DHAP + GAP .
Vtpi : DHAP <=> GAP .
Vgapdh : GAP <=> dPG + NADH .
Vpgk : dPG <=> 3PG + ATP .
Vpgm : 3PG <=> 2PG .
Veno : 2PG <=> PEP .
Vpk : PEP => ATP + pyr .
Vpepc : CO2 + PEP => OAA .
Vpdh : pyr => CO2 + AccoA + NADH .
Vcl : cit + ATP => OAA + AccoA .
Vcs : OAA + AccoA => cit .
Vaco : cit <=> icit .
Vidh : icit <=> aKG + CO2 + NADH .
Vkgdh : aKG <=> CO2 + NADH + succ .
Vsdh : succ <=> FADH2 + fum .
Vfum : fum <=> mal .
Vmdh : mal <=> NADH + OAA .
Vme : mal => CO2 + NADH + pyr .
Vg6pdh : G6P => CO2 + Ru5P + NADPH .
Vepi : Ru5P <=> X5P .
Vriso : Ru5P <=> Ri5P .
Vpgi_p : G6P_p <=> F6P_p .
Vg6pdh_p : G6P_p => CO2 + Ru5P_p + NADPH .
Vepi_p : Ru5P_p <=> X5P_p .
Vriso_p : Ru5P_p <=> Ri5P_p .
Vtk1_p : S7P_p + DHAP_p <=> Ri5P_p + X5P_p .
Vtk2_p : F6P_p + DHAP_p <=> E4P_p + X5P_p .
Vtald_p : S7P_p + DHAP_p <=> E4P_p + F6P_p .
Vpfk_p : ATP + F6P_p => F16bP_p .
Vald_p : F16bP_p <=> GAP_p + DHAP_p .

Vtpi_p : GAP_p <=> DHAP_p .
Vgapdh_p : DHAP_p => ATP + NADH + PEP_p .
Vpk_p : PEP_p => ATP + pyr_p .
Vrbco : Ru5P_p + CO2 => 2 DHAP_p .
Vpdh_p : pyr_p => CO2 + AccoA_p + NADH .
Tg6p : G6P <=> G6P_p .
Ttp : DHAP <=> DHAP_p .
Tpep : PEP <=> PEP_p .
Tx5p : X5P <=> X5P_p .
Vpglm_p : G1P_p <=> G6P_p .
Vat : G1P_p + ATP => ADPG .
Vss : ADPG => starch .
Vpglm : G1P <=> G6P .
Vut : G1P + UTP => UDPG .
Vsusy : Fru + UDPG <=> Suc .
Vinv : Suc => Fru + Glc .
Vsps : F6P + UDPG => SucP .
Vspase : SucP => Suc .
Vala : pyr + glu <=> ala + aKG .
Vasp : OAA + glu <=> asp + aKG .
Vgdh : aKG + NADH <=> glu .
Vgs : gln + aKG + NADPH => 2 glu .
Vgl_out : glu => gl_out .
Vasp_out : asp => asp_out .
Vala_out : ala => ala_out .
Vcw : UDPG => CellWall .
VFA16 : 8 AccoA_p + 7 ATP + 14 NADPH => FA16 .
VFA18 : AccoA + FA16 + ATP + NADPH => FA18 .
VFA : FA16 + 2 FA18 => 3 FA .
Vglyc3P : DHAP_p + NADH + ATP => glyc3P .
Vdag : glyc3P + ATP + 2 FA => DAG .
Vac_g : Glc => Glc_v .
Vac_f : Fru => Fru_v .
Vac_s : Suc => Suc_v .
Vac_m : mal => mal_v .
Vac_c : cit => cit_v .
NRJ1 : NADH => 2 ATP .
NRJ1b : 2 NADPH => 3 ATP .
NRJ2 : FADH2 => 2 ATP .
NRJ3 : ATP => UTP .

References

1. Poolman M, Miguet L, Sweetlove LJ, Fell D (2009) A genome-scale metabolic model of Arabidopsis and some of its properties. Plant Physiol 151:1570–1781
2. The Tomato Genome Consortium (2012) The tomato genome sequence provides insights into fleshy fruit evolution. Nature 485: 635–641

3. Francke C, Siezen RJ, Teusink B (2005) Reconstructing the metabolic network of a bacterium from its genome. Trends Microbiol 13(11):550–558
4. Junker BH, Schreiber F (eds) (2008) Analysis of biological networks. Global network properties. Wiley, Hoboken, pp 31–59
5. Fell DA, Wagner A (2000) The small-world of metabolism. Nat Biotechnol 18:1121–1122
6. Jeong H, Tombor B, Albert R, Oltvai ZN, Barabasi AL (2000) The large-scale organization of metabolic network. Nature 407: 651–654
7. Wagner A, Fell DA (2001) The small-world inside large metabolic networks. Proc Roy Soc Lond B 268:1803–1810
8. Barabasi AL, Oltvai ZN (2004) Network biology. Nat Rev Genet 5:101–113
9. Gillepsie DT (1977) Exact stochastic simulation of coupled chemical reactions. J Phys Chem 81(25):2340–2361
10. Srivatasta R, Peterson MS, Bentley WE (2001) Stochastic kinetic analysis of the *Escherichia coli* stress circuit using sigma(32)-targeted antisense. Biotechnol Bioeng 75 (1):120–129
11. Trinh CT, Wlaschin A, Srienc F (2009) Elementary mode analysis: a useful metabolic pathway analysis tool for characterizing cellular metabolism. Appl Microbiol Biotechnol 81:813–826
12. Seressiotis A, Bailey JE (1988) MPS: an artificially intelligent software system for the analysis and synthesis of metabolic pathways. Biotechnol Bioeng 31:587–602
13. Heinrich R, Schuster S (1996) The regulation of cellular systems. Springer, New York
14. Schuster S, Fell DA, Dandekar T (2000) A general definition of metabolic pathways useful for systematic organization and analysis of complex metabolic networks. Nat Biotechnol 18(3):326–32
15. Stelling J, Klamt S, Bettenbrock K, Schuster S, Gilles ED (2002) Metabolic network structure determines key aspects of functionality and regulation. Nature 420:190–193
16. Schuster S, Dandekar T, Fell DA (1999) Detection of elementary flux modes in biochemical networks: a promising tool for pathway analysis and metabolic engineering. Trends Biotechnol 17:53–60
17. Poolman MG, Fell DA, Raines CA (2003) Elementary modes analysis of photosynthate metabolism in the chloroplast stroma. Eur J Biochem 270:430–439
18. Rohwer JM, Botha FC (2001) Analysis of sucrose accumulation in the sugar cane culm on the basis of in vitro kinetic data. Biochem J 358:437–445
19. Poolman MG, Assmus HE, Fell DA (2004) Applications of metabolic modelling to plant metabolism. J Exp Bot 55:1177–1186
20. Klamt S, Gilles ED (2004) Minimal cut sets in biochemical reaction networks. Bioinformatics 20(2):226–234
21. Ballerstein K, von Kamp A, Klamt S, Haus U (2012) Minimal cut sets in a metabolic network are elementary modes in a dual network. Bioinformatics 28:381–387
22. Beurton-Aimar M, Beauvoit B, Monier A et al (2011) Comparison between elementary flux modes analysis and 13C-metabolic fluxes measured in bacterial and plant cells. BMC Syst Biol 5:95
23. Dieuaide-Noubhani M, Raffard G, Canioni P et al (1995) Quantification of compartmented metabolic fluxes in maize root-tips using isotope distribution from C13 labeled or C14 labeled glucose. J Biol Chem 270: 13147–13159
24. Rontein D, Dieuaide-Noubhani M, Dufourc EJ et al (2002) The metabolic architecture of plant cells. Stability of central metabolism and flexibility of anabolic pathways during the growth cycle of tomato cells. J Biol Chem 277:43948–43960
25. Schwender J, Ohlrogge JB, Shachar-Hill Y (2003) A flux model of glycolysis and the oxidative pentosephosphate pathway in developing Brassica napus embryos. J Biol Chem 278:29442–29453
26. Entwistle G, ap Rees T (1988) Enzymic capacities of amyloplasts from wheat (Triticum aestivum) endosperm. Biochem J 255:391–396
27. Borchert S, Harborth J, Schunemann D, Hoferichter P, Heldt HW (1993) Studies of the enzymic capacities and transport properties of pea root plastids. Plant Physiol 101: 303–312
28. Hill SA, ap Rees T (1994) Fluxes of carbohydrate metabolism in ripening bananas. Planta 192:52–60
29. Debnam PM, Emes MJ (1999) Subcellular distribution of enzymes of the oxidative pentose phosphate pathway in root and leaf tissues. J Exp Bot 50:1653–1661
30. Eicks M, Maurino V, Knappe S, Flügge UI, Fischer K (2002) The plastidic pentose phosphate translocator represents a link between the cytosolic and the plastidic pentose phosphate pathways in plants. Plant Physiol 128: 512–522
31. Kammerer B, Fischer K, Hilpert B et al (1998) Molecular characterization of a carbon

transporter in plastids from heterotrophic tissues: the glucose 6-phosphate/phosphate antiporter. Plant Cell 10:105–118

32. Fischer B, Kammerer M, Gutensohn B et al (1997) A new class of plastidic phosphate translocators: a putative link between primary and secondary metabolism by the phosphoenolpyruvate/phosphate antiporter. Plant Cell 9:453–462
33. Kang F, Rawsthorne S (1994) Starch and fatty acid synthesis in plastids from developing embryos of oilseed rape (*Brassica napus* L.). Plant J 6:795–805
34. Pfeiffer T, Sanchez-Valdenebro I, Nuno JC, Montero F, Schuster S (1999) METATOOL: for studying metabolic networks. Bioinformatics 15:251–257
35. Klamt S, Stelling J (2002) Combinatorial complexity of pathway analysis in metabolic networks. Mol Biol Rep 29:233–236
36. Junker BH, Lonien J, Heady LE, Rogers A, Schwender J (2007) Parallel determination of enzyme activities and in vivo fluxes in Brassica napus embryos grown on organic or inorganic nitrogen source. Phytochemistry 68: 2232–2242
37. de Oliveira Dal'Molin CG, Quek LE, Palfreyman RW, Brumbley SM, Nielsen LK (2010) AraGEM, a genome-scale reconstruction of the primary metabolic network in Arabidopsis. Plant Physiol 152:579–589
38. Radrich K, Tsuruoka Y, Dobson P, Gevorgyan A, Swainston N, Schwartz JM (2010) Reconstruction of an in silico metabolic model of Arabidopsis thaliana through database integration. BMC Syst Biol 4:114
39. Saha R, Suthers P, Maranas C (2011) Zea mays irs1563: a comprehensive genome-scale metabolic reconstruction of maize metabolism. PLoS ONE 6(7):e21784

Part I

Experimental Design and Acquistion of Labeling Data (Experimental Design and Data Acquisition for Isotopic Labeling Based Metabolic Flux Analysis)

Chapter 3

^{14}C Pulse Labeling to Estimate External Fluxes and Turnovers in Primary Metabolism

Katia Pianelli, Antoine Monier, Marie-Hélène Andrieu, Bertrand Beauvoit, and Martine Dieuaide-Noubhani

Abstract

Steady state ^{13}C-MFA is classically used to measure fluxes in complex metabolic networks. However, the modeling of steady state labeling allows the quantification of internal fluxes only and requires the estimation, by other methods, of the external fluxes, corresponding to substrate uptake (carbon input into the network) and to the production rate of compounds that accumulate within plant cells (network output). Additionally, it is not always possible to discriminate between different pathways that lead to the same label distribution. Methods to measure fluxes, based on direct measurements of pool size and on ^{14}C short-time labeling experiments, are described in this chapter. To illustrate this approach, we focus on the quantification of sucrose and starch turnovers.

Key words ^{14}C labeling, Substrate cycle, Sucrose turnover, Pulse labeling

1 Introduction

Plant heterotrophic cells depend, for their growth, on the import of carbon, mainly as sucrose provided by leaves. Sucrose is then converted to hexose-phosphates, a major metabolic crossroad toward the synthesis of structural (cell wall) or storage (starch) polysaccharides and to glycolysis and TCA cycle to produce ATP (respiration) and intermediates for biosyntheses (amino acids for proteins, fatty acids for lipids, etc.). If the reactions involved in the primary metabolism are quite well known, their regulation remains badly understood due to the lack of knowledge concerning their in vivo activities. Since the 1980s, methods to quantify fluxes in plant metabolic networks have been developed [1]. Today, steady state ^{13}C-MFA is probably the main method used. For instance, fluxes in primary metabolism of maize root tips [2, 3], tomato cells [4], Arabidopsis cells [5, 6], embryos [7–9], and seeds [10, 11] were successfully quantified by labeling cells with ^{13}C glucose (Glc).

Martine Dieuaide-Noubhani and Ana Paula Alonso (eds.), *Plant Metabolic Flux Analysis: Methods and Protocols*, Methods in Molecular Biology, vol. 1090, DOI 10.1007/978-1-62703-688-7_3, © Springer Science+Business Media New York 2014

With the assumptions of metabolic and isotopic steady state, the differential equations that describe the metabolic network can be simplified to algebraic equations, thus simplifying their resolution and reducing considerably the information required. However, steady state ^{13}C-MFA by itself does not allow to quantify all fluxes in a metabolic network, but intermediary fluxes only. Input and output fluxes (considered as external fluxes in the model) have to be determined independently. It concerns the flux of carbon uptake, often corresponding to the rate of sugar or amino acid import, and all the fluxes toward compounds that accumulate within cells, such as starch or cell wall polysaccharides. The other limitation of steady state modeling is that it is not possible to discriminate between pathways that produce similar dilution or rearrangement. For instance, when [1-^{13}C]-Glc is used as substrate, a label transfer from the C1 to the carbon C6 is often observed when regarding the internal glucose. This transfer is mainly due to the reversibility of the glycolysis that allows the resynthesis of hexose-P from triose-P [2, 12–14]. However, several routes can be considered to transfer the labeling from hexose-P to glucose: a direct dephosphorylation of hexose-phosphates via an hexose-phosphatase, the sucrose cycle (synthesis of sucrose from hexose-P and then, degradation to glucose and fructose via the invertase), and the starch cycle [15].

Besides the general MFA method dedicated to measure the whole combination of fluxes, more targeted methods exist to calculate individual fluxes and are detailed in this chapter. The first and basic method consists in measuring the variation of quantity of a compound with time. It is classically used to measure the accumulation of a specific compound when cells are growing [16]. Rigorously, this approach quantifies a net rate, corresponding to the difference between the synthesis and the degradation rates. Thus, the rate of synthesis can be underestimated if significant degradation occurs. In growing tissues, the rate of cell wall polysaccharides, lipids, and secondary metabolites accumulation can be approximated to their biosynthetic rates. The case of proteins, starch and sucrose is more complicated because these compounds are often turned over [2, 17–21].

The second method involves ^{14}C pulse labeling. In the past, this method has been used to identify the metabolic intermediates and their order in a pathway: the most labeled compound is the one closest to the initial substrate, and the succession of the compounds is defined according to the decrease in the specific radioactivity. This approach has been used to differentiate the C3 and C4 photosynthesis through the first labeled compound produced. Short-term labeling experiments are also used to quantify unidirectional fluxes of synthesis. As a complement to the pulse, the chase allows to calculate the degradation rate of the studied metabolite [22]. In the same order of idea, we present in this chapter, a protocol described by Stelling et al. [15] to estimate the turnover rate of sucrose and starch, based on (1) pulse labeling and (2) quantity measurements.

2 Materials

2.1 Culture, Labeling, and Sampling

1. Plant materials. Tomato cells of *Lycopersicon esculentum cerasiformae sweet 100.* Maize root tips, corresponding to the 3-mm tips of the primary roots excised from 3-day germinated Maize seeds.
2. Medium MS-Glc: Murashige and Skoog medium containing 166 mM glucose.
3. Medium S/4-Glc [23]: 0.7 mM KNO_3, 0.25 mM $Ca(NO_3)_2$, 0.25 mM KH_2PO_4, 0.16 mM $MgSO_4$, 0.75 mM NH_4NO_3, 6 μM H_3BO_4, 2.25 μM $MnSO_4$, 1 μM $ZnSO_4$, 25 nM $CuSO_4$, 10 nM $(NH_4)_6MoO_4$, 20 μM FeEDTA, 200 mM glucose and buffered to pH 5.8 by adding 20 mM Mes. Medium was aerated by bubbling 50 % O_2/N_2 (v/v).
4. [^{13}C]-Glucose (the choice between [1-^{13}C]-, [2-^{13}C]-, [6-^{13}C]-Glc depends on the metabolic network and the fluxes to be calculated) and [U-^{14}C]-glucose (10–120 dpm/nmol or 0.16–2 Bq/nmol).
5. 250 mL Erlenmeyer flasks.
6. Liquid nitrogen.

2.2 The External Flux Toward Proteins

1. Protein measurement Kit.
2. TCA 10 % (v/v).
3. 6 N HCl.
4. Centrifuge (capable of 10,000 × *g*).
5. Glass tubes to seal.
6. HPLC equipped with a fluorometer.
7. AccQ•Tag Column, 3.9 × 150 mm (Waters).
8. EDTA 1 mg/mL.
9. HPLC buffer A: for 2 L, weight 38 g sodium acetate,$3H_2O$, 1.4 g triethylamine and add 2 mL of 1 mg/L EDTA. Adjust pH to 5.7 with phosphoric acid.
10. HPLC buffer B: 60 % (v/v) acetonitrile (HPLC grade).

2.3 Quantification of Fluxes in Sucrose and Starch Cycles

2.3.1 Extraction, Purification, and Quantification of Sugars

1. Ethanol solutions for extraction: 80 % (v/v) and 50 % (v/v) ethanol in deionized water.
2. Cation exchange resin: Dowex 50-X8, Bio-Rad, hydrogen form; capacity 1 mequiv per mg resin.
3. Anion exchange resin: Dowex 1-X8, Bio-Rad, formate form or acetate form; capacity 1 mequiv per mg resin.
4. HPLC equipped with on-line refractometer.
5. ACCQ•Fluor Reagent™ Kit from Waters.

6. Aminex® HPX-87C Carbohydrate Column, 300 × 7.8 mm (BIO-RAD).
7. Amyloglucosidase (150 units/mL in 50 mM acetate buffer, pH 4.8).
8. Enzymatic Kit for glucose quantification.

2.3.2 Determination of the ^{14}C Labeling of Glucose-6-P

1. Phosphate buffer, 50 mM PiK (pH 7.5 with KOH).
2. KOH 2 % (w/v).
3. NADP 30 mM in phosphate buffer.
4. NADP-dependent Glc 6-P dehydrogenase 3,000 units/mL.
5. 3-Phosphogluconate dehydrogenase 2,000 units/mL.
6. 15-mL polyethylene vials, with push-in rubber stopper.

2.4 ^{14}C Counting

1. Liquid scintillation counter.
2. Liquid scintillation cocktail.
3. 15-mL polyethylene vials with polypropylene self-sealing cap.

3 Methods

3.1 Plant Tissue Culture and Labeling

1. Tomato cells were grown in the culture medium MS-Glc in a rotary shaker (150 rpm, 23 °C).
2. Maize root tips were grown at 25 °C in the culture medium S/4. Medium was aerated by bubbling 50 % O_2/N_2 (v/v).
3. Define the conditions for labeling experiments (*see* **Note 1**). All experiments are carried at the metabolic steady state: control that the concentration of the internal metabolites does not vary, whereas final products (considered as external metabolites) accumulate linearly. Then, define the time points of the kinetics to insure suitable conditions for short-time and steady state labeling experiments. If pulse labeling experiments are used to estimate fluxes toward metabolites that are partially degraded, such as sucrose or starch, the time has to be defined as short as possible to limit the loss of labeling by degradations of the metabolite concerned.
4. Incubate plant material with the stable [^{13}C]- or radioactive [^{14}C]-glucose: freshly excised maize root tips are directly transferred into the medium MS/4-Glc containing the labeled substrate (*see* **Note 2**); 10 mL of tomato cells in the exponential phase are diluted in 90 mL of fresh medium containing the labeled substrate in 250-mL Erlenmeyer flasks.
5. Stop the incubation: maize root tips are rinsed with cold deionized water and rapidly frozen in liquid nitrogen; tomato cells are filtered, washed with cold water, and frozen in liquid nitrogen.
6. Store the plant material at −20 °C (*see* **Note 3**).

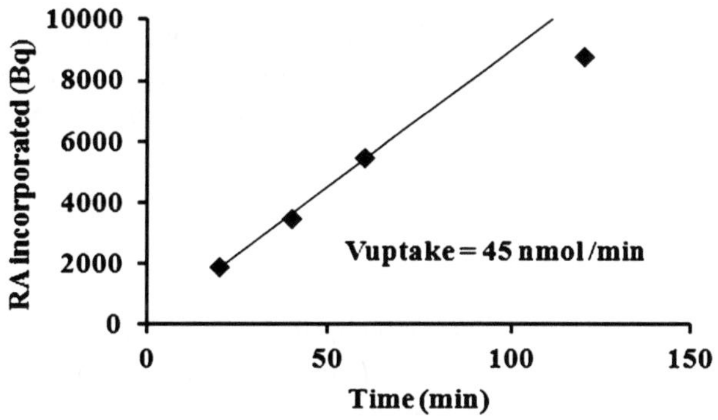

Fig. 1 Estimation of glucose uptake (a theoretical case). Plant cells or tissues were incubated with [U-^{14}C]-Glc (2 Bq /nmol) for 20, 40, 60, and 120 min. The incorporation of radioactivity (RA) into the plant material was measured and Vuptake measured from the initial slope of the curve and Eq. 2

3.2 Quantification of the External Fluxes

3.2.1 The Rate of the Substrate Uptake

1. Measure the decrease in the quantity (mol or gram) of the feeding substrate in the incubation medium with time:

$$V_{\text{uptake}} = \left|\frac{\Delta Q}{\Delta t}\right| \tag{1}$$

2. If the uptake rate is not high enough to detect differences in substrate concentration throughout the incubation time, make short-time labeling experiment:
   - Incubate cells or tissue with ^{14}C (or ^{13}C, depending on the availability) for short time periods.
   - Harvest cells or tissue.
   - Rinse with cold deionized water.
   - Measure the total radioactivity incorporated (*see* **Note 4**).

Figure 1 represents the increase in radioactivity with time. The slope of the line corresponds to the rate of the label incorporation (dR_A/dt). It can be converted to the rate of substrate uptake by dividing by the specific labeling (SR_S, that corresponds to the quantity of radioactivity per mole of compound) of the substrate S:

$$V_{\text{uptake}} = \frac{dR_A}{dt} \times \frac{1}{SR_S} \tag{2}$$

3.2.2 The Rate of Product Accumulation

The rate of product accumulation is often used to estimate the external fluxes in a metabolic network. Classically, it concerns the synthesis of final products such as cell structural components (cell wall polysaccharides, lipids, proteins) but also accumulation of starch, soluble sugars, secondary metabolites, etc.

1. Estimate the amount of each structural compound, expressed as in nmol of substrate equivalent, at the beginning and at the end of the incubation (*see* **Note 5**).
2. Calculate V_{accP}, the rate of accumulation of the product *P*:

$$V_{\text{accP}} = \frac{\Delta P}{\Delta t} \tag{3}$$

3. Alternatively, estimate the relative abundance (%*P*) of the product *P* in the tissue, expressed as in nmol of substrate equivalent, and the growth rate of the tissue or the cells.

$$V_{\text{accP}} = \%P \times \text{Growth rate} \tag{4}$$

3.2.3 The Particular Case of Proteins

Proteins synthesis does not correspond to a single flux of substrate utilization and has to be decomposed to the sum of the rate of synthesis of all amino acids, respectively, to their relative abundance into proteins.

1. Estimate the rate of protein accumulation
   - Grind the plant material in the protein extraction buffer.
   - Centrifuge 10 min at 10,000 × *g*.
   - Quantify protein content in the supernatant using a detergent compatible kit for protein assay. Use BSA as the external standard.
   - Calculate $V_{accProt}$, the rate of proteins accumulation (*see* Eq. 3).
2. Extract and hydrolyze proteins
   - Grind the plant material (tomato cells or maize root tips) in TCA 10 % (w/v).
   - Centrifuge 10 min at 10,000 × *g*.
   - Use a small aliquot to estimate the total protein amount (calculate Vprot).
   - Add HCl 6 N to the pellet and incubate 24 h at 110 °C in sealed tubes.
   - Remove HCl under nitrogen flux.
   - Solubilize amino acids in deionized water.
3. Estimate amino acid composition by HPCL.
   The method described in this chapter is based on the derivatization of amino acids with the reagent ACCQ•Fluor from Waters. Other methods can be used according to the disposable analytical material.

Table 1
Liquid chromatography gradient used for amino acids analysis

Time (min)	Buffer A (%)	Buffer B (%)
0	100	0
1	98	2
16	95	5
37	78	22
42	78	22
52	75	25
56	0	100
58	100	0
67	100	0

The flow rate is constant and equal to 1 mL/min

- Prepare the ACCQ•Fluor Reagent and derivative the amino acids (dilute before, if necessary) according to the instructions.
- Inject sample and elute derivatized amino acids using the gradient described in Table 1.
- Detect amino acids by fluorescence (λex: 250 nm, λem: 395 nm).
- Quantify the amino acid composition from the peak area and response factors determined by external standardization with the amino acid standard. Values thus obtained allow to estimate the percentage of each amino acid in the proteins (%AA).

4. Calculate the rate of synthesis for each proteogenic amino acid, V_{AA}.

$$V_{AA} = V_{accProt} \times \%AA \qquad (5)$$

3.3 Short-Time Labeling Experiments to Estimate Sucrose and Starch Turnover

Sucrose and starch are involved in substrate cycle and precautions have to be taken to estimate accurately their synthesis and degradation rates.

Plant material has been incubated for short times and harvested material stored at −20 °C as described in Subheading 3.1. A minimum of three points, if possible at regular interval, is required to estimate flux.

3.3.1 Extraction of Sucrose and Starch

Polar metabolites were extracted according to the protocols described by Stitt and ap Rees (1978) [24].

1. Add boiling 80 % ethanol (0.5 mL per 50 mg fresh weight) on the frozen material (*see* **Note 6**). Incubate at 80 °C for 15 min.
2. Transfer the supernatant S1 in a tube and add 50 % ethanol (0.5 mL per 50 mg fresh weight) on plant residue. Incubate at 80 °C for 15 min.
3. Add the supernatant to S1. Add water to plant residue and incubate at 80 °C for 15 min.
4. Add the third supernatant to the first two. The solution thus obtained contains all the polar metabolites, including sucrose.
5. Rinse the plant residue three times with deionized water to insure removing all the soluble sugars.
6. Heat 1 h at 120 °C.
7. Add 37 units of amyloglucosidase in 50 mM acetate buffer. Incubate 1 h at 37 °C.
8. Stop the reaction by heating 5 min at 90 °C.
9. Centrifuge 5 min at 3,700 × *g* and collect the supernatant containing glucose hydrolyzed from starch.

3.3.2 Sucrose Purification and Labeling Analysis

In this protocol, a preliminary treatment of the ethanolic extract is required to remove anions (mainly organic acids and phosphorylated intermediates) and cations (mainly amino acids) from free sugars prior their separation by HLPC.

1. Solubilize the ethanolic extract in water.
2. Prepare small columns (*see* **Note 7**) containing 200 μL cation exchange resin plus 200 μL anion exchange resin.
3. Pass the sample through column. Sugars, not retained on exchange resins, are collected. Rinse the resin with 2 mL of deionized water to assure a quantitative recovery of sugars.
4. Evaporate the neutral fraction in a rotative vacuum evaporator.
5. Solubilize the neutral fraction in water.
6. Separate the sugars by HPLC:
   - Heat the column at 80 °C.
   - Inject 20 μL of the neutral fraction (use an appropriate dilution).
   - Detect the sugars using a refractometer.
   - Collect the fraction corresponding to sucrose and count the radioactivity.
   - Estimate the sucrose amount from the peak area and using an external calibration with known solutions of sucrose.

3.3.3 Starch Analysis

1. Count the radioactivity incorporated into starch.
2. Estimate the starch amount using enzymatic assay.

3.3.4 Determination of the Specific Labeling of Glucose-6-Pi (SR_Glc6P)

1. Resuspend the ethanolic extract in 0.5 mL of phosphate buffer.
2. Determine spectrophotometrically the number of moles of glucose-6-Pi in 10 μL of extract, using the NADPH-linked Glc-6-P dehydrogenase assay.
3. Add unlabeled Glc-6-P in the extract up to a final concentration of 0.7 mM.
4. Add 5 μL of NADP and 1 μL of NADP-dependent Glc-6-P dehydrogenase and 3-phosphogluconate dehydrogenase.
5. Incubate the assay for 2 h at 25 °C in a hermetically closed 15-mL vial, with a central well inside containing a Whatman paper soaked with KOH.
6. Stop the reaction by injecting through the rubber stopper 50 μL of HCl to liberate CO_2.
7. Further incubate the assay for 30 min.
8. Open the vial and take out the Whatman paper, wash the central well with KOH and count the CO_2 radioactivity.
9. CO_2 recovery is measured in parallel experiments using a [U-^{14}C] Glc-6-P standardization. The Glc-6-P specific labeling, expressed as in disintegrations per min per nmol of Glc-6-P, is calculated from the CO_2 radioactivity, multiplied by 6 and divided by the yield of the reaction and by the amount of Glc-6-P in the assay (Fig. 2b).

3.3.5 Calculation of Synthesis and Degradation Fluxes

1. Calculate the accumulation rates of sucrose (Vacc-sucrose) and starch (Vacc-starch) (use Eq. 3).
2. Draw SR_Glc6P = *f*(*t*) (*see* Fig. 2a for a theoretical case).
3. Draw RA (incorporated into sucrose or starch) = *f*(*t*) (*see* Fig. 2b for a theoretical case).
4. Calculate the quantities of sucrose (Nsucrose) and starch (Nstarch) produced with time by using (Fig. 2c):

$$N_{t+1} = N_t + (\mathrm{RA}_{t+1} - \mathrm{RA}_t)/\mathrm{SR\_Glc6P}_t \quad (\textit{see}\ \textbf{Note 8}) \qquad (6)$$

5. Draw : Nsucrose = *f*(*t*) (*see* Fig. 2c for a theoretical case) and Nstarch = *f*(*t*) (*see* **Note 9**).
6. Calculate the initial slope of the curve to estimate the unidirectional rate of synthesis (Vsyn).
7. Calculation of the degradation rate (Vdeg).

$$\mathrm{V\,deg} = \mathrm{Vacc} - \mathrm{Vsyn} \qquad (7)$$

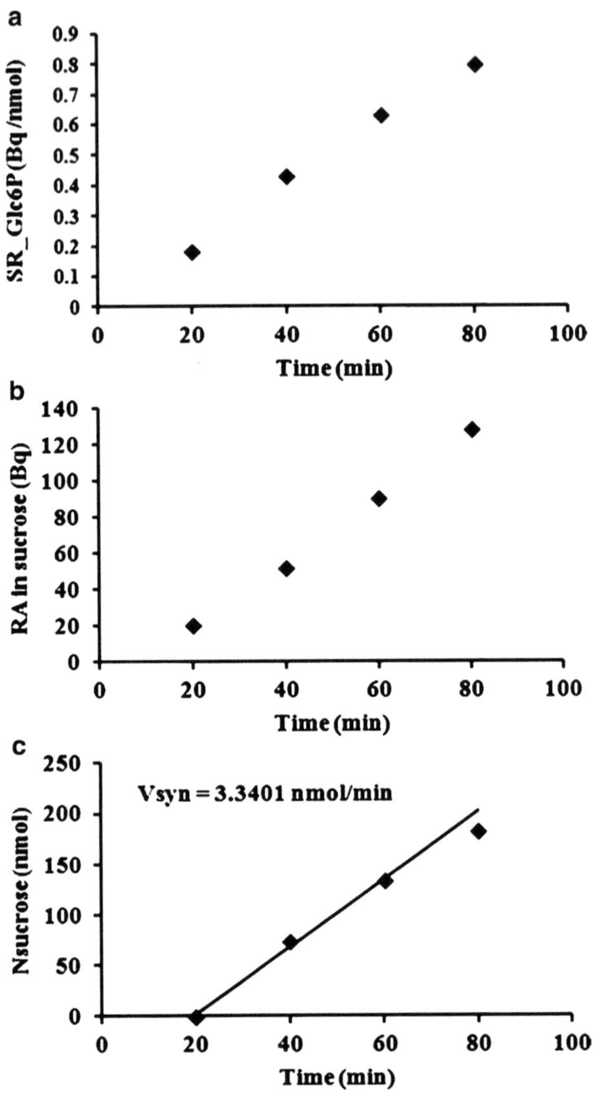

Fig. 2 Estimation of sucrose turnover (a theoretical case). Plant cells or tissues were incubated with [U-^{14}C]-Glc (2 Bq /nmol) for 20, 40, 60, and 80 min. The evolution of SR_Glc6P (**a**) and the incorporation of radioactivity (RA) into sucrose (**b**) was measured. The quantity of sucrose produced (Nsucrose) with time is deduced by using Eq. 6 (**c**). The sucrose synthesis rate (Vsyn) was calculated from the initial slope of the curve

4 Notes

1. Preliminary experiments are required to estimate the incubation time that allows the establishment of both the metabolic and isotopic steady states: plant cells or tissues are incubated in the presence of the tracer and the time-course evolution of the labeling in metabolic intermediates is analyzed.

2. The culture medium must be adapted to the tissue or organ studied. For example, in the case of maize root tips, 200 mM glucose was added as carbon source. This obviously high concentration was necessary to keep the respiratory rate constant and to allow root growth [23].
3. ^{13}C labeled samples will be analyzed by NMR or LC-MS (*see* Chapters 5, 6, and 9 of that volume) and data thus obtained used to estimate internal fluxes of the metabolic network using formalism of steady state labeling.
4. To count the radioactivity incorporated into cells, it can be done by adding directly liquid scintillation cocktail on cells. For tissues, it is necessary to grind and incubate the material in the presence of a tissue solubilizer for 24 h, prior the addition of liquid scintillation cocktail.
5. The variation of quantity over time can be converted to a flux provided that the rate of accumulation is constant, i.e. the biological system is at the metabolic steady state.
6. For ethanolic extraction, in the case of cell-suspension or maize root tips, boiling ethanol can be added directly on the frozen material. Alternatively, plant material can be ground, and eventually dried in a lyophilizator prior extraction.
7. Cation and anion exchange columns are formed by using pipet tips filled with the resin, retained by glass cotton.
8. The estimation of the sucrose and starch synthesis rates using the specific radioactivity of Glc-6-P relies on the hypothesis that hexose-phosphates are in rapid isotopic exchange, so that the precursors of sucrose and starch (Glc-6-P, Fru-6-P, UDP-Glc and ADP-Glc) are similarly labeled.
9. For accurate determination, a straight line must be obtained during a sufficient time-domain. A decrease in the rate of sucrose synthesis would be an indicator that radioactivity is lost in the reaction of degradation. In this case, reduce the incubation time.

References

1. Salon C, Raymond P, Pradet A (1988) Quantification of carbon fluxes through the tricarboxylic acid cycle in early germinating lettuce embryos. J Biol Chem 263:12278–12287
2. Dieuaide-Noubhani M, Raffard G, Canioni P, Pradet A, Raymond P (1995) Quantification of compartmented metabolic fluxes in maize root tips using isotope distribution from 13C- or 14C-labeled glucose. J Biol Chem 270:13147–13159
3. Alonso AP, Raymond P, Hernould M et al (2007) A metabolic flux analysis to study the role of sucrose synthase in the regulation of the carbon partitioning in central metabolism in maize root tips. Metab Eng 9:419–432
4. Rontein D, Dieuaide-Noubhani M, Dufourc EJ, Raymond P, Rolin D (2002) The metabolic architecture of plant cells. Stability of central metabolism and flexibility of anabolic pathway during the growth cycle of tomato cells. J Biol Chem 277:43948–43960
5. Williams TCR, Miguet L, Masakapalli SK et al (2008) Metabolic network fluxes in heterotrophic Arabidopsis cells: stability of the flux distribution under different oxygenation conditions. Plant Physiol 148:704–718

6. Masakapalli SK, Kruger NJ, Ratcliffe RG (2013) The metabolic flux phenotype of heterotrophic Arabidopsis cells reveals a complex response to changes in nitrogen supply. Plant J. doi:10.1111/tpj.12142
7. Sriram G, Fulton DB, Iyer VV et al (2004) Quantification of compartmented metabolic fluxes in developing soybean embryos by employing biosynthetically directed fractional 13C labeling, two dimensional [13C, 1H] nuclear magnetic resonance, and comprehensive isotopomer balancing. Plant Physiol 136:3043–3057
8. Lonien J, Schwender J (2009) Analysis of metabolic flux phenotypes for two Arabidopsis mutants with severe impairment in seed storage lipid synthesis. Plant Physiol 151:1617–1634
9. Alonso AP, Dale VL, Shachar-Hill Y (2010) Understanding fatty acid synthesis in developing maize embryos using metabolic flux analysis. Metab Eng 12:488–497
10. Spielbauer G, Margl L, Hannah LC et al (2006) Robustness of central carbohydrate metabolism in developing maize kernels. Phytochemistry 67:1460–1475
11. Alonso AP, Val DL, Shachar-Hill Y (2011) Central metabolic fluxes in the endosperm of developing maize seeds and their implications for metabolic engineering. Metab Eng 13:96–107
12. Keeling PL, Wood JR, Tyson RH, Bridges IG (1988) Starch biosynthesis in developing wheat grain. Evidence against the direct involvement of triose phosphates in the metabolic pathway. Plant Physiol 87:311–319
13. Hatzfeld WD, Stitt M (1990) A study of the rate of recycling of triose phosphates in heterotrophic *Chenopodium rubrum* cells, potato tubers and maize endosperm. Planta 180:198–204
14. Viola R, Davies HV, Chudeck AR (1991) Pathways of starch and sucrose biosynthesis in developing tubers of potato (*Solanum tuberosum* L.) and seeds of faba bean (*Vicia faba* L.). Planta 183:202–208
15. Alonso AP, Vigeolas H, Raymond P, Rolin D, Dieuaide-Noubhani M (2005) A new substrate cycle in plants. Evidence for a high glucose phosphate-to-glucose turnover from in vivo steady-state and pulse labeling experiments with [13C]glucose and [14C]glucose. Plant Physiol 138:2220–2232
16. Marx A, de Graaf AA, Wiechert W et al (1996) Determination of the fluxes in the central metabolism of *Corynebacterium glutamicum* by nuclear magnetic resonance spectroscopy combined with metabolite balancing. Biotechnol Bioeng 49:111–129
17. Hargreaves JA, ap Rees T (1988) Turnover of starch and sucrose in roots of *Pisum sativum*. Phytochemistry 27:1627–1629
18. Dancer J, Hatzfeld WD, Stitt M (1990) Cytosolic cycle regulate the turnover of sucrose in heterotrophic cell-suspension cultures of *Chenopodium rubrum* L. Planta 182:223–231
19. Geigenberger P, Stitt M (1991) A "futile" cycle of glucose synthesis and degradation is involved in regulating partitioning between sucrose, starch and respiration in cotyledons of germinating *Ricinus communis* L. seedlings when phloem transport is inhibited. Planta 185:81–90
20. Hill SA, ap Rees T (1994) Fluxes of carbohydrate metabolism in ripening bananas. Planta 192:52–60
21. N'tchobo H, Dali N, Nguyen-Quoc B et al (1999) Starch synthesis in tomato remains constant throughout fruit development and is dependent on sucrose supply and sucrose synthase activity. J Exp Bot 50:1457–1463
22. Trethewey RN, Riesmeier JW, Willmitzer L, Stitt M, Geigenberger P (1999) Tuber-specific expression of a yeast invertase and a bacterial glucokinase in potato leads to an activation of sucrose phosphate synthase and the creation of a sucrose futile cycle. Planta 208:227–238
23. Saglio PH, Pradet A (1980) Soluble sugar, respiration, and energy charge during aging of excised maize root tips. Plant Physiol 66:516–519
24. Stit M, ap Rees T (1978) Pathways of carbohydrate oxidation in leaves of *Pisum sativum* and *Triticum aestivum*. Phytochemistry 18: 1905–1911

Chapter 4

Optimization of Steady-State ^{13}C-Labeling Experiments for Metabolic Flux Analysis

Nicholas J. Kruger, Shyam K. Masakapalli, and R. George Ratcliffe

Abstract

While steady-state ^{13}C metabolic flux analysis is a powerful method for deducing multiple fluxes in the central metabolic network of heterotrophic and mixotrophic plant tissues, it is also time-consuming and technically challenging. Key steps in the design and interpretation of steady-state ^{13}C labeling experiments are illustrated with a generic protocol based on applications to plant cell suspension cultures.

Key words Arabidopsis, Cell culture, Metabolic flux analysis, Heterotrophic carbon metabolism, Isotope labeling, Metabolic modeling, Network analysis, Primary metabolism

1 Introduction

Statistically robust protocols have been developed that allow steady-state ^{13}C metabolic flux analysis (MFA) to be applied to heterotrophic and mixotrophic plant tissues [1, 2]. While steady-state ^{13}C MFA is currently the method of choice for measuring multiple fluxes within the compartmented network of primary metabolism, the implementation of the method is challenging and it requires a considerable commitment of time and effort by the researcher. It follows that it is important to optimize the design, execution, and interpretation of the labeling experiments on which MFA depends in order to obtain the best possible estimates of the fluxes in the metabolic network.

A steady-state ^{13}C MFA investigation can be broken down into several steps (*see* Fig. 1). First, the tissue of interest is incubated with a ^{13}C-labeled substrate for an extended period, often several days or longer. Cell suspensions and developing oil seed embryos are the most commonly used tissues, but in principle the method is applicable to any heterotrophic or mixotrophic system that can be maintained at a metabolic steady state. Secondly, the labeling of metabolic intermediates and end-products is analyzed when the

Martine Dieuaide-Noubhani and Ana Paula Alonso (eds.), *Plant Metabolic Flux Analysis: Methods and Protocols*, Methods in Molecular Biology, vol. 1090, DOI 10.1007/978-1-62703-688-7_4, © Springer Science+Business Media New York 2014

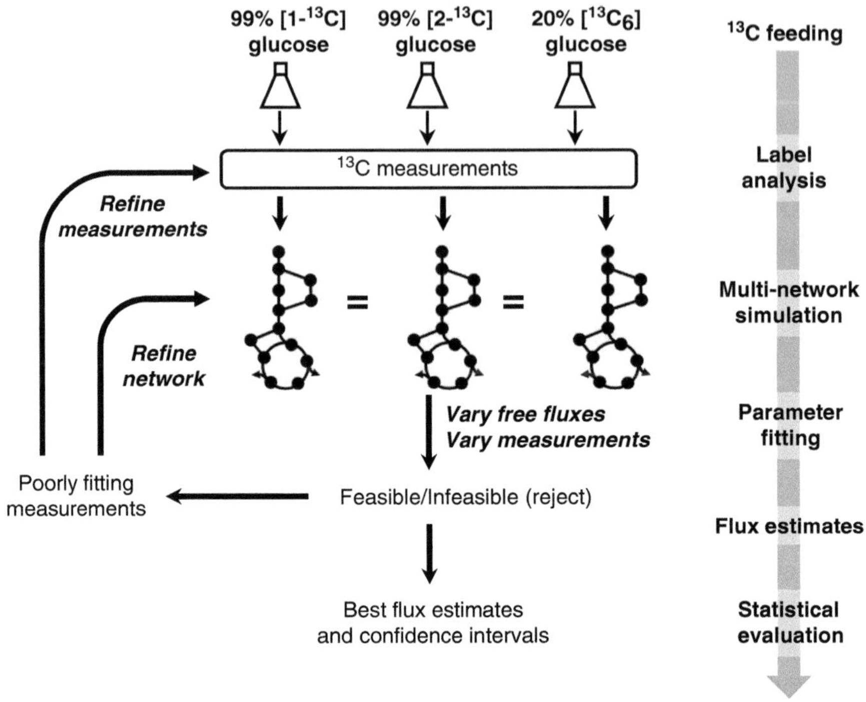

Fig. 1 Outline strategy for steady-state MFA of heterotrophic metabolic networks in plants. First, replicate samples of the tissue of interest are incubated with different forms of a ^{13}C-labeled substrate. Secondly, after the system has reached an isotopic steady state, the redistribution of label within metabolic intermediates and end-products is analyzed. Thirdly, a metabolic model is constructed which consists of multiple copies of the network that defines the carbon atom relationships between the reactants of the successive metabolic steps that generate cellular biomass. Fourth, iterative numerical analysis is used to deduce the values of the flux parameters that are consistent with the labeling data from all the feeding experiments as well as any other metabolic constraints. Fifth, the fluxes that best fit the metabolic model are determined. Finally, the quality of the fit and the reliability of the deduced fluxes are established

system has reached an isotopic steady state. Gas-chromatography/mass-spectrometry (GC-MS), which detects mass isotopomers, and nuclear magnetic resonance (NMR) spectroscopy, which detects positional isotopomers, are the two most commonly used analytical techniques for quantifying the redistribution of label within the tissue [3, 4]. Thirdly, a metabolic network model is constructed that defines the carbon atom relationships between successive metabolites during the conversion of the system inputs, including the ^{13}C-labeled substrate, into the system outputs, typically biomass components such as protein, cell wall, lipids, and carbohydrate storage forms. Fourth, the metabolic fluxes that explain the redistribution of the label are deduced by fitting the model to the labeling data and biomass constraints. Typically this is an iterative process that requires critical evaluation of both the experimental data and the solution space in which the actual fluxes might lie.

Finally, the analysis converges on the best estimate of the fluxes, and the quality of the fit and the reliability of the deduced fluxes are established.

The implementation of the procedure outlined in the preceding paragraph is still far from routine for plant tissues. Steps are being taken to increase the throughput of the analysis by streamlining the workflow [5], but the quality of the final flux map is still very dependent on the precise way in which the individual steps are carried out. With this in mind this protocol describes key steps in the design and interpretation of steady-state ^{13}C labeling experiments within the framework of the widely used 13C-FLUX software package [6]. It is based on extensive experience with plant cell cultures, but the guidance provided on the optimization of the ^{13}C MFA experiment is broadly applicable, and analogous procedures can be used in other heterotrophic and mixotrophic plant systems.

2 Materials

1. A heterotrophic or mixotrophic plant cell suspension culture growing under defined conditions.
2. Carbon sources for growth labeled with ^{13}C, for example [1-^{13}C]-, [2-^{13}C]- and [$^{13}C_6$]glucose.
3. GC-MS and NMR equipment for measuring mass isotopomers and positional isotopomers, respectively.
4. Software for the analysis of steady-state stable isotope labeling experiments, for example 13C-FLUX.

3 Methods

3.1 Labeling Strategy, Experimental Design, and Analytical Measurements

1. Set up triplicate experiments for each labeled substrate by transferring replicate samples of cell suspension into growth medium containing 99 % [1-^{13}C]glucose, 99 % [2-^{13}C]glucose, or 20 % [$^{13}C_6$]glucose (*see* **Note 1**).
2. Incubate samples under defined growth conditions for sufficient time to allow the metabolic pools to achieve isotopic steady state and to permit a measurable accumulation of label in biosynthetic end-products (*see* **Note 2**).
3. Harvest the cells and freeze in liquid nitrogen.
4. Extract metabolites, retaining both the soluble, low molecular weight compounds and the insoluble, high molecular weight biopolymers (*see* **Note 3**).
5. Obtain data on the labeling of selected soluble metabolites and derived biosynthetic products using procedures described elsewhere in this volume.

3.2 Verification of the Isotopic Steady State

Use either or both of the following approaches to establish the extent to which the compounds in the metabolic network are at isotopic steady state (*see* **Note 4**).

3.2.1 Comparison of Time Points

1. Obtain labeling data from triplicate cultures growing under identical conditions and harvested at two or more time points, preferably at least 24 h apart.
2. For each metabolite, compare the relative abundance of the isotopomers between time points using the following two approaches.
3. Analysis of variance (ANOVA)—the labeling patterns of metabolites of cultures at isotopic steady state should yield no significant difference between the samples harvested at different time points.
4. Principal component analysis—the isotopomer composition of samples at isotopic steady state should produce a heterogeneous cluster of points on a scores plot of the individual samples with no obvious separation between samples obtained from different time points (*see* **Note 5**).

3.2.2 Quantification of Isotopic Enrichment

1. For cultures incubated in 20 % [$^{13}C_6$]glucose, determine the average ^{13}C fractional abundance (F) of a metabolite or metabolite fragment using the equation:

$$F = \frac{1}{n}\sum_{j=1}^{n} (F_j \times j)$$

   where F_j is the fractional abundance of the mass isotopomer containing j ^{13}C atoms, and n is the total number of carbon atoms in the molecule.
2. Compare the calculated fractional ^{13}C abundance of each metabolite with that of the substrate used to label the culture. Metabolic intermediates that are at isotopic steady state will be labeled to an average ^{13}C abundance equivalent to that of the substrate in which all carbon positions are equally labeled (*see* **Note 6**).

3.3 Compensating for Incomplete Isotopic Equilibration

Storage products generated as outputs from the network of central carbon metabolism often fail to reach isotopic steady state because of the presence of material synthesized prior to labeling.

Use a combination of the following approaches to compensate for the contribution of these preexisting pools to the labeling patterns of material that has not reached an isotopic steady state (*see* **Note 7**).

3.3.1 Correction by Scaling Adjustment (See Note 8)

1. Using the ^{13}C fractional abundances of a reporter metabolite (S), that has reached an isotopic steady state, and a product (P), calculate the correction factor (Δ):

$$\Delta = \frac{F_S - F_P}{F_S - 0.0113}$$

where 0.0113, the natural abundance of ^{13}C, is the expected value of F for an unlabeled metabolite. A value of 0 for Δ implies that P and S are in the same isotopic steady state; a value of 1 implies that P is derived entirely from a preexisting unlabeled pool.

2. Estimate the contribution of the unlabeled pool to the observed mass isotopomers by multiplying the fractional abundance of each of the mass isotopomers of P by Δ.
3. Subtract the resulting values from the measured mass isotopomer values for P and rescale to sum to 1 to generate the corrected mass isotopomer measurements for P.
4. Use the corrected mass isotopomer values for P when fitting the flux parameters of the model.

3.3.2 Correction Using Proxy Reactions (See **Note 9***)*

1. For each product requiring correction of the measured isotopomer values, define the isotopomer composition of the preexisting unlabeled component based on ^{13}C natural abundance and include this metabolite as an additional substrate in the network.
2. For each compound define inputs into the relevant metabolite pool in the network from (1) the precursor metabolite present in the network, and (2) the unlabeled substrate pool. Also, specify the output from the metabolite pool to end-product formation (i.e., an output from the network).
3. Use the estimate of the rate of product formation to constrain the net flux of the step converting the precursor metabolite to product, but allow the net flux from unlabeled input and the network output from the metabolite pool to vary during the subsequent fitting procedure (*see* **Note 10**).

3.4 Analyzing Parallel Experiments (See **Note 11***)*

1. Construct a metabolic model containing replicate copies of the stoichiometry and carbon atom transitions associated with each step in the network, including input and output reactions. This produces a model consisting of several structurally identical networks that will allow the data from each labeling experiment to be fitted simultaneously (*see* **Note 12**).
2. Assign unique identifiers, corresponding to the individual labeling experiments, to each of the equivalent steps and pathway intermediates in the replicated networks.
3. Define the isotopic composition of the substrate inputs for the model (*see* **Note 13**).
4. Constrain the flux through equivalent steps in the replicate networks to be equal. This ensures that the flux distribution through each network in the model is identical (*see* **Note 14**).

5. Assign the isotopomer measurements from each labeling experiment to the appropriate network.

3.5 Parameter Fitting (See Note 15*)*

1. Define values of any input and output fluxes that are available to constrain the fitting procedure (*see* **Note 16**).
2. Identify a set of free fluxes that can be varied during the fitting procedure and assign initial values to these fluxes expressed as arbitrary proportions of the input flux (*see* **Note 17**).
3. Use a numerical optimization tool to determine the values of the free fluxes by least squares fitting of experimental and predicted measurements (*see* **Note 18**).
4. Perform Monte Carlo simulations with bootstrap sampling of the isotopomer measurements, i.e., repeat the preceding step multiple times with statistical sampling of the labeling data as defined by the values and variance of each isotopomer measurement (*see* **Note 19**).
5. Exclude flux solutions that generate infeasible flux values (*see* **Note 20**).
6. Exclude from the remaining feasible flux solutions those with high residuum values that reflect poorly fitting local minima (*see* **Note 21**).
7. Calculate the mean value for each of the free fluxes in the network from the remaining low-residuum feasible flux solutions, and use these as the initial values for the free flux parameters of the network model.
8. Refit the model to the original measured isotopomer dataset using the mean flux values derived in the preceding step to obtain the flux solution with the lowest residuum (*see* **Note 22**).
9. Perform a χ^2 test on the flux solution from the preceding step by comparing the residuum with the χ^2 distribution for the relevant degrees of freedom (*see* **Note 23**).
10. If the residuum is lower than the appropriate critical χ^2 value (i.e., p is greater than the chosen significance level) then the model provides an adequate description of the labeling patterns and the flux map is acceptable; otherwise proceed to the next step (*see* **Note 24**).
11. Identify the isotopomer measurements with the largest variance-weighted deviation from their predicted values.
12. Reassess the identity, value, and estimate of error of each poorly fitting isotopomer and make corrections as appropriate.
13. Consider altering the structure of the network by the addition or deletion of particular reactions (*see* **Note 12**).
14. Repeat **steps 3–9** to obtain an optimized flux map with a residuum that satisfies the χ^2 criterion (*see* **Note 25**).

3.6 Determination of Confidence Intervals

Use either or both of the following methods to determine the confidence limits for the fluxes deduced from the parameter fitting procedure (*see* **Note 26**).

3.6.1 Monte Carlo Analysis

1. Obtain a set of feasible flux distributions from a large number of Monte Carlo simulations as described in Subheading 3.5 (*see* **Note 27**).
2. Sort the flux estimates for each reaction separately into rank order (*see* **Note 28**).
3. Select the range of values appropriate for the desired confidence interval. For example, the 95 % confidence interval is defined by the range remaining after discarding the largest and smallest 2.5 % of values, whereas the 68 % confidence interval is obtained from the range of values remaining after eliminating the smallest and largest 16 % of the flux estimates.

3.6.2 Network Sensitivity Analysis

1. Set all free fluxes to their best estimates.
2. Constrain the value of a particular free flux, refit the model, and note the residuum.
3. Vary the value of the single constrained flux in a stepwise manner and refit the model to obtain the residuum at each value.
4. Plot the residuum against the value of the varied flux.
5. Identify the flux values at which the increase in residuum exceeds the appropriate critical χ^2 value for 1 degree of freedom—these define the confidence limits of the selected flux. The threshold values of χ^2 for 68 %, 90 %, 95 %, and 99 % confidence limits are 0.99, 2.71, 3.84, and 6.63, respectively.
6. Repeat **steps 2–5** for each free flux in turn.

4 Notes

1. The precision with which a particular flux can be determined depends on the structure of the network, the fluxes that flow through it, the labeling of the substrate, and the isotopomer measurements of intermediates and end-products. Thus no single ^{13}C-substrate is likely to report fully on all fluxes under all conditions, and the optimum labeling strategy is likely to vary between cell cultures, growth conditions, and the fluxes of interest. If the network fluxes are known, then it is possible to determine the optimum combination of ^{13}C-substrates needed to define a specific flux or combination of fluxes through in silico analysis [7, 8]. Although this approach can be useful in refining estimates of specific fluxes, at the outset of analysis the flux distribution is not known, and so the labeling strategy must be largely heuristic.

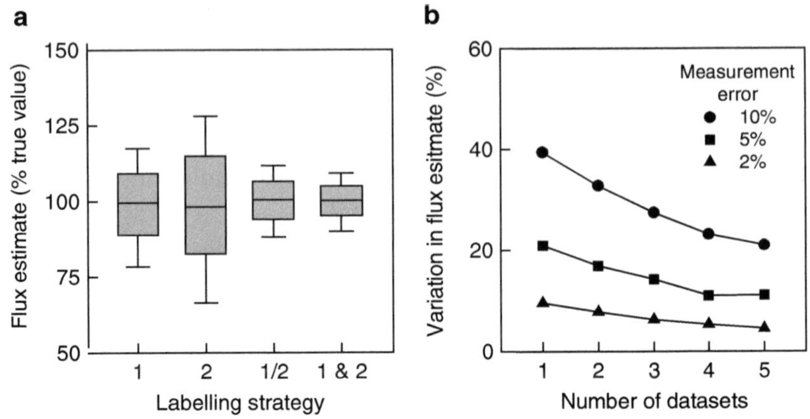

Fig. 2 Factors influencing the reliability of flux estimates. (**a**) The reliability of the flux estimates of a typical reaction is dependent on the ^{13}C-labeling pattern of the substrate and is often improved by combining data from different ^{13}C-labeling studies. Here provision of two differently labelled forms of the substrate in combination (1/2) yields a better flux estimate than either labelling strategy (1 or 2) in isolation. Even greater improvement is achieved by supplying the two substrates separately to replicate tissue samples and combining the data in a single analysis (1 & 2). (**b**) The reliability of flux estimates is also generally improved by analyzing replicate datasets and by increasing the precision of the measurements of isotopic composition of metabolites. Figure based on data obtained from in silico analysis of a metabolic network [1]

Usually ^{13}C is supplied in the form of the major metabolic substrate to ensure adequate labeling of the system, and for heterotrophic plant cells this normally means a ^{13}C-labeled sugar, with [^{13}C]glucose being the most useful because of the availability of a wide range of different labeled forms at comparatively low cost and the ability of glucose to substitute for sucrose as the major respiratory substrate in many systems. Typically, we employ a combination of [1-^{13}C]-, [2-^{13}C]-, and [$^{13}C_6$]glucose which provides good coverage of a broad range of fluxes throughout the network of central metabolism in heterotrophic plant cells. Supplying these ^{13}C-substrates separately provides more information than labeling strategies involving mixtures of two or more isotopic forms of the substrate (*see* Fig. 2a) [9, 10]; and the precision of the flux estimates is further improved by triplicating each ^{13}C-stubstrate feeding experiment (*see* Fig. 2b). This experimental design generates nine datasets, three replicates for each of three feeding strategies. Data from the different feeding strategies must be modeled in three networks since the redistribution of label from each of the ^{13}C-substrates will differ. Replicate feeding studies conducted using the same labeling strategy may be treated in a similar manner. However, this is only required if the labeling patterns of metabolites obtained from the replicate studies differ substantially. In such circumstances there may be a correlation between the relative isotopomer abundances of different metabolites, and it may be important to ensure that data from the same sample are considered as a group. Normally this is not an issue and data from separate samples may be considered as replicate measurements for the relevant metabolite and included in the same network. This is particularly helpful to limit the size of the model that would otherwise contain nine networks.

2. The duration of the labeling experiment depends on the system under investigation. For cell suspension cultures we usually begin labeling immediately after subculture and harvest the cells at mid-log phase of the culture cycle— for an Arabidopsis cell culture this is at days 4–5. Typically most intermediates in the central network of carbon metabolism are labeled to isotopic steady state within a few hours. However, pools of end-products including soluble sugars and organic acids can take much longer [11]. It is possible to correct for the contribution of preexisting material to the isotopomer composition of these pools of metabolites (*see* **Notes 8** and **9**). Even so, having a greater proportion of labeled material in a pool that turns over slowly improves the accuracy of this correction thus favoring longer labeling times. Against this, the longer the labeling period, the more difficult it becomes to maintain the system in a metabolic steady state. Ultimately, the incubation time is a compromise between these two conflicting factors.

3. Various techniques may be used to extract the intracellular metabolites, and common solvents include perchloric acid [12], chloroform/methanol [10], and ethanol [13]. The critical considerations are rapid inactivation of enzyme activity, and extraction of the metabolites in a form compatible with the subsequent ^{13}C analysis.

4. These methods will generally show that the biomass components of the network do not reach a true isotopic steady state because of the presence of material formed before labeling of the system.

5. The extent of clustering of the sample scores obtained from principal component analysis is usually gauged subjectively, although it may be more critically assessed using ANOSIM, a nonparametric permutation based nearest-neighbor test [13].

6. This approach is only useful for metabolites that are synthesized exclusively from the major metabolic substrate. Since intracellular CO_2 frequently fails to achieve the same ^{13}C fractional abundance as the labeled substrate it is not possible to predict the steady state average ^{13}C abundance of metabolites formed by reactions involving CO_2 and/or methylene-tetrahydrofolate as a substrate.

Table 1
Reporter metabolites used in the scaling correction for incomplete isotopic equilibration of amino acids

Compound	Reporter metabolite
Aspartate	Malate/oxaloacetate
Glutamate	2-Oxoglutarate OR Malate/oxaloacetate and citrate
Isoleucine	
Fragment C1-6	Alanine and 2× malate/oxaloacetate
Fragment C2-6	Alanine and 1.5× malate/oxaloacetate
Leucine	Alanine
Lysine	5× Alanine and 7× malate/oxaloacetate
Phenylalanine	Alanine
Proline	2-Oxoglutarate OR Malate/oxaloacetate and citrate
Threonine	Malate/oxaloacetate
Tyrosine	Alanine
Valine	Alanine

The average ^{13}C abundance of the fraction of each compound formed during labeling with [$^{13}C_6$]glucose should be the same as that of the reporter metabolite. For compounds formed from more than one source, the predicted average labeling of the compound is based on the weighted average of the relative contributions of the various reporter metabolites to the carbon atoms of the compound (or specific fragment) analyzed

7. Incomplete isotopic equilibration in network products is frequently encountered in labeling studies of plant tissue, even when network intermediates are labeled to isotopic steady state. This is revealed by a difference in the average fractional ^{13}C abundance of substrates and products of the steps in the network. The disparity in isotope abundance arises principally because an appreciable fraction of the product in the sample was present before labeling began and was not degraded or subject to turnover during the period of labeling. The compounds that are most affected by this problem are soluble metabolites that accumulate in the vacuole, proteins, starch granules, and cell wall polysaccharides.
8. The scaling adjustment for P hinges on identifying a measurable pool of S that reaches an isotopic steady state in, typically, a 20 % [$^{13}C_6$]glucose labeling experiment. It is more difficult to apply to metabolites derived from intermediates of the TCA cycle because assimilation of atmospheric CO_2 by PEP carboxylase will dilute the ^{13}C derived from the labeled glucose. In this instance the fractional abundance of the precursor can be estimated from that of a reporter metabolite that is synthesized

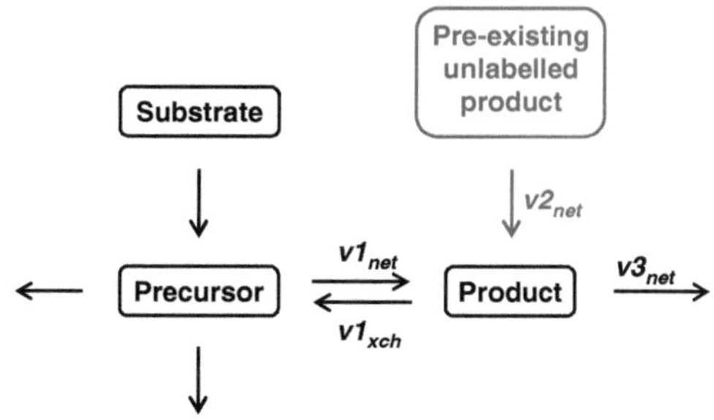

Fig. 3 Use of proxy reactions to accommodate preexisting (unlabeled) pools of product in steady-state MFA. The figure models a fragment of the metabolic network in which a product is reversibly synthesized from a metabolic precursor by a reaction with net and exchange fluxes v_{1net} and v_{1xch}, respectively. Preexisting unlabeled product is modeled (in *grey*) as an additional input (v_{2net}). The output flux (v_{3net}) is defined by v_{1net} and v_{2net} which reflect the relative contributions of newly synthesized and preexisting product to the isotopomer measurements. Each proxy reaction introduces one additional free flux into the model

from the same precursor and is believed to have reached isotopic equilibrium. Table 1 provides a list of reporter metabolites for amino acids which can be used for this approach.

9. When scaling adjustments are not possible because of the lack of adequate estimates of average ^{13}C fractional abundance in a suitable reporter metabolite, then proxy reactions (*see* Fig. 3) can be used to account for isotopic disequilibrium [14, 15]. In principle, this approach can be also used in studies based on positionally labeled substrates.

   The introduction of proxy reactions to the network adds an additional free flux to the metabolic model for each unlabeled pool. However, models produced using the experimental design described in this chapter are typically massively overdetermined and as a consequence these extra free fluxes normally have only a minor effect on the fitting procedure. Nevertheless, when analyzing models containing large numbers of free fluxes, it is helpful initially to constrain the values of the proxy fluxes during Monte Carlo analysis of the model and only to set them as free in the final optimization to find the best fit. This procedure avoids generating an unacceptably high proportion of infeasible flux solutions during the Monte Carlo analysis. The approach must be modified for proxy reactions carrying negligible fluxes since 13C-FLUX does not permit an input flux to be constrained below 0.0001. Removing such reactions from the network avoids this limitation and helps in parameter fitting by minimizing the total number of free fluxes in the model. This has little practical impact on the final outcome and is justified because a very low flux through a proxy reaction implies that the contribution of the preexisting material to the metabolite pool is negligible and thus the need for compensation is minimal.

10. This defines the flux to product during the labeling period, without constraining the relative contribution of preexisting material to the isotopomer composition of the pool of metabolite analyzed. Note that the estimates of flux from unlabeled input (v_{2net}) and to the network output (v_{3net}) from the metabolite pool have no kinetic significance and simply account for the proportion of preformed metabolite contributing to the product pool. The relative values of v_{2net} and v_{1net} reflect the contributions of preexisting and de novo material to the product pool. Corrections deduced from a [$^{13}C_6$]glucose labeling experiment can be applied to measurements obtained in experiments with other positionally labeled substrates provided all the experiments are conducted in parallel on the same batch of cells.

11. Analyzing data obtained from multiple feeding experiments has several advantages (Fig. 2). First, it improves the precision of flux estimates throughout the network by allowing the amalgamation of data from separate labeling strategies that inevitably differ in the reliability with which they define specific steps. Secondly, combining multiple independent datasets using the same labeling strategy also improves precision. Thirdly, it overcomes difficulties in combining the results obtained from fitting individual datasets separately. Flux estimates obtained by averaging are likely to be inaccurate because different labeling strategies differ in the reliability with which they define specific steps and because there is not a proportional relationship between fluxes and the labeling patterns of metabolites in the network.

12. Formulation of the metabolic network that best explains the experimental data is likely to be an iterative process involving the following steps: (1) construction of the *largest possible* network based on the available biochemical information on the relevant pathways and their compartmentation; (2) definition of a subset of free fluxes to be varied during the parameter fitting procedure to account for the measured redistribution of label in the network; (3) identification and elimination of structurally indeterminable fluxes to simplify the network; (4) use of the resulting network to deduce the best-fit flux estimates from a labeling dataset; (5) identification and potential elimination of all well-defined zero fluxes; and (6) identification and possible elimination of all poorly defined fluxes if they cannot be improved by further data collection. This procedure will lead to a network in which all the fluxes are well defined. This network can then be fixed and replicated in the model used to analyze the parallel labeling experiments.

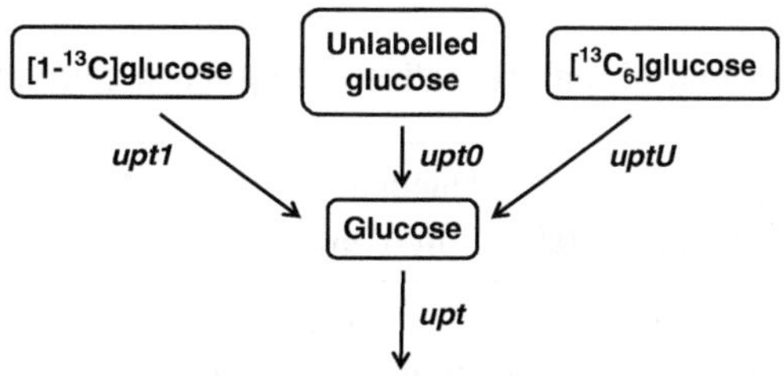

Fig. 4 Modeling inputs from combinations of differently labeled substrate. This scheme allows the composition of the substrate supplied to the network to be varied by defining the relative values of the uptake fluxes (*upt0*, *upt1*, and *uptU*) for three separate ^{13}C-labeled sources, while the rate of glucose consumption is determined by the value of input flux (*upt*) from the common intermediate pool of glucose. Individual labeled substrates can be excluded from a network by constraining the corresponding uptake flux (*upt**) to zero

13. When defining the isotopic composition of the substrate, it is often convenient to include reactions from each labeled form of the substrate that converge on a common pool of the substrate that enters the network (*see* Fig. 4). The isotopic composition of the substrate can then be selected by defining the relative fluxes from the different sources. This arrangement allows the absolute rate of substrate consumption to be constrained while permitting flexibility in the precise proportion of the alternative isotopomers utilized by the network to account for the labeling pattern of the metabolites. This is particularly useful when analyzing experiments involving metabolism of a mixture of differently labeled substrates (such as [$^{13}C_6$]glucose and unlabeled glucose containing ^{13}C at natural abundance) especially if the exact proportions of the labeled sources is uncertain due to carryover of unlabeled medium during subculturing of tissues or the changing composition that occurs during introduction of labeled substrate in a continuous flow culture system. Additionally, this approach avoids the need to recalculate the isotopomer composition of the substrate formed by combining varying proportions of differently labeled sources, and it simplifies the organization of models containing two or more structurally identical networks each supplied with a differently labeled substrate. Note also that each labeled substrate is typically supplied at 98–99 % ^{13}C purity and is therefore a mixture of different isotopomers that must be correctly specified in the model in order to accurately reflect the isotopic composition of the input. Errors in the definition of the label inputs can result in poorly fitting, biased flux solutions. The isotopic composition of representative [^{13}C]glucose sources is provided in Table 2.
14. In 13C-FLUX this is achieved by defining equality constraints in the ftbl file describing the model. Variable steps that are free or constrained in one network are defined as determined in

Table 2
Isotopic composition of representative [^{13}C]glucose sources used in steady-state labeling studies

Unlabeled glucose (natural abundance)		[$^{13}C_6$]glucose (99 % abundance)		[1-^{13}C]glucose (99 % abundance)		[1,2-$^{13}C_2$]glucose (99 % abundance)	
Isotopomer	Fractional abundance	Isotopomer	Fractional abundance	Isotopomer	Fractional abundance	Isotopomer	Fractional abundance
000000	0.934	111111	0.940	100000	0.935	110000	0.936
100000	0.011	011111	0.010	110000	0.011	111000	0.011
010000	0.011	101111	0.010	101000	0.011	110100	0.011
001000	0.011	110111	0.010	100100	0.011	110010	0.011
000100	0.011	111011	0.010	100010	0.011	110001	0.011
000010	0.011	111101	0.010	100001	0.011	100000	0.010
000001	0.011	111110	0.010	000000	0.010	010000	0.010

The distribution of label in each isotopomer is indicated by 0 and 1 for ^{12}C and ^{13}C, respectively, in each C atom position. Only isotopomers with a fractional abundance greater than 0.001 are considered. The precision of ^{13}C measurement is typically in the order of 1 %. Therefore, although each source contains further isotopic impurities, isotopomers contributing less than 0.1 % of the labeled substrate are likely to have a negligible impact on the deduced flux distribution and can be ignored. For computational purposes, it is important that the fractional abundances of the isotopomers considered sum to 1

subsequent iterations of the network in the model and are set to be equal to the flux through the equivalent step in the first network; the only exception being the initial uptake steps that are independent and are defined individually to match the labeling strategy associated with the isotopomer measurements being analyzed in each specific network of the model.

15. Although parameter fitting may be performed in general mathematical programs such as MatLab, there is a range of dedicated software packages available for flux analysis. Currently we favor 13C-FLUX although in future we are likely to migrate to its successor, 13C-FLUX2 [16]. This software package has the advantages of being: (1) able to utilize both mass isotopomer and positional isotopomer data; (2) sufficiently adaptable to accommodate the complexities and unusual features of the plant network; and (3) capable of handling the large networks of reactions associated with the simultaneous analysis of isotopomer patterns obtained from multiple independently supplied ^{13}C-substrates that forms the core of our approach.
16. Fluxes may be expressed in any convenient units. However, note that the numerical values of fluxes will influence the maximum computable limit of exchange fluxes relative to the corresponding net fluxes, and thus the estimate of the

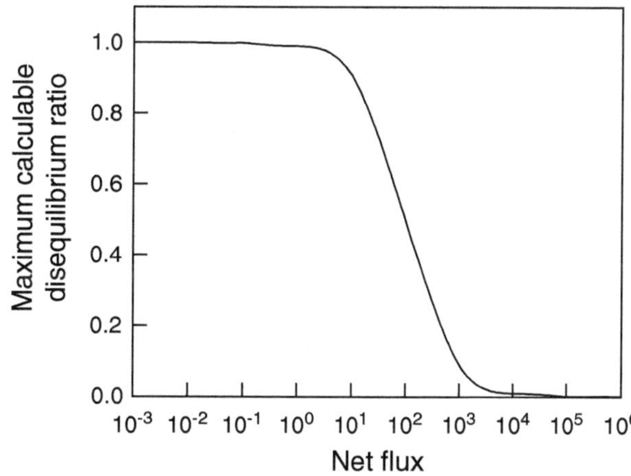

Fig. 5 Influence of flux scaling on the limit of estimated reaction reversibility. The maximum exchange flux obtained from the largest computable value of $v_{xch[0,1]}$ in 13C-FLUX was calculated from the relationship $v_{xch} = v_{xch[0,1]}/(1 - v_{xch[0,1]})$ for various values of the net (forward) flux (v_{net}) as indicated. The ratio of reverse (v_r) and forward (v_f) fluxes was determined from $v_r/v_f = v_{xch}/(v_{net} + v_{xch})$. This is equivalent to the disequilibrium ratio which varies from 0 for an irreversible reaction to 1 for a reaction at equilibrium. High absolute values for v_{net} constrain the degree of displacement from equilibrium that can be achieved by reactions in the flux network determined using 13C-FLUX

reversibility of the reaction (v_r/v_f). This is because 13C-FLUX employs a hyperbolic normalized scaling function, $v_{xch[0,1]} = v_{xch}/(1 + v_{xch})$, which transforms all exchange fluxes (which in principle can range from zero to infinity) to the finite range 0-1, and also sets the maximum permissible value of $v_{xch[0,1]}$ as 0.99 [17]. This limits the maximum possible value of v_{xch} that can be obtained in the fitting procedure, thereby constraining the extent of reversibility that can be calculated (*see* Fig. 5). When using the transformation function, we recommend scaling all fluxes to an input flux for the main substrate of unity. This allows good sensitivity over the range of flux reversibility likely to be encountered in metabolic networks and ensures comparability between different flux maps.

17. Free fluxes provide a set of parameters that are sufficient to define all fluxes in the network. Realistic networks of central carbon metabolism in plants are unlikely to contain a unique set of free fluxes, and the precise combination of which fluxes are free can be determined, in part, by the researcher. However, this should not affect the ultimate distribution of fluxes in the network.

18. A range of different algorithms are available for deducing fluxes in 13C-FLUX, including an evolutionary algorithm, CooolE-voAlpha, and gradient-based local optimization tools such as

CooolBFGSAlpha. Currently we favor using Donlp2, a sequential quadratic programing method, which provides fits much faster than the other approaches. Donlp2 is implemented within 13C-FLUX using the command "Donlp2 *.ftbl" where "*.ftbl" is the network model in ftbl format.

19. The number of Monte Carlo simulations performed by Donlp2 is defined by the value applied to the option "–m". Each simulation starts from a different set of random initial free flux parameters. Specifying the option "–bootstrap" varies the isotopomer values within the normal distribution defined by the value and variance of each measurement for each successive Monte Carlo simulation. We normally aim to generate at least 50 low-residuum feasible flux solutions at this stage to provide an adequate reflection of the flux space defined by the isotopomer measurements.
20. Infeasible flux solutions are those that fail to satisfy the constraints defined in the model. For simple networks containing only a few free fluxes every simulation may yield a feasible flux solution, but for larger models up to 90 % of the flux solutions may be infeasible depending on the size and complexity of the network and the number of constraints imposed on the model. Infeasible flux values can be identified in the output file generated by the Donlp2 routine in 13C-FLUX. Such a flux solution is designated as an "infeasible point" in the defined output file, and this label may be used to identify and remove the corresponding series of free flux values from the "*.ftbl.bs_sto" summary file of best-fit solutions generated automatically by the Donlp2 routine.
21. The feasible flux solutions generated by Donlp2 typically yield a population of closely related solutions with low residua and a smaller proportion of more variable, poorly fitting solutions that are taken to represent locally optimal fits to the labeling data. We normally discriminate between the two classes of solutions by principal component analysis of the flux values for each best-fit solution, with the low residua solutions forming a tight cluster and the higher residua solutions being more diffusely scattered distribution in a plot of the scores values. There is usually a distinct separation in the residuum between the two classes.
22. Apply the option "–m 0" when using Donlp2 to obtain a single best-fit solution based on the measured isotopomer values, starting from the initial free flux values entered in the model. The best-fit values of free fluxes which in combination with constrained input and output fluxes are sufficient to define the flux network are contained in the output file specified when implementing the Donlp2 routine. Additionally, the best-fit

values of all fluxes in the network may be obtained from the "out.fimprove.donlp2.txt" file generated automatically when running Donlp2.

23. The χ^2 value is used to assess the probability of the sum of the normalized differences between the observed and predicted isotopomer measurements occurring by chance assuming that the model correctly describes the metabolism of the substrate. Thus a χ^2 value greater than the critical value for $p = 0.05$ means that a discrepancy between the observed and predicted measurements of this size would be expected in less than 5 % of the analyses.

24. The degrees of freedom (df) are determined by:

$$\mathrm{df} = m_{\mathrm{ind}} - p_{\mathrm{tot}}$$

in which m_{ind} is the number of independent measurements and p_{tot} is the number of parameters estimated in the model. The data set may contain both mass isotopomer (m^{MS}) and positional isotopomer (m^{NMR}) measurements. Thus:

$$m_{\mathrm{ind}} = m_{\mathrm{ind}}^{\mathrm{MS}} + m_{\mathrm{ind}}^{\mathrm{NMR}}$$

Mass isotopomer data are usually expressed as fractional abundances. For a specific chemical fragment, the number of independent measurements is typically 1 less than the total number of mass isotopomers (m_{tot}), and so for a data set containing a number of fragments (m_{frag}):

$$m_{\mathrm{ind}}^{\mathrm{MS}} = m_{\mathrm{tot}}^{\mathrm{MS}} - m_{\mathrm{frag}}^{\mathrm{MS}}$$

Similarly, positional isotopomer data obtained from ^{13}C NMR measurements normally provide only the relative abundance of different groups of isotopomers which are scaled during the fitting procedure. For a data set in which the NMR measurements are considered as a number of discrete isotopomer scaling groups (m_{grp}) then:

$$m_{\mathrm{ind}}^{\mathrm{NMR}} = m_{\mathrm{tot}}^{\mathrm{NMR}} - m_{\mathrm{grp}}^{\mathrm{NMR}}$$

25. Two further checks can be used to assess the validity of the parameter optimization procedure. First, if the flux model is correct, then the standard deviation-weighted residuals will be normally distributed with a mean of 0 and standard deviation of 1. This may be assessed using a normal probability plot. Alternatively, the Kolmogorov–Smirnov or Shapiro–Wilks test can be used to test that the residuals are normally distributed. If the data fail this test or the standard deviation of the weighted residuals is significantly different from one, then either the residuals have been improperly weighted, or a subset of the isotopomers have been incorrectly assigned, either of which

may result in biased flux estimates. Secondly, a jack-knife analysis may be used to check that the flux estimates are not unduly influenced by a single aberrant data set. In this analysis, the Monte Carlo parameter fitting procedure is repeated omitting the data obtained from each labeled culture, in turn, to test for any bias in the derived flux solutions.

26. The two techniques describe in this section produce equivalent results that are preferable to the confidence intervals calculated using the EstimateStat routine in 13C-FLUX. The latter is based on local linear approximations of the sensitivity, which can lead to over- and underestimation of the reliability of flux estimates [18]. The approach based on Monte Carlo simulations has the advantage of allowing the confidence intervals of all fluxes to be determined from one set of simulations. However, the disadvantage is that reliable estimates of the confidence limits require a large number of simulations (*see* **Note 27**). The network sensitivity approach [18] generally requires far fewer simulations, although in the 13C-FLUX implementation described here it can only used to estimate the confidence limits of the free fluxes. In principle, this could be circumvented by defining a different set of free fluxes and repeating the analysis, although this is likely to be extremely time-consuming. However, the network sensitivity approach can be used to estimate the confidence limits for all fluxes in some other software packages such as Metran.

27. The number of Monte Carlo solutions needed to obtain robust estimates of the confidence interval of the flux values is considerably greater than required to obtain the mean value for each of the free fluxes during parameter determination. As a broad approximation, the 95 % limit of accuracy of a confidence interval (a) is defined by:

$$a \approx 1/\sqrt{n}$$

where n is the number of Monte Carlo simulations [19]. Thus the confidence interval determined from100 solutions will have an accuracy of about 10 % and the number of solutions needs to be increased to 10,000 to achieve an accuracy of 1 %. The disproportionate increase in effort needed to get the higher degree of accuracy is probably not worthwhile given the likely degree of biological variability, and we generally compromise by using about 400 simulations (nominally 5 % accuracy). In the future this constraint may be relaxed without massive investment in additional computer hardware by exploiting cloud computing to enable the simulations to be run in parallel on multiple virtual machines, thereby greatly decreasing the time needed to obtain very large numbers of solutions [20].

28. When using the Donlp2 routine in 13C-FLUX, the complete set of flux estimates for the best-fit solution of each Monte Carlo simulation can be extracted from the "out.fimprove.donlp2.txt" file using the value of the residuum to identify the relevant flux solution.

Acknowledgments

S.K.M. acknowledges financial support from the University of Oxford (Clarendon Scholarship), Exeter College (Mr Krishna Pathak Scholarship) and a UK Overseas Research Student award. We thank W. Wiechert (Forschungszüentrum Jülich GmbH) for permission to use 13C-FLUX and P. Spelluci (Fachbereich Mathematik, Technische Universität Darmstadt, Germany) for developing the Donlp2 algorithm.

References

1. Kruger NJ, Masakapalli SK, Ratcliffe RG (2012) Strategies for investigating the plant metabolic network with steady-state metabolic flux analysis: lessons from an *Arabidopsis* cell culture and other systems. J Exp Bot 63:2309–2323
2. O'Grady J, Schwender J, Shachar-Hill Y, Morgan JA (2012) Metabolic cartography: experimental quantification of metabolic fluxes from isotopic labelling studies. J Exp Bot 63:2293–2308
3. Ratcliffe RG, Shachar-Hill Y (2006) Measuring multiple fluxes through plant metabolic networks. Plant J 45:490–511
4. Allen DK, Ratcliffe RG (2009) Quantification of isotope label. In: Schwender J (ed) Plant metabolic networks. Springer, New York, pp 105–149
5. Huege J, Poskar CH, Franke M, Junker BH (2012) Towards high throughput metabolic flux analysis in plants. Mol Biosyst 8:2466–2469
6. Wiechert W, Möllney M, Petersen S, de Graaf AA (2001) A universal framework for ^{13}C metabolic flux analysis. Metab Eng 3:265–283
7. Möllney M, Wiechert W, Kownatzki D, de Graaf AA (1999) Bidirectional reaction steps in metabolic networks. IV. Optimal design of isotopomer labeling experiments. Biotechnol Bioeng 66:86–103
8. Libourel IGL, Gehan JP, Shachar-Hill Y (2007) Design of substrate label for steady state flux measurements in plant systems using the metabolic network of *Brassica napus* embryos. Phytochemistry 68:2211–2221
9. Schwender J, Shachar-Hill Y, Ohlrogge JB (2006) Mitochondrial metabolism in developing embryos of *Brassica napus*. J Biol Chem 281:34040–34047
10. Masakapalli SK, Le Lay P, Huddleston JE et al (2010) Subcellular flux analysis of central metabolism in a heterotrophic *Arabidopsis thaliana* cell suspension using steady-state stable isotope labeling. Plant Physiol 152:602–619
11. Troufflard S, Roscher A, Thomasset B et al (2007) *In vivo* ^{13}C NMR determines metabolic fluxes and steady state in linseed embryos. Phytochemistry 68:2341–2350
12. Kruger NJ, Troncoso-Ponce MA, Ratcliffe RG (2008) ^{1}H NMR of metabolite fingerprinting and metabolomic analysis of perchloric acid extracts from plant tissues. Nat Protoc 3:1001–1012
13. Kruger NJ, Huddleston JE, Le Lay P et al (2007) Network flux analysis: impact of ^{13}C-substrates on metabolism in *Arabidopsis thaliana* cell suspension cultures. Phytochemistry 68:2176–2188
14. Lonien J, Schwender J (2009) Analysis of metabolic flux phenotypes for two Arabidopsis mutants with severe impairment in seed storage lipid synthesis. Plant Physiol 151: 1617–1634

15. Williams TCR, Poolman MG, Howden AJ et al (2010) A genome-scale metabolic model accurately predicts fluxes in central carbon metabolism under stress conditions. Plant Physiol 154:311–323
16. Weitzel M, Nöh K, Dalman T et al (2013) 13CFLUX2—high-performance software suite for 13C-metabolic flux analysis. Bioinformatics 29:143–145
17. Wiechert W, de Graaf AA (1997) Bidirectional reaction steps in metabolic networks. I. Modeling and simulation of carbon isotope labeling experiments. Biotechnol Bioeng 55:101–117
18. Antoniewicz MR, Kelleher JK, Stephanopoulos G (2006) Determination of confidence intervals of metabolic fluxes estimated from stable isotope measurements. Metab Eng 8:324–337
19. Chernick MR (2007) Bootstrap methods: a guide for practitioners and researchers, 2nd edn. Wiley, Hoboken, NJ
20. Dalman T, Dörnemann T, Juhnke E et al (2013) Cloud MapReduce for Monte Carlo bootstrap applied to metabolic flux analysis. Future Gen Comp Syst 29:582–590

Chapter 5

Quantification of ^{13}C Enrichments and Isotopomer Abundances for Metabolic Flux Analysis Using 1D NMR Spectroscopy

Shyam K. Masakapalli, R. George Ratcliffe, and Thomas C.R. Williams

Abstract

The analysis of stable isotope incorporation following feeding of ^{13}C-labeled precursors to plant tissues provides the constraints necessary for metabolic flux analysis. This protocol describes the use of one-dimensional ^{1}H and ^{13}C nuclear magnetic resonance spectroscopy for the quantification of ^{13}C enrichments and isotopomer abundances in mixtures of metabolites or hydrolyzed biomass components.

Key words NMR spectroscopy, Flux analysis, Stable isotope labeling, Plant metabolism, Isotopomers, ^{13}C enrichments

1 Introduction

Metabolic flux analysis provides a framework for the simultaneous quantification of multiple fluxes in plant metabolism [1, 2]. The approach is based on the construction of a metabolic model that can explain the extent to which stable isotope label, most often ^{13}C, is incorporated into intermediates and end products of metabolism during feeding of a metabolically active tissue with an isotopically labeled precursor.

Isotope incorporation can be described in various ways [3, 4], with each type of measurement providing different amounts of information about the location and extent of labeling within metabolites, and hence about the metabolic fluxes responsible for the label distribution. Fractional enrichments, for example, indicate the proportion of atoms within a population of molecules that are labeled (*see* Fig. 1a). Positional enrichments provide additional detail about the location of isotope incorporation, in that they represent the proportion of molecules incorporating stable isotope in a specific atom (*see* Fig. 1b). Isotope isomers (isotopomers) are molecules differing only in the presence or absence of stable isotope

Martine Dieuaide-Noubhani and Ana Paula Alonso (eds.), *Plant Metabolic Flux Analysis: Methods and Protocols*, Methods in Molecular Biology, vol. 1090, DOI 10.1007/978-1-62703-688-7_5,

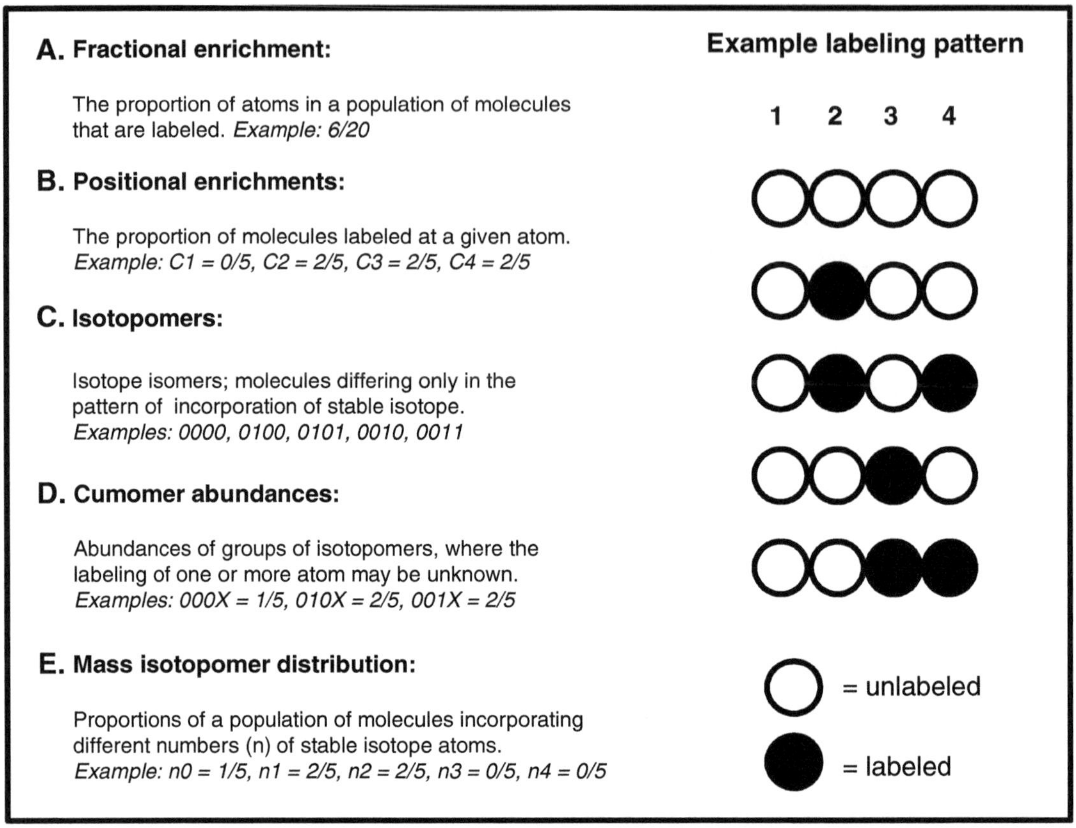

Fig. 1 Types and examples of labeling data. Definitions of the different types of labeling data are given together with example values determined for the population of 5 four-carbon molecules illustrated. For isotopomers and cumomers, 1 represents a labeled (^{13}C) atom, 0 represents an unlabeled atom (^{12}C), and X represents an atom where the labeling is either unknown or not specified (C4 in this example). Mass isotopomers (E) are also often identified using the M+X notation, where X is the number of stable isotope atoms

label in each atom, and for any compound labeled with ^{13}C there are therefore 2^n isotopomers [4] where n is the number of carbon atoms in the compound (*see* Fig. 1c). Quantification of isotopomer abundances describes the pattern of incorporation of isotope label within a compound and hence can provide constraints for the determination of metabolic fluxes. It is, however, technically demanding to measure all isotopomers for any metabolite with more than three carbon atoms. The cumomer (cumulative isotopomer) representation [5] of labeling addresses this problem by placing different isotopomers into groups where the labeling state of one or more atoms within a compound may be unknown (*see* Fig. 1d). Finally, mass isotopomer distributions (groups of isotopomers differing only in mass) indicate the proportion of a population of molecules, or molecular fragments, incorporating a given number of stable isotope atoms, and therefore indicate the extent to which a metabolite is labeled (*see* Fig. 1e). Which of these types of measurement can be obtained depends to a large extent upon the technique used for the analysis of stable isotope incorporation.

The most commonly used stable isotope for metabolic flux analysis, ^{13}C, has a nuclear magnetic moment, and so nuclear magnetic resonance (NMR) spectroscopy is an important tool for quantifying the pattern of label incorporation [3, 4, 6]. Whilst mass spectrometry can be readily used to obtain fractional enrichments and mass isotopomer distributions [3], the fact that different atoms within a compound give rise to signals at different positions in NMR spectra means that NMR spectroscopy can be used to measure positional enrichments [3, 6]. This approach is particularly informative when labeling experiments are carried out using positionally labeled substrates such as [1-^{13}C] glucose (*see* for example, 7). However, through the detection of spin–spin coupling between magnetically inequivalent ^{13}C atoms NMR can also be used to detect when isotope has been incorporated into adjacent atoms, thus making it suitable for the quantification of isotopomer and cumomer abundances [8, 9] and for use with samples from labeling experiments that generate molecules containing multiple label atoms [10–13].

NMR spectroscopy can be used in a variety of ways to determine isotopomer abundances and fractional enrichments, and the NMR experiment of choice will depend on the type of measurement required, the tissue under study and the labeling experiment that was performed, as well as on the configuration of the available instrument. However, a crucial choice to be made is whether to employ relatively straightforward one-dimensional (1D) NMR, where all the labeling information is presented in a single frequency dimension, or more complex multidimensional NMR experiments where this information is dispersed into two or three dimensions. Two-dimensional (2D) NMR methods offer high sensitivity and less extensive peak overlap, and they have been employed with great success for the analysis of stable isotope labeling in the metabolic flux analysis of plant tissues [10, 13]. Fast 2D and three-dimensional NMR methods have also been developed recently for the measurement of positional enrichments that promise to increase even further the number of signals that can be separated [14, 15] whilst decreasing the analysis time. However, 1D NMR experiments are preferred under certain conditions. Firstly, achieving optimal sensitivity for 2D NMR may not be straightforward, typically requiring the use of specialized hardware (e.g., pulsed field gradient generators) that may not be available, the use of carefully designed broadband decoupling pulses to avoid significant sample heating and the modification of pulse sequences to permit the acquisition of quantitative data [16, 17]. Secondly, the quantitative analysis of spectra obtained from 2D NMR experiments is nontrivial, normally necessitating the averaging of portions of the spectra to produce pseudo-1D spectra [10], and the confirmation that all peaks in the F1 dimension (typically ^{1}H for homonuclear or ^{13}C for heteronuclear experiments) contain sufficient data points to be reliably quantified. Thirdly, and perhaps most importantly, due to the presence of bond-specific parameters such as transfer times in most heteronuclear 2D NMR pulse sequences, the peak intensities of ^{13}C multiplets cannot be compared [18]. Whilst this problem can be partially circumvented by the use of J-resolved 2D NMR techniques [19, 20] or modification of standard pulse sequences [18] it has so far limited the use of heteronuclear 2D NMR to labeling experiments that produce significant quantities of multiply labeled molecules, typically those carried out with uniformly labeled substrates [10, 13]. However these experiments are not necessarily the most informative for metabolic flux analysis [9, 21]. In contrast, 1D NMR experiments can be performed with almost any NMR spectrometer, and providing suitable precautions are taken, informative data for metabolic flux analysis can be obtained from extracts of tissue subject to almost any type of labeling strategy.

This protocol covers the measurement of positional isotope enrichments and isotopomer abundances using 1D ^{1}H and ^{13}C NMR [7, 8, 22]. Samples obtained from stable isotope labeled tissue are dissolved in a deuterated solvent containing a chemical shift reference and, in the case of aqueous samples, phosphate buffer. The chelating agent EDTA is also included to reduce the line-broadening effect of paramagnetic cations and/or calcium ions that may be present in the sample. ^{1}H spectra are recorded using a standard 90° pulse sequence with pre-saturation to remove the signal from residual water. ^{13}C spectra are recorded using a 90° pulse sequence with ^{1}H decoupling. In both pulse sequences particular care must be paid to the relaxation delay to ensure that signals throughout the entire spectrum are directly proportional to the amount of ^{13}C or ^{1}H present. Signals from these spectra are quantified using line fitting or deconvolution, and the peak areas obtained are used to calculate positional enrichments and isotopomer abundances that can then be used for metabolic flux analysis.

2 Materials

1. Dried extracts of tissue fed with ^{13}C labeled precursor (*see* **Note 1**).
2. For 1D ^{13}C NMR: 25 mM 1,4-dioxane, 10 mM KH_2PO_4/K_2HPO_4 (pH 7.5), 10 mM EDTA in D_2O (*see* **Note 2**).
3. For 1D ^{1}H NMR: 0.3 mM 3-(trimethylsilyl) propionic-2, 2, 3, 3-d_4 acid (TSP), 17 mM KH_2PO_4/K_2HPO_4 (pH 7.5), 0.3 mM EDTA in D_2O (*see* **Note 2**).

3 Methods

3.1 Sample Preparation

1. Dissolve the labeled plant extract in the appropriate NMR solvent (*see* Subheading 2): typically 0.8 ml for 5 mm probes and 4 ml for 10 mm probes.
2. Centrifuge the sample (10 min, 10,000 × *g*) or pass sample through a 0.22 μm syringe filter to remove any insoluble particles.
3. Transfer the supernatant to a clean 5 mm or 10 mm NMR tube as appropriate (*see* **Note 3**).

3.2 Instrument Setup

1. Install the ^{1}H or broadband (^{13}C) probe according to the manufacturer's instructions and set the probe temperature to 20 °C.
2. Introduce the sample into the NMR spectrometer and allow several minutes for the temperature of the sample to stabilize.
3. Load the appropriate standard NMR pulse sequence: 1D ^{13}C with ^{1}H decoupling for ^{13}C NMR, or 1D ^{1}H with a pre-saturation pulse for ^{1}H NMR.
4. Tune and match the ^{1}H and broadband channels (*see* **Note 4**).
5. Optimize the lock signal and turn on lock regulation (*see* **Note 5**).
6. Shim the sample to optimize field homogeneity, which will typically correspond to an intrinsic linewidth of less than 1 Hz for TSP (^{1}H NMR) or 2 Hz for 1,4-dioxane (^{13}C NMR) (*see* **Note 6**).

3.3 Pulse Sequence Setup and Data Acquisition

Isotopomer and Cumomer Abundances Using ^{13}C NMR

1. Input the ^{13}C 90° pulse calibration details for the installed probe and set the detection period to a suitable value, typically 1 s (*see* **Note 7**).
2. Input the appropriate relaxation delay and decoupling channel configuration; 19 s without induction of the nuclear Overhauser effect (NOE) for acquisition of fully relaxed spectra or 6 s with low power ^{1}H decoupling for induction of the NOE when full relaxation is not required (*see* **Note 8**).
3. Input the ^{1}H 90° pulse calibration details for the installed probe and set up low-power Waltz16, or an equivalent sequence, for ^{1}H decoupling during the detection period.
4. Set number of scans to 1024/block (*see* **Note 9**) and acquire data.
5. Confirm that the lock signal level remains constant over the first few minutes of acquisition, that the linewidth of the

internal standard is satisfactory (*see* **Note 6**) and that ^{1}H decoupling is complete.

Positional Enrichments Using 1H NMR

1. Input the ^{1}H 90° pulse calibration details for the installed probe and set the detection period to 4 s.
2. Set the relaxation delay to 2 s (*see* **Note 8**).
3. Turn on pre-saturation during the relaxation delay and adjust the pre-saturation offset and power parameters to minimize the residual water signal (*see* **Note 10**).
4. Set the number of scans to at least 64 and acquire data (*see* **Note 9**).
5. Confirm that the lock level remains constant over the first few minutes of acquisition and that the linewidth of the internal standard is satisfactory (*see* **Note 6**).

3.4 Spectral Processing

1. Import data into processing software (*see* **Note 11**) and carry out summation of blocks if applicable.
2. Zero fill the data and apply a baseline offset correction to the free induction decay if necessary.
3. Apply line-broadening (typically exponential multiplication) of 1 Hz for ^{1}H spectra and 2.5 Hz for ^{13}C spectra.
4. Apply the Fourier transformation.
5. Apply automatic phase correction to the spectrum and then adjust the phase manually if necessary.
6. Flatten the spectral baseline using a linear regression method, thereby removing baseline oscillations.
7. Set the chemical shift reference: 67.3 ppm for 1,4-dioxane (^{13}C spectra), or 0 ppm for TSP (^{1}H spectra).

3.5 Peak Assignment and Quantification

1. Assign peaks based on comparison with literature values, mixtures of prepared standards or 2D NMR spectra (*see* **Note 12**).
2. Assign peaks arising from spin–spin coupling (*see* **Note 13**).
3. Carry out line fitting to deconvolute overlapping peaks and determine peak areas (*see* **Note 14** and Fig. 2). Use a Lorentzian–Gaussian peak model and allow the software to adjust peak height, width and the fractional Lorentzian contribution to minimize the difference between the real and simulated spectra.
4. Assign peaks to cumomers (*see* Fig. 1d) using 0 to denote ^{12}C, 1 to denote ^{13}C and X to indicate where labeling is unknown (*see* **Notes 13** and **15**, and Figs. 2 and 3).

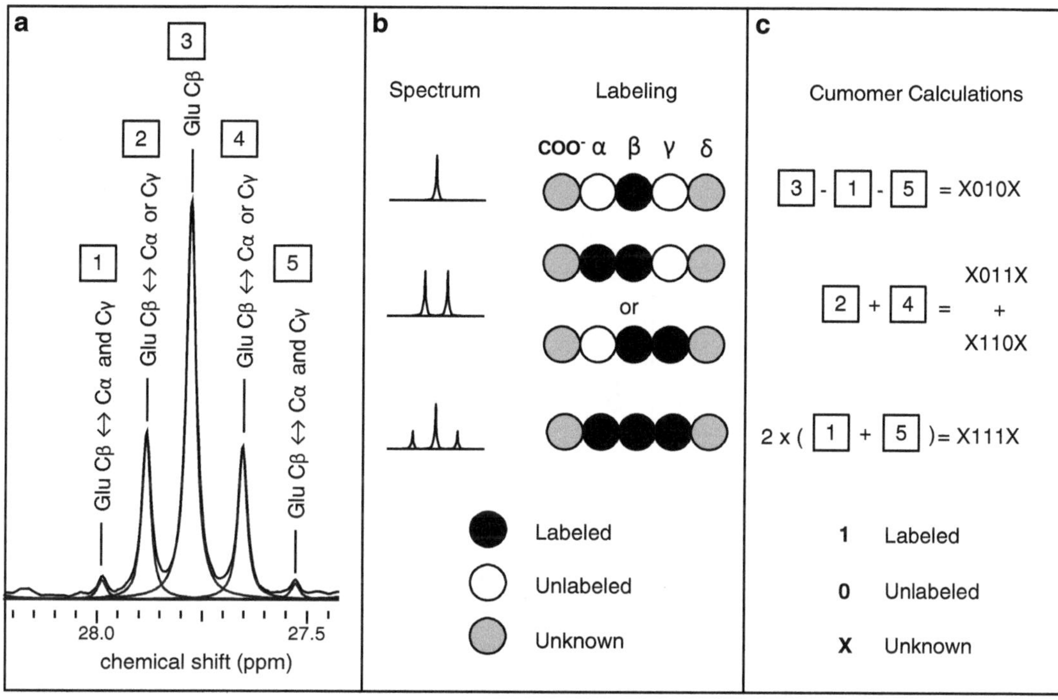

Fig. 2 Peak assignment, line fitting, and calculation of cumomer abundances in a representative 1D ^{13}C NMR spectrum. (**a**) Assignment and line fitting of signals arising from the β carbon of the amino acid glutamate. Double headed arrows (↔) indicate signals arising from spin–spin coupling between the β carbon atom and the α/γ carbon atoms. The spectrum was recorded from a mixture of amino acids obtained after hydrolyzing the protein from a heterotrophic Arabidopsis cell suspension culture fed with [1-^{13}C]glucose for 5 days. (**b**) Interpretation of coupling patterns. The NMR spectra on the left result from the labeled molecules that are illustrated on the right. Isotopomers labeled at Cβ but not at Cα or Cγ give rise to the singlet peak 3. Isotopomers labeled at Cβ and Cα, or Cβ and Cγ produce a doublet with a 1:1 ratio; signals 2 and 4 in the spectrum arise from these molecules. Isotopomers labeled at Cβ, Cα, and Cγ produce a triplet with the ratio 1:2:1. The two smaller signals correspond to peaks 1 and 5 in the spectrum, whilst the central peak of the triplet contributes to peak 3. The spectrum does not contain details about the labeling of the COO$^-$ group or Cδ, and thus all peaks illustrated here arise from mixtures of isotopomers (cumomers) where the labeling of these atoms is unknown. (**c**) Calculation of cumomer abundances using the peaks areas determined in (**a**) and the relationships between the coupling patterns and labeling illustrated in **b**

5. Sum the peak areas arising from the same isotopomer or cumomer (*see* **Note 15** and Fig. 2).
6. For ^{1}H spectra calculate positional enrichments directly as the ratio of ^{13}C-^{1}H peak areas to total peak areas for a specific ^{1}H signal (*see* **Notes 15** and **16** and Figs. 3 and 4).
7. Calculate measurement errors as a fixed or variable percentage of the measured value (*see* **Note 17**).

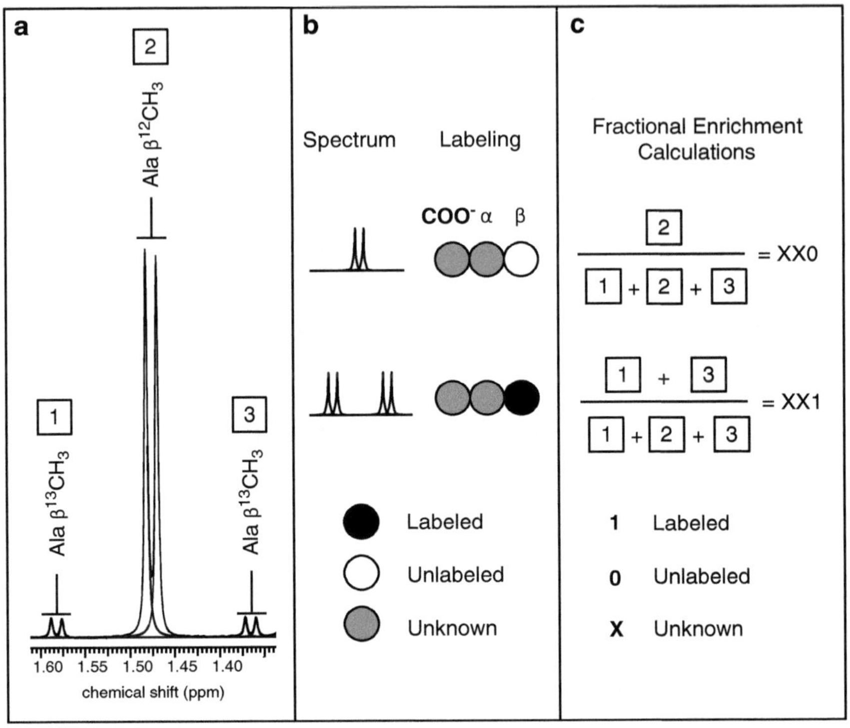

Fig. 3 Peak assignment, line fitting, and calculation of cumomer abundances in a representative 1D ^{1}H NMR spectrum. (**a**) Assignment and line fitting of signals arising from the protons attached to the β carbon of the amino acid alanine. The spectrum was obtained from a perchloric acid extract of a heterotrophic Arabidopsis cell suspension culture fed with [1-^{13}C]glucose for 5 days. (**b**) Interpretation of coupling patterns. The ^{1}H NMR spectra on the left result from the labeled molecules illustrated on the right. Isotopomers labeled at Cβ give rise to the pair of doublets 1 and 3. Isotopomers unlabeled at Cβ produce the doublet 2. The doublets arise from ^{1}H coupling between the Cβ and Cα H atoms. (**c**) Calculation of fractional enrichments. The fractional enrichment of Cβ can be calculated using the peaks areas determined in (**a**) and the relationships between the coupling patterns and labeling illustrated in **b**

4 Notes

1. The methods described here can be applied to aqueous or organic extracts of plant biomass [7–9, 22], hydrolyzed biomass constituents including protein [8, 10, 13] and starch [23], and growth media [24]. 1D ^{1}H NMR spectroscopy can typically detect metabolites in the micromolar concentration range in the sample, whilst 1D ^{13}C NMR spectroscopy can detect metabolites at millimolar concentrations. Extraction of several grams fresh weight of labeled tissue is therefore normally required to obtain a sufficient signal-to-noise ratio for reliable quantification of stable isotope incorporation. The success of these methods will also depend upon the labeling strategy employed. Labeling experiments carried out using precursors

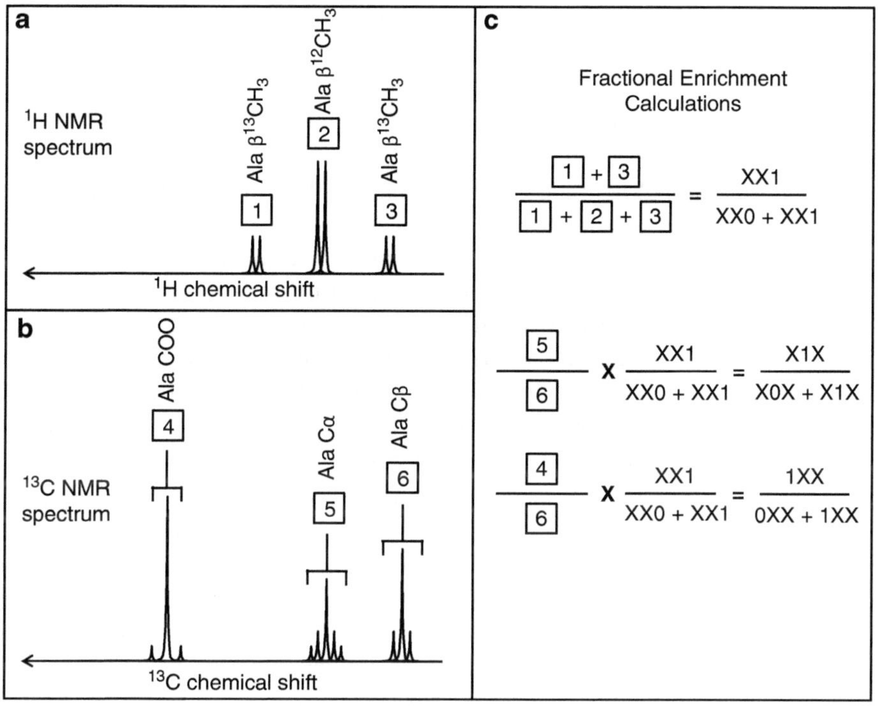

Fig. 4 Calculation of fractional enrichments using ^{1}H and ^{13}C NMR spectra. (**a**) Portion of an illustrative ^{1}H NMR spectrum of the amino acid alanine indicating peaks arising from isotopomers with labeled and unlabeled Cβ atoms. (**b**) Illustrative fully relaxed ^{13}C NMR spectrum of alanine, showing peaks arising from the presence of ^{13}C in each of the three carbon atoms. Peak areas 4, 5, and 6 represent the sum of all the signals indicated. (**c**) Calculation of fractional enrichments for atoms not observed using ^{1}H NMR. The fractional enrichment of Cβ determined from spectrum A together with the peak areas 4, 5, and 6 from spectrum B can be used to calculate the fractional enrichments of Cα and COO

with low ^{13}C enrichment may not lead to sufficient incorporation of ^{13}C in metabolites and biomass constituents for subsequent analysis by ^{1}H or ^{13}C NMR spectroscopy. Given the variation in field strength and probe sensitivity between instruments as well as biological variation it is not possible to give definitive information about the amount of labeled tissue that must be processed to obtain a sample that will produce a good quality spectrum. It is therefore strongly recommended to carry out tests using unlabeled samples before carrying out a labeling experiment that will be analyzed using NMR spectroscopy. Signals that can be detected at natural abundance ^{13}C levels, either as singlets in 1D ^{13}C NMR spectra or as peaks arising from ^{1}H-^{13}C coupling in ^{1}H spectra, will certainly be detectable in labeled extracts.

2. If samples are to be analyzed by ^{13}C NMR only, then it is only necessary to add sufficient D_2O to obtain a lock signal (10 % v/v). Samples to be analyzed by ^{1}H NMR should be

dissolved in 100 % D_2O, and stocks of concentrated EDTA and phosphate buffer also made up in D_2O. 1,4-dioxane is highly toxic and flammable and must be handled and disposed of appropriately. Deuterated chloroform can be used to dissolve samples soluble in organic solvents.

3. We have typically used a 10 mm broadband probe for the acquisition of 1D ^{13}C data and a 5 mm triple resonance probe for 1D ^{1}H experiments, though in principle this same 5 mm probe could also be used for direct detection of ^{13}C, albeit with reduced sensitivity. The sample volume should be sufficiently large to avoid end effects during shimming. Using the same volume for each sample will reduce the need for shimming between samples. After analysis samples may be recovered and stored at −80 °C for future reanalysis if necessary.
4. Samples containing high concentrations of inorganic salts, for example those obtained from perchloric acid extractions [25], may prove difficult to tune and match. Dilution of the sample or removal of salt using ion exchange chromatography is a possible solution. Furthermore, certain probes offer increased tolerance to high salt concentrations.
5. Lock signal: 1D ^{13}C NMR experiments for the determination of isotopomer abundances are typically long (several hours at least), making the use of a lock signal for correction of instrument/field drift indispensible.
6. Automatic optimization of the Z shims and the low level non-spinning shims is usually sufficient to obtain spectra suitable for integration or peak deconvolution. Sample spinning is not usually necessary. Linewidths of the internal standards should be measured after a few scans have been recorded and the sample re-shimmed if the appropriate linewidth is not obtained. The linewidths of several analyte peaks can also be checked. Excessively broad peaks reduce sensitivity and complicate the integration and line fitting of the signals. If linewidths cannot be improved by further shimming this may indicate the presence of high salt concentrations or paramagnetic ions.
7. Optimal signal to noise will only be obtained with calibrated 90° pulses. Follow the manufacturer's instructions for hard pulse calibration for the installed probe. The detection period must be long enough for the signal to decay to zero before the application of zero filling and line-broadening to prevent the introduction of artifacts into the final spectrum, and can be extended if necessary. Confirm that receiver overload is not occurring, and decrease the receiver gain if necessary.
8. In order for all signals throughout the spectrum (^{1}H or ^{13}C) to be directly proportional to each other complete relaxation must take place between each 90° pulse. An appropriate relaxation

time can be determined empirically by comparing the signal to noise obtained for specific signals of interest after a fixed number of scans with different relaxation delays; signal to noise will be reduced if relaxation is not complete. Increasing the relaxation delay will inevitably reduce the signal to noise that can be obtained in a given length of time. For certain labeling experiments, in particular those using fully labeled precursors, the relative intensities of 1D ^{13}C signals arising from different carbon atoms within a molecule contain little information about metabolic fluxes. In this case a short relaxation delay (6 s) can be used and the NOE induced, and only relative signals within multiplets considered without significant loss of information [9]. A relaxation delay of 2 s has been used successfully for metabolite fingerprinting using ^{1}H NMR [25].

9. The precision of the flux estimates that can be obtained using MFA depends on the measurement precision of the enrichments and isotopomer abundances used to constrain the flux distribution. NMR spectra with high signal to noise must therefore be obtained. The number of scans required to obtain sufficient signal to noise for MFA will depend upon the equipment, the concentration of metabolites in the extract and the amount of ^{13}C incorporated during the labeling experiment. This parameter must therefore be determined empirically for each type of sample; however as a guide we aim for an RMS signal to noise of at least 100 for the most intense peaks originating from labeled compounds in 1D ^{13}C spectra. This corresponds to an estimated error (*see* **Note 17**) of approximately 3 %. For long ^{13}C experiments consider splitting the experiment into blocks of smaller numbers of scans. This will allow any changes in the sample that occur during the acquisition to be detected.

10. Adjust the pre-saturation power and offset to reduce the residual water signal to as close to zero as possible without distorting signals at similar chemical shifts (typically ^{1}H attached to the anomeric carbons of sugars). This can be carried out empirically by carrying out short experiments with different values for these parameters.

11. The majority of software packages for analysis of NMR data are capable of the processing steps detailed here. Consult the documentation supplied with the software for additional information on how to implement these steps.

12. Simple sugars, organic acids and amino acids are typically responsible for the majority of signals in 1D ^{13}C NMR spectra of plant extracts that are of sufficient intensity to be used for the measurement of stable isotope incorporation. For this reason assignment of peaks is normally straightforward. 1D ^{1}H NMR

spectra on the other hand are often highly crowded and assignment may be more challenging; 2D experiments such as ^{1}H-^{1}H COSY performed on representative samples may be a useful aid to assignment in these cases.

13. Signals arising from spin–spin coupling in 1D ^{13}C spectra can be assigned by comparison of 1-bond coupling constants with values available in the literature and public databases. When values are unavailable, knowledge of the structure of the molecule can narrow down the list of isotopomers that may be responsible for the signal. Where signals arising from adjacent positions within the same compound are present in the spectrum these can be used to cross-check the assignment of peaks arising from spin–spin coupling. *See* Fig. 2 for an example of peak assignment.
14. If fully resolved signals have been obtained they may be integrated directly. If signals overlap then it is necessary to use a line fitting (deconvolution) procedure to determine the relative contribution of the individual peaks. Spectral resolution can, in the case of spectra with good signal to noise, be improved by applying a Lorentzian–Gaussian resolution enhancement window function prior to Fourier transformation. The process of line fitting or integration is more challenging for spectra containing signals from many metabolites or metabolites giving rise to extensive spin–spin coupling. Samples can be simplified by chemical fractionation (for example ion exchange chromatography) or by derivatization to constrain the conformation of sugars [23]. Multidimensional NMR experiments also represent a possible solution in this instance [14–16, 18, 19]. The use of labeling strategies expected to produce less spin–spin coupling in analytes (e.g., positional labeling strategies) may also help simplify spectra.
15. Positional labeling patterns can be indicated conveniently using the 1, 0, X nomenclature, indicating labeled, unlabeled and unknown positions respectively within a compound (*see* Figs. 1, 2, and 3). For some software this method of cumomer identification may be carried through to the metabolic model used to interpret the labeling data [26]. In this case it is crucial that the peak annotation reflects the requirements of the model exactly.
16. If fully relaxed ^{1}H and ^{13}C spectra have been obtained for the same extract, fractional enrichments of atoms for which signals were not detected by 1D ^{1}H NMR can also be calculated using one experimentally determined enrichment from ^{1}H NMR and the incorporation of ^{13}C in other positions within the same molecule from the 1D ^{13}C spectrum (*see* Fig. 4).
17. Estimates of measurement errors are required for parameter estimation during MFA, and are typically calculated as a

percentage of the fractional enrichment or isotopomer abundance value [26]. This percentage may be determined using an empirically determined relationship between signal to noise and measurement precision [8] or fixed for all measurements (for example, 7). If a fixed value is used particular care must be taken to obtain high signal to noise for all peaks, thereby avoiding assigning unrealistically low errors to small peaks.

Acknowledgements

TCRW was supported by a Petrobras (Petroleo Brasileiro S.A.) postdoctoral grant. SKM acknowledges financial support from the University of Oxford (Clarendon Scholarship), Exeter College (Mr Krishna Pathak Scholarship), and a UK Overseas Research Student award.

References

1. Ratcliffe RG, Shachar-Hill Y (2006) Measuring multiple fluxes through plant metabolic networks. Plant J 45:490–511
2. O'Grady J, Schwender J, Shachar-Hill Y et al (2012) Metabolic cartography: experimental quantification of metabolic fluxes from isotopic labelling studies. J Exp Bot 63:2293–2308
3. Allen DK, Ratcliffe RG (2009) Quantification of isotope label. In: Schwender J (ed) Plant Metabolic Networks. Springer, New York, pp 105–149
4. Wiechert W (2001) ^{13}C Metabolic Flux Analysis. Metab Eng 3:196–206
5. Wiechert W, Möllney M, Isermann N et al (1999) Bidirectional reaction steps in metabolic networks: III. Explicit solution and analysis of isotopomer labeling systems. Biotechnol Bioen 66:69–85
6. Ratcliffe RG, Shachar-Hill Y (2005) Revealing metabolic phenotypes in plants: inputs from NMR analysis. Biol Rev Camb Philos Soc 80:27–43
7. Alonso AP, Goffman FD, Ohlrogge JB et al (2007) Carbon conversion efficiency and central metabolic fluxes in developing sunflower (*Helianthus annuus L.*) embryos. Plant J 52:296–308
8. Williams TC, Miguet L, Masakapalli SK et al (2008) Metabolic network fluxes in heterotrophic Arabidopsis cells: stability of the flux distribution under different oxygenation conditions. Plant Physiol 148:704–718
9. Masakapalli SK, Le Lay P, Huddleston JE et al (2010) Subcellular flux analysis of central metabolism in a heterotrophic Arabidopsis cell suspension using steady-state stable isotope labeling. Plant Physiol 152:602–619
10. Sriram G, Fulton DB, Iyer VV et al (2004) Quantification of compartmented metabolic fluxes in developing soybean embryos by employing biosynthetically directed fractional ^{13}C labeling, two-dimensional [^{13}C, ^{1}H] nuclear magnetic resonance, and comprehensive isotopomer balancing. Plant Physiol 136:3043–3057
11. Ettenhuber C, Radykewicz T, Kofer W et al (2005) Metabolic flux analysis in complex isotopolog space. Recycling of glucose in tobacco plants. Phytochemistry 66:323–335
12. Ettenhuber C, Spielbauer G, Margl L et al (2005) Changes in flux pattern of the central carbohydrate metabolism during kernel development in maize. Phytochemistry 66:2632–2642
13. Sriram G, Fulton DB, Shanks JV (2007) Flux quantification in central carbon metabolism of Catharanthus roseus hairy roots by ^{13}C labeling and comprehensive bondomer balancing. Phytochemistry 68:2243–2257
14. Giraudeau P, Massou S, Robin Y et al (2011) Ultrafast quantitative 2D NMR: an efficient tool for the measurement of specific isotopic enrichments in complex biological mixtures. Anal Chem 15:3112–3119
15. Giraudeau P, Cahoreau E, Massou S et al (2012) UFJCOSY: A Fast 3D NMR Method for Measuring Isotopic Enrichments in Complex Samples. Chemphyschem 13:3098–3101

16. Massou S, Nicolas N, Letisse F et al (2007) Application of 2D-TOCSY NMR to the measurement of specific ^{13}C-enrichments in complex mixtures of ^{13}C-labeled metabolites. Metab Eng 9:252–257
17. Thrippleton MJ, Keeler J (2003) Elimination of Zero-Quantum interference in two-dimensional NMR spectra. Angew Chem Int Ed 42:3938–3941
18. Peterson DJ, Loening NM (2007) QQ-HSQC: a quick, quantitative heteronuclear correlation experiment for NMR spectroscopy. Magn Reson Chem 45:937–941
19. Cahoreau E, Peyriga L, Hubert J et al (2012) Isotopic profiling of ^{13}C-labeled biological samples by two-dimensional heteronuclear J-resolved nuclear magnetic resonance spectroscopy. Anal Biochem 427:158–163
20. Ludwig C, Viant MR (2010) Two-dimensional J-resolved NMR spectroscopy: review of a key methodology in the metabolomics toolbox. Phytochem Anal 21:22–32
21. Libourel IGL, Gehan JP, Shachar-Hill Y (2007) Design of substrate label for steady state flux measurements in plant systems using the metabolic network of *Brassica napus* embryos. Phytochemistry 68:2211–2221
22. Rontein D, Dieuaide-Noubhani M, Dufoure EJ et al (2002) The metabolic architecture of plant cells: stability of central metabolism and flexibility of anabolic pathways during the growth cycle of tomato cells. J Biochem 46:43948–43960
23. Allen DK, Shachar-Hill Y, Ohlrogge JB (2007) Compartment-specific labeling information in ^{13}C metabolic flux analysis of plants. Phytochemistry 68:2197–2210
24. Chen X, Alonso AP, Allen DK et al (2011) Synergy between ^{13}C-metabolic flux analysis and flux balance analysis for understanding metabolic adaptation to anaerobiosis in *E. coli*. Metab Eng 13:38–48
25. Kruger NJ, Troncoso-Ponce MA, Ratcliffe RG (2008) ^{1}H NMR of metabolite fingerprinting and metabolomic analysis of perchloric acid extracts from plant tissues. Nat Protocols 3:1001–1012
26. Wiechert W, Möllney M, Petersen S et al (2001) A universal framework for ^{13}C metabolic flux analysis. Metab Eng 3:265–283

Chapter 6

Analysis of Proteinogenic Amino Acid and Starch Labeling by 2D NMR

Quyen Truong and Jacqueline V. Shanks

Abstract

Comprehensive analysis of isotopic labeling patterns of metabolites in proteinogenic amino acids and starch for plant systems lay in the powerful tool of 2-Dimensional [^{1}H, ^{13}C] Nuclear Magnetic Resonance (2D NMR) spectroscopy. From ^{13}C-labeling experiments, 2D NMR provides information on the labeling of particular carbon positions, which contributes to the quantification of positional isotope isomers (isotopomer). 2D Heteronuclear Single Quantum Correlation (HSQC) NMR distinguishes particularly between the labeling patterns of adjacent carbon atoms, and leads to a characteristic enrichment of each carbon atom of amino acids and glucosyl and mannosyl units present in hydrolysates of glycosylated protein. Furthermore, this technique can quantitatively classify differences in glucosyl units of starch hydrolysate and of protein hydrolysate of plant biomass. Therefore, the 2D HSQC NMR method uses proteinogenic amino acids and starch to provide an understanding of carbon distribution of compartmentalization in the plant system. NMR has the advantage of minimal sample handle without separate individual compounds prior to analysis, for example multiple isotopomers can be detected, and their distribution extracted quantitatively from a single 2D HSQC NMR spectrum. The peak structure obtained from the HSQC experiment show multiplet patterns, which are directly related to isotopomer balancing. These abundances can be translated to maximum information on the metabolic flux analysis. Detailed methods for the extractions of protein, oil, soluble sugars, and starch, hydrolysis of proteinogenic amino acid and starch, and NMR preparation using soybean embryos cultured in vitro as a model plant systems are reported in this text. In addition, this chapter includes procedures to obtain the relative intensity of 16 amino acids and glucosyl units from protein hydrolysate and the glucosyl units of starch hydrolysate of soybean embryos in 2D HSQC NMR spectra.

Key words Isotopomer measurement, 2D HSQC NMR, Metabolic flux analysis, Proteinogenic amino acid hydrolysis, Starch hydrolysis, Chemical shifts

1 Introduction

Nuclear Magnetic Resonance (NMR) is a well-known analytical method for the structural identification of newly isolated compounds that have been structurally preserved. In recent years, NMR has become a notable biotechnology for the elucidation of plant metabolomics, such as biosynthetic pathways and metabolite

Martine Dieuaide-Noubhani and Ana Paula Alonso (eds.), *Plant Metabolic Flux Analysis: Methods and Protocols*, Methods in Molecular Biology, vol. 1090, DOI 10.1007/978-1-62703-688-7_6,

flux via the quantitative assessment of relative abundances of isotope isomers (isotopomers). NMR utilizes a stable isotopic tracer of carbon, nitrogen, phosphorus, sulfur, and hydrogen to yield information about metabolites in the plant system. Especially with carbon, the ^{13}C-labeling experiment can pinpoint the distribution of carbon fluxes to identify pathways in complex metabolite networks such as parallel metabolic, metabolic cycles, reversible or bidirectional reaction, and compartmentalization pathways of plant system [1]. The greatest features of ^{13}C-labeling NMR methods are: (a) flux analysis information such as carbon enrichments (the amount of ^{13}C label that can be integrated in an individual carbon position of a metabolite), (b) quantification of positional isotopomers (the various isotope isomer patterns that can form the same metabolite), and (c) labeling of adjacent carbon atoms (provide information on carbon-to-carbon bond formation and cleavage [2, 3]). The isotopomer distribution is determined from the analysis of ^{13}C-^{13}C coupling patterns in labeled metabolites(s) in 1-Dimensional (1D) ^{13}C or 2-Dimensional (2D) [^{1}H,^{13}C] NMR spectra [4]. The advantage of 2D NMR techniques is to provide a solution for signal overlap from different metabolites by extending into second dimension [5–7]. Further, 2D ^{13}C-heteronuclear single quantum correlation (HSQC) provides an indirect detection technique and increases the sensitivity of experiments [8]. Therefore, in this chapter, we focus on the application of 2D [^{1}H, ^{13}C] HSQC NMR.

2D [^{1}H, ^{13}C] HSQC utilizes detailed information about the coupling of the carbon atoms. 2D HSQC detects labeling patterns of the target carbon atoms, as well as the adjacent carbon atoms. In addition to the labeling pattern of carbon atoms, 2D HSQC analysis can identify the protons that are attached to either ^{12}C or^{13}C, which provides detailed information about the neighboring carbon of the molecule [4]. The NMR spectrum can provide information on the subcellular compartmentalization of metabolites [9]. In recent studies, HSQC analysis can distinguish between the glucosyl units on the protein and starch hydrolysates from soybean tissue, which provides crucial information about different pools of glucose-6-phosphate between the cytosolic and plastidic compartments within plant cells [10, 11]. This important information about compartmentalization in plants was modeled by the *Arabidopsis* cell [12–14], which concluded that starch is exclusively synthesized from the plastidic glucose 6-phosphate pool. The peak structure obtained from the HSQC experiment showed multiplet patterns proportional to the isotopomer abundances. These abundances can be translated to flux information.

2D NMR spectrum is useful for plant metabolic flux analysis because it contributes summarized information about the quantities and identities of the metabolites present in vivo. ^{13}C labeling experiment and 2D NMR application in plant metabolic engineering have been well examined in comprehensive reviews by many researchers over the last decade [1, 3, 15–18]. In recent years, elevated interest of omics-based research and plant models for improved crop production of food and feed industries has found the application of carbon labeled metabolic flux analysis (^{13}C-MFA) to be a practical tool for systems biology. ^{13}C-MFA generated metabolic flux maps of central metabolic pathways with subcellular compartmentation provide a detailed picture of observable in vivo fluxes in plants. Metabolic flux maps of soybean [11, 19, 20], maize [21, 22], sunflower [23], and rapeseed oil [24–26] are a few of crop metabolic flux map studies. Hence, we are using soybean embryos cultured in vitro as the model plant system for the application of 2D [^{1}H,^{13}C] HSQC NMR analysis, with the focus on proteinogenic amino acid and starch labeling.

In the methods section, we describe an ideal protocol of extraction of the biomass components of proteinogenic amino acid and starch labeling from soybean embryo culture. Generating sufficient NMR samples can have a high cost due to the labeling substrates. Moreover, for each NMR sample, approximately 20–50 mg of material is required to obtain an adequately resolved 2D HSQC NMR spectrum with a sufficient signal-to-noise ratio for quantification. Therefore, proper techniques and careful planning for NMR sample preparation are presented. Furthermore, the section provides summary of the 2D NMR analysis, which includes the techniques used to obtain the relative intensities of the 16 amino acids and glucosyl units measured from protein fraction, as well as the determination of the latter in the starch fraction of soybean embryos.

2 Materials

1. Freeze dry system (Labconco, Kansas City, MO; cat. no. 77500-00).
2. RapidVap evaporator (Labconco, Kansas City, MO; cat. no. 7900002).
3. Hydrolysis tube 20-ml (Pierce, Rockford, IL; cat. no. 29564).
4. Hydrochloric acid (constant boiling) 6 N Sequencing Grade (Pierce, Rockford, IL; cat. no. 24308).
5. Deuterium oxide, D_2O, (Sigma-Aldrich, St. Louis, MO; cat. no. CAS 7789-20-0).
6. Deuterium chloride, DCl, (Sigma-Aldrich, St. Louis, MO; cat. no. CAS 7698-05-7).
7. Dialysis Cassette (2,000 MWCO) (Pierce, Rockford, IL; cat. no. 66212).
8. Heating block (Pierce, Rockford, IL; cat. no. 18870).

9. 3-(Trimethylsilyl)-1-propanesulfonic acid sodium salt (TMS) (Sigma-Aldrich, St. Louis, MO; cat. no. CAS 2039-96-5).
10. NMR tubes (Kimble-Chase, Vineland, NJ; Kontes Article no. 897240-0000).
11. Coomassie Plus (Bradford) Protein Assay (Pierce, Rockford, IL; cat. no. 23236).
12. β-Mercaptoethanol (Bio-Rad, Hercules, CA; cat. no. 1610710).
13. Amyloglucosidase from *Aspergillus niger* (Sigma-Aldrich, St. Louis, MO; cat. no. CAS 9032-08-0).
14. Glucose Assay Kit (Sigma-Aldrich, St. Louis, MO; cat. no. GAGO20).
15. Soda Lime (VWR, Radnor, PA; cat. no. CAS 8006-28-8).
16. 50-ml plastic tubes with screw caps (Falcon™ Conical Centrifuge Tubes, BD Biosciences, San Jose, CA).
17. Vacuum Pump (model RV5 115/230 V, 1-ph, 50/60 Hz) (Edwards, Crawley Sussex, England).
18. −80 °C Freezer (So-Low Environmental Equipment, Cincinnati, OH).
19. Spin-X® Centrifuge Tube Filter (0.22 μm nylon, Costar, Corning, NY; cat. no. 8169).
20. Refrigerated Microcentrifuge Model 5415R (Eppendorf).
21. Precision Water Bath Model 25 (Thermo Scientific, cat. no. 51221080).
22. Water bath sonicator (Fisher Scientific FS 110H Ultrasonic Cleaner).
23. Geno/Grinder® (SPEX SamplePrep, Metuchen NJ, USA).

3 Methods

The powerful NMR technique does not require any pretreatment method, such as derivatization, prior to analysis, which makes standardization straightforward. However, NMR samples have to have a sufficient amount of materials for the analysis of intracellular fluxes, which are the extraction of protein and starch fractions for 2D NMR analysis. One challenge in the extraction protocol is to have sufficient amount of materials for NMR analysis. Our extraction method is unique because oil, protein, soluble sugars, and starch can sequentially in a separation process scheme, be extracted from one biomass sample (Fig. 1).

The biomass fractionation of soybean embryos provides a way to extract individual soluble metabolites separately. This prevents the complication of mixing crude extracts from the tissue, which

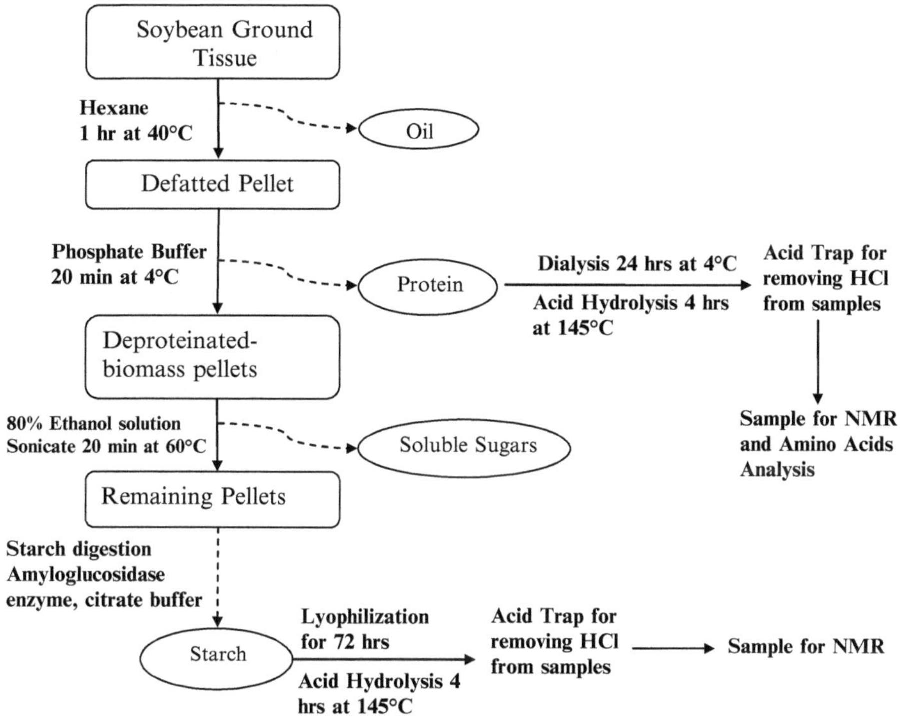

Fig. 1 Schematic diagram of biomass fractionation of soybean dry tissue and preparation of protein and starch hydrolysates for NMR analysis

could potentially interfere with the analysis of carbon coupling patterns in 2D HSQC NMR spectra. Furthermore, the proteinogenic amino acids observed in the spectra of protein hydrolysates illustrates a unique method that is able to distinguish the intracellular compartmentation within the tissue. One of many examples, the biosynthesis pathway of the amino acid histidine is located inside the plastid [27]; therefore, histidine plays an important role in compartment identification between plastid and cytosol. Another unique aspect of our method is that the glucosyl units observed in the 2D HSQC NMR spectra of protein and of starch hydrolysate directly reflect glucose-6-phosphate pools in the cytosol and the plastid compartments within the cell, respectively. As mentioned in Sriram et al. [10], acid hydrolysate of soybean protein contains hydrolysis products of glucose and mannose. Starch is a glucose polymer; therefore, acid hydrolysis of soybean starch yields glucose monomers.

The analysis of the carbon coupling patterns from the signals of 2D HSQC NMR spectra of the proteinogenic amino acids and starch materials from the extract requires the highest available purity. Therefore, consider autoclaving or sterilizing all of the glass bottles that will contain buffer and reagents materials before proceeding with the protocol.

3.1 Biomass Extraction, Hydrolysis, Amino Acid Quantification

1. Soybean in vitro cultures are harvested by pouring the contents of each culture into a sieve, rinsing with 100 ml distilled water, blotting dry, and weighing and freezing in liquid nitrogen prior to lyophilization at −50 °C and 0.0158 mbar for 72 h. The lyophilized embryos are finely grounded using a Geno/Grinder® prior to further analysis.
2. Approximately 100 mg of dry powdered soybean embryo sample is extracted with 1 ml of *n*-hexane at 40 °C for 1 h; the extracts are centrifuged at high speed (16,100 × *g*) for 10 min at room temperature. The process is repeated five times and after each extraction, the solvent containing the lipids is pooled into a pre-weighed glass tube (approximately 4.8 ml final volume) and dried for four days in a hood at room temperature. The mass of lipids after solvent evaporation is measured gravimetrically.
3. The remaining hexane-extracted biomass is dried and further extracted for protein in 600 μl of 200 mM phosphate buffer (pH = 7.2) containing 14 mM β-mercaptoethanol at 4 °C for 20 min to dissolve and suspend the protein in solution. Then the solution is centrifuged at high speed (16,100 × *g*) at 4 °C for 15 min and the supernatant is transferred into a microcentrifuge tube. The extraction is performed on the pellet two more times with 400 μl of the buffer (per extraction) and the supernatants are pooled with that from the initial extraction (approximately 1.2 ml final volume). Protein contents of the extract are measured by Bradford Assay (28, *see* **Note 1**).
4. Residual soluble sugars in the defatted/deproteinated-biomass pellets are extracted into 1 ml of 80 % aqueous ethanol in a water bath sonicator at 60 °C for 20 min, vortexed every 5 min. The extraction is repeated four times and after each extraction, the solvent containing the soluble sugars are pooled into a pre-weighed glass tube and dried for 3 days in a 40 °C oven, and the soluble sugar content in these fractions is measured gravimetrically.
5. The remaining pellets of each sample, after soluble sugars extraction in **step 4**, are placed in 1.5 ml of Distilled De-Ionized (DDI) water, covered with aluminum foil, and autoclaved (liquid cycle at 121 °C and a pressure of 15 psi) for 30 min prior to starch digestion and extraction (*see* **Note 2**).
6. Starch is digested in 1.5 ml of 100 mM citrate buffer (pH 5.0) containing amyloglucosidase at a ratio of 0.025 mg enzyme/1 mg of tissue dry weight (approximately 2 μl of amyloglucosidase or 0.6 U of enzyme are added to each sample). Samples are incubated in a 30 °C water bath overnight. The extractions from **step 5** are centrifuged at high speed for 30 min at room temperature. Starch content is quantified using a glucose assay kit (*see* Subheading 2).

7. Extracted starch sample is frozen in a −80 °C freezer for 4 h and lyophilized at −50 °C and 0.0158 mbar for 72 h (*see* **Note 3**). Then, starch powder is dissolved in DDI water.
8. Protein extract is dialyzed using dialysis cassettes in 10 mM phosphate buffer (pH 7.2) at 4 °C for 24 h to remove any soluble sugar residues in the sample (*see* **Note 4**).
9. The protein and starch samples are vacuum hydrolyzed in hydrolysis tubes using 6 N constant boiling hydrochloric acid (HCl). The acid is added in the ratio of 6 ml HCl to 10 mg protein/starch.
10. The hydrolysis tubes of protein and starch samples are evacuated multiple times purged with nitrogen to remove residual air, and re-evacuated to prevent oxidation during hydrolysis.
11. The protein and starch samples in the hydrolysis tubes are hydrolyzed for 4 h at 140–145 °C in a heat block (*see* **Note 5**).
12. After hydrolysis, a RapidVap evaporator is used to evaporate the hydrochloric acid for 2 h at 35 % speed, 40 bar pressure, and a temperature of 45 °C (*see* **Note 6**).
13. Each dried hydrolysate (one for protein and one for starch) is reconstituted in 2 ml of de-ionized water using a RapidVap evaporator for 30 min at 70 % speed, atmospheric pressure, and room temperature (*see* **Note 7**).
14. The reconstituted samples are filtered using centrifuge and Spin-X® centrifuge tube with filter (*see* **Note 8**). The filters are discarded, the protein and starch samples tubes are tightly sealed with parafilm, and holes are poked for vaporization of water during freeze-drying.
15. The protein and starch samples are frozen in a −80 °C freezer for 4 h and lyophilized at −50 °C and 0.0158 mbar for 72 h or until the samples are completely dried.
16. The protein and starch samples are reconstituted in 400 μl of deuterium oxide and vortexed. Then, a small portion of protein sample is diluted to 20–50 μl using DDI water and amino acids are analyzed by HPLC. 16 Amino Acids can be analyzed by using pre-column derivatization with 6-Aminoquinolyl-*N*-Hydroxysuccinimidyl carbamate [29].

3.2 NMR Sample Preparation (Protein and Starch)

Protein and starch samples are reconstituted with deuterium oxide as described in the previous section. Next, adjust the pH of the sample to 0.8–1.0 using DCl/D_2O mixture if required; this step is crucial due to chemical shifts of the carbon and hydrogen is highly depending on pH of the samples. A pinch of TMS standard flakes are added, vortexed well, and then transferred to clean NMR tubes.

3.3 2-D HSQC Spectral Analysis of Amino Acids

1. 2-D [^{1}H, ^{13}C] Heteronuclear Single Quantum Correlation (HSQC) spectroscopy of hydrolyzed protein and starch samples are performed on a Bruker Avance DRX 500 MHz spectrometer at 298 K. The HSQC spectra were acquired using a modified version of the INEPT spectra (insensitive nuclei enhanced by polarization transfer) pulse sequence [8]. TMS is used as an internal standard: the reference zero p.p.m. is set by the methyl signal of TMS. The magnetic resonance frequency of ^{13}C (Fl dimension) and ^{1}H (F2 dimension) are 125.7 MHz and 499 MHz, respectively. The spectral widths along the Fl and F2 dimensions are 5028.05 Hz and 5482.26 Hz, respectively. Peak aliasing is applied to minimize the sweep width in the Fl dimension. The number of complex data points is 900 (^{13}C) and 1024 (^{1}H). The number of scans ranges from 16 to 32.
2. NMR spectra are acquired using Xwinnmr (Bruker) software and analyzed by using the free software NMRview [30]. NMRview is available in both Linux and Window versions.
3. Cross peaks of s, d, and t can be processed directly to obtain the raw signal intensities by using the NMRview software.
4. Both aliphatic and aromatic carbon atoms of 16 amino acids in the spectra are identified from hydrolysate of the soybean protein samples, as shown in Figs. 2 and 3, respectively [11].
5. Each carbon of the amino acid listed in the figures is detectable by its unique ^{13}C and ^{1}H chemical shifts [31] as well as distinctive coupling patterns and J-coupling constants (Jcc) [32]. The chemical shifts and the J-coupling-constant (Jcc) values obtained from [33] are organized in Table 1 (*see* **Note 9**).
6. Table 1 displays the relationship of the ^{13}C labeling pattern in the multiplet to carbon atoms. A multiplet can contain one or any combination of the following: a singlet "s", a doublet "d", a doublet split by a smaller scalar coupling "d1", a doublet split by a larger scalar coupling "d2", a doublet of doublets "dd", and a triplet "t" [4]. The metabolic precursors to the given metabolites measured are included in Table 1 as well.
7. Furthermore, Table 1 includes the multiplet component of the cross-peak to the specific isotopomer. For example, singlet "s" under "Cross-Peak" corresponds to the first isotopomer listed under "Isotopomer". The notation for each of these isotopomers indicates the labeling state of the observed carbon "Carbon number" and its neighbor(s). The boldfaced carbon atom(s) in Isotopomer indicates labeled carbon atom(s), normal font indicates unlabeled carbon atom(s), and x indicates an unknown labeling state of the carbon atom(s).

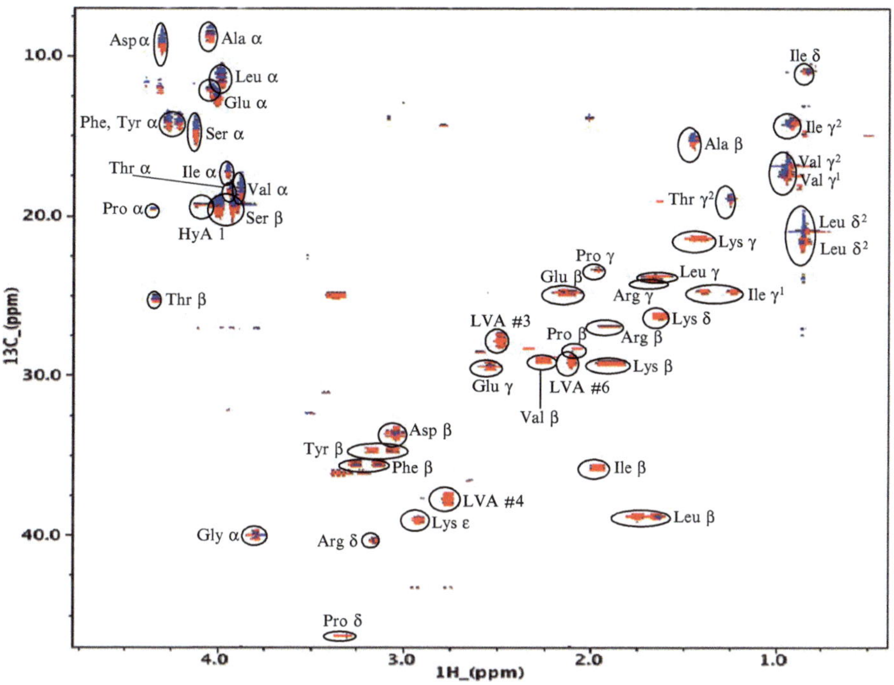

Fig. 2 Detailed scheme for analyzing the carbon coupling patterns in 2D [^{1}H, ^{13}C] HSQC spectrum of 16 proteinogenic amino acids and glucosyl units of soybean (*Glycine max)* zygotic embryo *cv.* Jack [11]

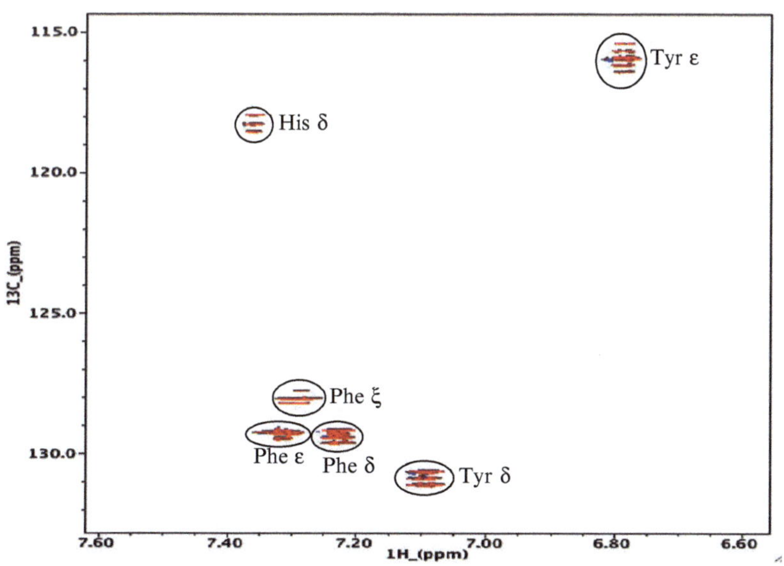

Fig. 3 2D HSQC spectrum of aromatic rings of protein extracted from soybean zygotic embryo *cv. Jack* [11]

8. To quantify overlapping multiplets [(s, d1, d2, dd), (s, d1, d2, d3, dd1, dd2, dd3, qd)] on the HSQC spectra, peak deconvolution software was developed in our lab, based on a spectral model originally proposed by van Winder et al. [36].

Table 1
The cross-peaks to carbon atoms, chemical shift values for the 16 amino acids, and their J-coupling constant from a 2D HSQC NMR spectrum of soybean embryos

Amino acid	Carbon position	Carbon number	Cross-peaks	1-H	13-C	Jcc (d1, d2)	Precursor	Isotopomer
Ala	α	2	s, d1, d2, dd	4.08, 4.12	11.54	59.6, 34.3	Pyruvate	[123], [123], [123], [123]
	β	3	s, d	1.59	17.98	34.3		[x23], [x23]
Arg	α	2	s, d1, d2, dd	4.18	14.69	53.4, 33.7	Glutamate	
	β	3	s, d, t	1.97	29.69	33.7, 34.1		[x234x], [x234x]+[x234x], [x234x]
	γ	4	s, d, t	1.67	26.52	34.1, 35.8		
	δ	5	s, d	3.23	43.07	35.8		[xxx45], [xxx45]
Asp	α	2	s, d1, d2, dd	4.34	12.15	59.8, 37.6	Oxaloacetate	[123x], [123x], [123x], [123x]
	β	3	s, d1, d2, dd	3.1	36.5	37.6, 55.5		[x234], [x234], [x234], [x234]
Glu	α	2	s, d1, d2, dd	4.49	14.94	59.6, 33.7	Glutamate	
	β	3	s, d, t	2.21	27.6	33.7, 36.3		[x234x], [x234x]+[x234x], [x234x]
	γ	4	s, d1, d2, dd	2.61	32.18	36.3, 54.7		[xx345], [xx345], [xx345], [xx345]
Gly	α	2	s, d	3.85	42.84	59.6	3-phosphoglycerate	[12x], [12x]
His	α	2	s, d1, d2, dd	4.4	14.93	59.8, 34.6	Ribose-5-phosphate (Plastid)	[xx345], [xx345], [xx345], [xx345]
	β	3	s, d1, d2, dd	3.42	27.86	51, 34.6		[x234x], [x234x], [x234x], [x234x]
	δ2	4	s, d	7.55	40.87	74.5		[12xxx], [12xxx]
Ile	α	2	s, d1, d2, dd	3.91	21.18	59.2, 32.5	Pyruvate (Plastid)	[12xx]·[x2x], [12xx]·[x2x], [12xx]·[x2x], [12xx]·[x2x]
	β	3	s, d1, d2, dd	2.04	38.52	32.5, 33.9		

	γ1	4	s, d, t	1.26, 1.46	27.46	33, 34.9		[x2x]·[xx**34**], [x**2**x]·[xx34] + [x2x]·[xx**34**], [x**2**x]·[xx**34**]
	γ2	5	s, d	1	16.92	33.9		[x23], [x**23**]
	δ	6	s, d	0.92	13.67	34.9		[xx3**4**], [xx**34**]
Leu	α	2	s, d1, d2, dd	4.04	14.29	59.6, 32.8	Pyruvate (Plastid)	[12]·[x2x], [**12**]·[x2x], [12]·[x**2**x], [**12**]·[x**2**x]
	β	3	s, d, t	1.72, 1.83	41.63	32.8, 32.4		[*x*2]·[x2x]·[x2x], [*x*2]·[x**2**x]·[x2x] + [*x*2]·[x2x]·[x2x], [*x*2]·[x**2**x]·[x**2**x]
	γ	4	s, d1, d2, dd	1.75	26.72	32.4, 34.3		
	δ1	5	s, d	0.94	24.36	34.3		[x23], [x**23**]
	δ2	6	s, d	0.93	23.65	34.3		[x2x]·[xx**3**], [x**2**x]·[xx**3**]
Lys	α	2	s, d1, d2, dd	4.05	15.54	53.6, 34.2	Oxaloacetate/ Pyruvate	
	β	3	s, d, t	1.93, 1.99	32.02	34.2, 34.5		½{[x2**34**] + [x**23**]·[xxx4]}, ½{[x**234**]+[x2**34**] + [x**23**]·[xxx4] + [x2**3**]·[xxx**4**]}, ½{[x**234**] +[x**23**]·[xxx**4**]
	γ	4	s, d, t	1.48	24.16	34.5, 34		[xx3**4**]·[xx3], [xx**34**]·[xx3] + [xx**34**]·[xx**3**], [xx**34**]·[xx**3**]
	δ	5	s, d, t	1.69	29.03	34, 35.4		½{[x2**34**] + [x**23**]·[xxx4]}, ½{[x**234**]+[x2**34**] + [x**23**]·[xxx4] + [x2**3**]·[xxx**4**]}, ½{[x**234**] + [x**23**]·[xxx**4**]}
	ε	6	s, d	2.99	41.75	35.4		½{[x**23**]+[x23x]}, ½{[x**23**]+[x**23**x]}
Met	α	2	s, d1, d2, dd	3.97	21.28		Oxaloacetate (Plastid)	
	β	3	s, d1, d2, dd	2.25	31.63			
	γ	4	s, d	2.6	31.29			[xx3**4**], [xx**34**]
	ε	5	s, d	2.1	16.62			
Phe	α	2	s, d1, d2, dd	4.32	16.93	59.9, 32.3	Phosphoenolpyruvate	[123], [**1**23], [1**23**], [**123**]

(continued)

Table 1
(continued)

Amino acid	Carbon position	Carbon number	Cross-peaks	1-H	13-C	Jcc (d1, d2)	Precursor	Isotopomer
	β	3	s, d1, d2, dd	3.34, 3.21	38.32	32.3, 43.7		[x2**3**]·[x2x], [x**23**]·[x2x], [x2**3**]·[x**2**x], [x**23**]·[x**2**x]
	γ	4	s, d, t	6.99, 7.36	25.08	43.7		
	ε	5	s, d, t	6.66, 7.07	20.03			
Pro	α	2	s, d1, d2, dd	4.39	22.53		Glutamate (Plastid)	
	β	3	s, d, t	2.15, 2.41	31.07			
	γ	4	s, d, t	2.03	26.12			[xx3**4**5], [xx**34**5]+[xx3**45**], [xx**345**]
	δ	5	s, d	3.39	48.97			[xxx4**5**], [xxx**45**]
Ser	α	2	s, d1, d2, dd	4.31	15.05	59.2, 36.4	Serine	[1**2**3], [**12**3], [1**23**], [**123**]
	β	3	s, d	3.98, 4.06	22.1	36.4		[x2**3**], [x**23**]
Thr	α	2	s, d1, d2, dd	4.17	17.56	58.8, 36.3	Oxaloacetate (Plastid)	[1**2**3x], [**12**3x], [1**23**x], [**123**x]
	β	3	s, d, t	4.34, 4.39, 4.44	28	36.6, 37.5		
	γ2	4	s, d	1.32	21.67	37.5		[xx3**4**], [xx**34**]
Tyr	α	2	s, d1, d2, dd	4.26	17.03	53.2, 33	Phosphoenolpyruvate	[1**2**3], [**12**3], [1**23**], [**123**]
	β	3	s, d, t	3.25, 3.13	37.5	33, 33.7		[x2**3**]·[x2x], [x**23**]·[x2x] + [x2**3**]·[x**2**x], [x**23**]·[x**2**x]
	δ1	5	s, d, t	6.99, 7.34	13.6	57.5, 58		[x2**3**]·[1xxx], [x**23**]·[1xxx]+[x2**3**]·[**1**xxx], [x**23**]·[**1**xxx]
	δ2	5	s, d, t	6.99, 7.34	13.6	57.5, 58		[x2x]·[**4**3xx], [x**2**x]·[**4**3xx]+[x2x]·[**43**xx], [x**2**x]·[**43**xx]

	ε1	6	s, d, t	6.77, 6.85, 6.96	38.69	58, 60.5		[xx3]·[**1**2xx], [xx**3**]·[**1**2xx] + [xx3]·[**12**xx], [xx**3**]·[**12**xx]
Val	α	2	s, d1, d2, dd	4	20.14	59.1, 32.2	Pyruvate (Plastid)	[1**2**x]·[x2x], [**12**x]·[x2x], [1**2**x]·[x**2**x], [**12**x]·[x**2**x]
	β	3	s, d1, d2, dd	2.25	31.77	34.4, 33.7		
	γ1	4	s, d	1.04	20.14	34.4		[x2**3**], [x**23**]
	γ2	5	s, d	1.02	19.56	33.7		[x2x]·[xx**3**], [x**2**x]·[xx**3**]

The notation of the cross-peak multiplet: s indicates singlet, d indicates doublet, d1 and d2 indicate the first and second doublet, dd indicates double doublet, and t indicates triplet. The precursor corresponding to each amino acid is indicated and its subcellular localization is referred in parentheses. The boldfaced carbon atom(s) in isotopomer indicates the labeled carbon atom(s), normal font indicates unlabeled carbon atom(s), and x indicates unknown labeling state of the carbon atom(s)

Table 2
Equations to determine each fraction intensity from multiplet s, d, t, and noise fraction from 2D HSQC NMR spectrum

Multiplet type	Singlet	Doublet	Triplet
s, d	$\frac{s \text{ intensity}}{\text{total intensity}}$	$\frac{d \text{ intensity} + d \text{ intensity}}{\text{total intensity}}$	
s, d, t	$\frac{s \text{ intensity} - (t \text{ intensity} + t \text{ intensity})}{total \text{ intensity}}$	$\frac{d \text{ intensity} + d \text{ intensity}}{\text{total intensity}}$	$\frac{2 * (t \text{ intensity} + t \text{ intensity})}{\text{total intensity}}$
Noise fraction	$\frac{n \text{ estimated intensity}}{\text{total intensity}}$		

9. Since the peaks of the aromatic amino acids and some of the glucosyl units are crowded, in order to separate the resonances additional 2D HSQC spectra are acquired that are J-scaled along the ^{13}C dimension, by integrating pulse sequences reported previously [34, 35]. J-scaling increases multiplet separation by an even integral factor J and eliminates multiplet overlap. J-scaling factors of six are employed in processing aromatic amino acid and some of glycosyl units and will be described below [10].
10. The multiplet intensities (s, d, t) provided by the NMRView program are raw intensities. However, NMR2Flux [11], the simulation program used to calculate the metabolic fluxes, requires intensity fractions.
11. Table 2 shows the equation that converts the raw intensities of the carbon atom of the amino acids and their noise data to the relatively intensities of the carbon atoms amino acids and noise (*see* **Note 10**)

3.4 2-D HSQC Spectra Analysis of Glucosyl Units

1. Figure 2 displays the 2D HSQC spectrum of 16 amino acids and glucosyl units, hydroxyacetone (HyA) and levulinic acid (LVA), from soybean protein hydrolysate (*see* **Note 11**).
2. The reaction products of the hydrolysis of the sugar residues in the glycoprotein are HyA and LVA (Fig. 4) [10].
3. Table 3 shows glucosyl units, carbon number, cross-peak, precursors, and isotopomers of the protein and starch hydrolysates of soybean embryo culture.
4. The HyA1 cross peak multiplet (s, d) of both protein and starch hydrolysates can be quantified by NMRview software on 2D HSQC spectrum to obtain raw intensities.
5. To quantify the overlapping multiplets (LVA 3, LVA 4, LVA 6 of protein and starch hydrolysates) in the HSQC spectra, a peak deconvolution software is written based on a spectral model originally proposed by van Winden et al. [36].

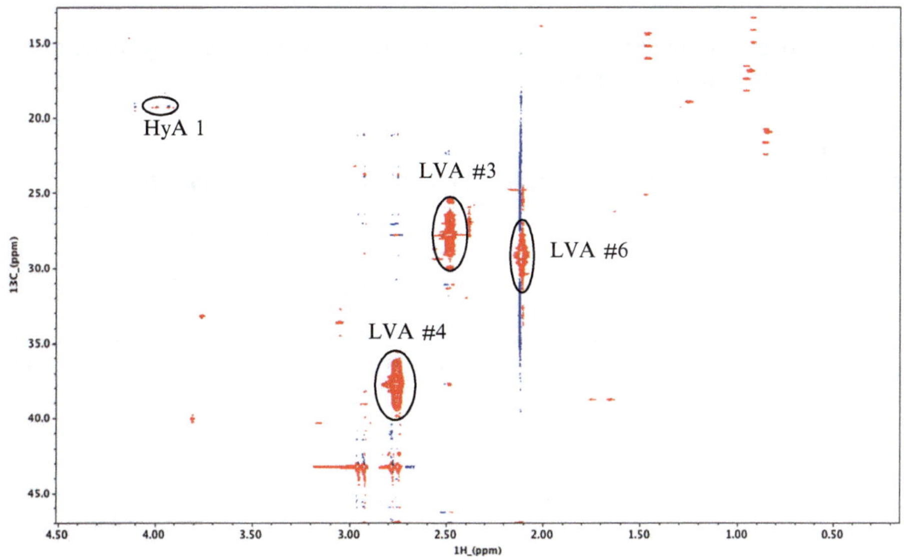

Fig. 4 2D HSQC spectrum of hydrolyzed starch samples from soybean zygotic embryo *cv.* Jack

4 Notes

1. The total soluble protein and starch contents of the cultures are estimated at 50 mg each, which should be more than sufficient for NMR analysis. As a result, a total of ~300 mg of dry powdered soybean embryo will need to be processed in multiple extractions of 100 mg each, and the content pooled for analysis, rather than scaling up the extraction process for one-step. As expected, the Bradford Assay result from extractable protein is lower than the method of determining the total protein content of the embryos via elemental analysis. Therefore, it is crucial to measure the amount of protein that can be extracted from the tissue.
2. Besides autoclaving starch samples, it is helpful to have positive/negative starch control samples. Prepare positive starch control samples, simply by placing 10–15 mg waxy corn starch in 3–12 × 75 mm glass tubes (weigh tube accurately), add 1.5 ml DDI water to each of the three tubes, vortex samples until dissolved, cover with aluminum foil. For negative starch control samples, just place 1.5 ml DDI water to each of the 3–12 × 75 mm glass tubes and cover with aluminum foil.
3. The purpose to lyophilize starch sample is to concentration and reduce the volume prior to acid hydrolyze treatment.

Table 3
Carbon number, cross-peak multiplet, precursors, and Isotopomers of glucosyl units of protein and starch hydrolysate samples from soybean culture

Hydrolysate	Glucosyl unit	Carbon number	Cross-peak	Precursors	Isotopomer
Protein	HyA 1	6	s, d	Glucose-6-phosphate/ Fructose-6-phosphate (Cytosol)	[**1**2xxxx], [**12**xxxx]
Protein	LVA 3	6	s, d1, d2, dd	Glucose-6- phosphate/ Fructose-6-phosphate (Cytosol)	[x2**3**4xx], [x**23**4xx], [x2**34**xx], [x**234**xx]
Protein	LVA 4	6	s, d1, d2, d3, dd1, dd2, dd3, qd	Glucose-6- phosphate/ Fructose-6-phosphate (Cytosol)	[xx3**4**56], [xx**34**56], [xx3**45**6], [xx3**4**5**6**], [xx**345**6], [xx**34**5**6**], [xx3**456**], [xx**3456**]
Protein	LVA 6	6	s, d1,d2, dd	Glucose-6- phosphate / Fructose-6-phosphate (Cytosol)	[xxx45**6**], [xxx4**56**], [xxx**4**5**6**], [xxx**456**]
Starch	HyA 1	6	s, d	Glucose-6-phosphate (Plastid)	[**1**2xxxx], [**12**xxxx]
Starch	LVA 3	6	s, d1, d2, dd	Glucose-6-phosphate (Plastid)	[x2**3**4xx], [x**23**4xx], [x2**34**xx], [x**234**xx]
Starch	LVA 4	6	s, d1, d2, d3, dd1, dd2, dd3, qd	Glucose-6-phosphate (Plastid)	[xx3**4**56], [xx**34**56], [xx3**45**6], [xx3**4**5**6**], [xx**345**6], [xx**34**5**6**], [xx3**456**], [xx**3456**]
Starch	LVA 6	6	s, d1,d2, dd	Glucose-6-phosphate (Plastid)	[xxx45**6**], [xxx4**56**], [xxx**4**5**6**], [xxx**456**]

4. Dialysis of the protein extract is necessary to eliminate sugar molecules that may have been co-extracted with the protein. The dialysis cassette model that was listed in the material section could hold up between 3 and 12 ml in volume; however, during the injection process of sample to the cassette, air can easily be trapped in the cassette. Therefore, a maximum of 6 ml in volume per cassette should be placed. Place the 6 ml-protein-cassette into the 2 l of 10 mM phosphate buffer with pH 7.2 at 4 °C; replace the buffer every 12 h.

5. Turn on the heating block sometime in advance so that the temperature is 140–145 °C before hydrolyzing your protein and starch samples. Also, before closing the hydrolysis tube make sure that no acid sticks to the top of the tubes; it can damage the Teflon in the cap. A maximum volume of the

sample to be stored in hydrolysis tube is 6 ml (1/3rd of the tube size) since the vacuum and heat cause the level to rise, which possibility causes the content to pull up.

6. After 4 h of incubation the protein and starch samples are removed from the heating block and let at room temperature. Then, place samples on ice or in the refrigerator before opening caps. Be extremely careful before opening the cap since the tube samples are under considerable pressure. Soda lime can be used to trap acid from the protein and starch acid hydrolyzed samples. If the RapidVap step is not being done on the same day, store the hydrolysis tube in the refrigerator without releasing the vacuum.
7. After the protein and starch samples have been placed in the high temperature RapidVap evaporator, these liquid samples are nearly evaporated. Therefore, add DDI water to reconstitute sample thoroughly.
8. The maximum volume capacity of the Spin-X® microcentrifuge tube with filter is less than 0.8 ml due to the insertion of the filter. Therefore, make sure not to add more than 0.8 ml of protein/starch acid hydrolysis to the filter insertion of the microcentrifuge tube.
9. Table 1 lists each carbon of the 16 amino acids; however, some of the carbon atoms of the amino acid provide a similar information regarding the precursor of the metabolic pathway. Therefore, the extra measurements of the amino acid can be used to check the validity of that particular flux in the metabolic pathway.
10. Taking a noise measurement is necessary for each peak which has its intensity quantified. For this purpose, the intensity of about 10 noise peaks that are closed to the target peak is averaged. Please verify that the sum of the fractional intensities of s, d, and t add up to 1. The raw intensity of the cross-peak is then input in the Excel spreadsheet form.
11. For clarity, hydrolyzed protein extract contains HyA and LVA peaks (located in the cytosol compartment) and hydrolyzed starch extract also contains HyA and LVA peaks (located in the plastid compartment).

References

1. Wiechert W (2001) C-13 metabolic flux analysis. Metab Eng 3(3):195–206
2. Massou S, Nicolas C, Letisse F, Portais JC (2007) NMR-based fluxomics: Quantitative 2D NMR methods for isotopomers analysis. Phytochemistry 68:2330–2340
3. Kruger NJ, Masakapalli SK, Ratcliffe RG (2012) Strategies for investigating the plant metabolic network with steady-state metabolic flux analysis: lessons from an Arabidopsis cell culture and other systems. J Exp Bot 63:2309–2323
4. Szyperski T (1995) Biosynthetically directed fractional C-13-labeling of proteinogenic amino-acids—an efficient analytical tool to investigate intermediary metabolism. Eur J Biochem 232:433–448

5. Schmidt K, Nielsen J, Villadsen J (1999) Quantitative analysis of metabolic fluxes in escherichia coli, using two-dimensional NMR spectroscopy and complete isotopomer models. J Biotechnol 71(1–3):175–189
6. van Winden WA, Heijnen JJ, Verheijen PJT (2002) Cumulative bondomers: a new concept in flux analysis from 2D C-13, H-1 COSYNMR data. Biotechnol Bioeng 80(7):731–745
7. Yang C, Hua Q, Shimizu K (2002) Quantitative analysis of intracellular metabolic fluxes using GC-MS and two-dimensional NMR spectroscopy. J Biosci Bioeng 93(1):78–87
8. Bodenhausen G, Ruben DJ (1980) Natural abundance nitrogen-15 NMR by enhanced heteronuclear spectroscopy. Chemical Physics Letters 69(1):185–189
9. Last RL, Jones AD, Shachar-Hill Y (2007) Towards the plant metabolome and beyond. Nat Rev Mol Cell Biol 8(2):167–174
10. Sriram G, Iyer VV, Fulton DB, Shanks JV (2007) Identification of hexose hydrolysis products in metabolic flux analytes: a case study of levulinic acid in plant protein hydrolysate. Metab Eng 9:442–451
11. Sriram G, Fulton DB, Iyer VV, Peterson JM, Zhou RL, Westgate ME, Spalding MH, Shanks JV (2004) Quantification of compartmented metabolic fluxes in developing soybean embryos by employing Biosynthetic ally directed fractional C-13 labeling, C-13, H-1 two-dimensional nuclear magnetic resonance, and comprehensive isotopomer balancing. Plant Physiol 136:3043–3057
12. Masakapalli SK, Le Lay P, Huddleston JE, Pollock NL, Kruger NJ, Ratcliffe RG (2010) Subcellular flux analysis of central metabolism in a heterotrophic arabidopsis cell suspension using steady-state stable isotope labeling. Plant Physiol 152:602–619
13. Williams TCR, Miguet L, Masakapalli SK, Kruger NJ, Sweetlove LJ, Ratcliffe RG (2008) Metabolic network fluxes in heterotrophic arabidopsis cells: stability of the flux distribution under different oxygenation conditions. Plant Physiol 148(2):704–718
14. Lonien J, Schwender J (2009) Analysis of metabolic flux phenotypes for Two arabidopsis mutants with severe impairment in seed storage lipid synthesis. Plant Physiol 151(3):1617–1634
15. Libourel IGL, Shachar-Hill Y (2008) Metabolic flux analysis in plants: from intelligent design to rational engineering. Annu Rev Plant Biol 59:625–650
16. Stephanopoulos G, Vallino JJ (1991) Network rigidity and metabolic engineering in metabolite overproduction. Science 252(5013):1675–1681
17. Ratcliffe RG, Shachar-Hill Y (2006) Measuring multiple fluxes through plant metabolic networks. Plant J 45(4):490–511
18. Schwender J (2008) Metabolic flux analysis as a tool in metabolic engineering of plants. Curr Opin Biotechnol 19(2):131–137
19. Allen DK, Ohlrogge JB, Shachar-Hill Y (2009) The role of light in soybean seed filling metabolism. Plant J 58(2):220–234
20. Iyer VV, Sriram G, Fulton DB, Zhou R, Westgate ME, Shanks JV (2008) Metabolic flux maps comparing the effect of temperature on protein and oil biosynthesis in developing soybean cotyledons. Plant Cell Environ 31(4): 506–517
21. Alonso AP, Dale VL, Shachar-Hill Y (2010) Understanding fatty acid synthesis in developing maize embryos using metabolic flux analysis. Metab Eng 12(5):488–497
22. Alonso AP, Val DL, Shachar-Hill Y (2011) Central metabolic fluxes in the endosperm of developing maize seeds and their implications for metabolic engineering. Metab Eng 13(1): 96–107
23. Alonso AP, Goffman FD, Ohlrogge JB, Shachar-Hill Y (2007) Carbon conversion efficiency and central metabolic fluxes in developing sunflower (Helianthus annuus L.) embryos. Plant J 52(2):296–308
24. Hay J, Schwender J (2011) Computational analysis of storage synthesis in developing Brassica napus L. (oilseed rape) embryos: flux variability analysis in relation to (13)C metabolic flux analysis. Plant J 67(3):513–525
25. Schwender J, Ohlrogge JB, Shachar-Hill Y (2003) A flux model of glycolysis and the oxidative pentosephosphate pathway in developing Brassica napus embryos. J Biol Chem 278 (32):29442–29453
26. Schwender J, Shachar-Hill Y, Ohlrogge JB (2006) Mitochondrial metabolism in developing embryos of Brassica napus. J Biol Chem 281(45):34040–34047
27. Stepansky A, Leustek T (2006) Histidine biosynthesis in plants. Amino Acids 30(2): 127–142
28. Bradford MM (1976) Rapid and sensitive method for quantitation of microgram quantities of protein utilizing principle of protein-Dye binding. Anal Biochem 72(1–2): 248–254
29. Cohen SA (2000) Amino acid analysis using precolumn derivatization with 6-aminoquinolyl-N-hydroxysuccinimidyl carbamate. Methods Mol Biol 159:39–47
30. Johnson BA, Blevins RA (1994) NMR view—a computure-program for the visualization and

analysis of NMR data. J Biomol Nmr 4(5): 603–614

31. Wuthrich K (1976) NMR in biological research: peptides and proteins. North Holland, Amsterdam
32. Harris RK (1983) Nuclear magnetic resonance spectroscopy: a physiochemical view. Pitman Books, London
33. Krivdin LB, Kalabin GA (1989) Structural applications of One-bond carbon-carbon spin-spin coupling-constants. Prog Nucl Magn Reson Spectrosc 21:293–448
34. Brown LR (1984) Differential scaling along omega-1 in COSY experiments. J Magn Reson 57(3):513–518
35. Willker W, Flogel U, Leibfritz D (1997) Ultra-high-resolved HSQC spectra of multiple-C-13-labeled biofluids. J Magn Reson 125 (1):216–219
36. van Winden W, Schipper D, Verheijen P, Heijnen J (2001) Innovations in generation and analysis of 2D C-13, H-1 COSYNMR spectra for metabolic flux analysis purposes. Metab Eng 3(4):322–343

Chapter 7

Analysis of Kinetic Labeling of Amino Acids and Organic Acids by GC-MS

Wagner L. Araújo, Takayuki Tohge, Adriano Nunes-Nesi, Toshihiro Obata, and Alisdair R. Fernie

Abstract

Plant metabolic pathways and the molecular and atomic fluxes through them can be deduced using stable isotopically labeled substrates. To this end one prerequisite is accurate measurement of the labeling pattern of targeted metabolites. Experiments are generally limited to the use of single-element isotopes, mainly ^{13}C. Here, we summarize the application of gas chromatography-time of flight mass spectrometry (GC-TOF-MS) for metabolic studies using differently labeled elemental isotopes applied to both intact organelles and whole plant tissue. This method allows quantitative evaluation of a broad range of metabolic pathways without the need for laborious (and potentially inaccurate) chemical fractionation procedures commonly used in the estimation of fluxes following incubation in radiolabeled substrates. We focus herein on the determination of isotope labeling in organic and amino acids.

Key words Metabolic flux analyses, Gas chromatography–mass spectrometry, Stable isotope labeling, Kinetic flux profiling, ^{13}C labeling

1 Introduction

Most likely the best characterized network within biological systems is that of metabolism. Given that it supports both plant growth and development experiments which provide insights into metabolic flux, as well as its regulation and control, can be invaluable tools to both improve our understanding of biological systems and ultimately aid in the discovery of gene function. Stable isotopic labeling experiments have been applied to investigate particular aspects of plant metabolism both in isolated organelles [1–3] and in whole plants or plant organs [4–7]. To this end feeding experiments using stable isotopes (e.g., ^{13}C, ^{14}C, and ^{15}N-labeled precursors) have been extensively used to investigate metabolic flux. Such experiments offer high sensitivity, and fractionation of labeled metabolites and biomass components can readily indicate the fate of metabolized radiolabel. For instance, feeding experiments with

Martine Dieuaide-Noubhani and Ana Paula Alonso (eds.), *Plant Metabolic Flux Analysis: Methods and Protocols*, Methods in Molecular Biology, vol. 1090, DOI 10.1007/978-1-62703-688-7_7,

$^{14}CO_2$ demonstrated the effect of reduced malate dehydrogenase activity on photosynthetic carbon assimilation in tomato leaves, and revealed that ascorbate feeding led to increased photosynthesis and altered assimilate partitioning in these transgenic lines [4]. Quantifying the release of $^{14}CO_2$ from the metabolism of different isotopomers of ^{14}C-glucose can also provide information about the relative activity of different metabolic pathways [4]. However, following the incubation with ^{13}C- or ^{15}N-labeled substrates and the use of gas chromatography-time of flight mass spectrometry (GC-TOF-MS) offers the possibility to determine higher resolution metabolic information and although this method still requires considerable computational time [8] it is far less laborious than simple chromatographic separation and radiolabel counting procedures and as such offers an important complement other post-genomic strategies such as metabolite profiling. Although here we discuss only the application of ^{13}C-labeled substrate, this method could be readily adapted to focus in detail on a certain branch of metabolism [8] by utilizing the wide range of commercially available stably labeled substrates. Successful applications of ^{13}C-labeling and GC-TOF-MS have been reported for a range of species and tissues, including: *Escherichia coli* [9, 10], *Saccharomyces cerevisiae* [11], photoautotrophic cyanobacteria, *Synechocystis* sp. [12], *Arabidopsis thaliana* and rice [13], potato tubers [8, 14], and tomato leaves [6, 15] and fruits [16].

The metabolism of a ^{13}C-labeled substrate through different pathways originates distinct labeling patterns and we concentrate here on examples of labeling of organic and amino acids. This method has been used to demonstrate the kinetic labeling of several compounds of the primary metabolism including amino acids, organic acids and glycolytic intermediates in potato tubers following incubation in presence of U-^{13}C-glucose [8]. A similar approach was used to investigate the fate of ^{13}C labeled glutamate in leaves of tomato plants with decreased succinyl-CoA ligase, and revealed that tricarboxylic acid (TCA) cycle flux was maintained in these plants by diversion of carbon through the GABA shunt [15]. Additionally by tracing the metabolism of ^{13}C-pyruvate by mitochondria isolated from *Arabidopsis* plants with reduced manganese superoxide dismutase activity it was possible to demonstrate that the decreased TCA cycle flux occurs most likely as a result of decreased TCA cycle flux caused by oxidative damage [1]. Furthermore, as a complement to the use of genetic approaches, isotope tracer experimentation using [U-^{13}C]-lysine or [U-^{13}C]-valine revealed that isovaleryl-CoA dehydrogenase is involved in degradation of the branched-chain amino acids, phytol, and Lysine, while 2-hydroxyglutarate dehydrogenase is involved exclusively in lysine degradation [17]. Such experiments with ^{13}C ring labeled phenylalanine, when coupled with LC-MS analysis of phenylpropanoids, additionally provide important information concerning the nexus of primary and secondary metabolism [16]. When considered together these studies clearly suggest that this approach has clear utility in the definition of gene function.

The data obtained can then be further analyzed mathematically either in simple or complex manners. Simple modeling was crucial in the proper that the glycolytic enzyme which associate to the outer mitochondrial membrane are functionally active [2, 3] whilst more complex modeling was recently used to determine the quantitative contributions of alternative pathways to the plant mitochondrial electron transport chain [18]. Actually several different approaches are possible for modeling data resulting from such studies. However, these are beyond the scope of this article so the interested reader is rather referred to several excellent recent reviews [19–23]. Additionally labeling experiments and complex metabolic modeling have demonstrated that developing green embryos are able to decrease and decreased losses of fixed carbon during oil synthesis by both refixing CO_2 released by respiration [7], and reducing the need for TCA cycle flux through the use of photosynthesis to meet energy demands [24]. More complex methods have also been widely applied for the discovery of gene function in microorganisms [25, 26], and their feasibility in plants has been demonstrated by experiments using *Arabidopsis* embryos [27] which revealed how flux can be rearranged in embryos deficient in two plastidial pyruvate kinase isoforms.

There is a considerable body of work demonstrating the use of nuclear magnetic resonance (NMR) for the quantification of intracellular fluxes in plant cells following incubations in stably labeled isotopes [28–33]. It should be noted, however, that in several instances such studies were augmented by the use of more sensitive GC-MS-based methods [2, 29, 34]. For example, the combination of GC-MS and NMR techniques has been also used to facilitate the construction of a model of glycolysis and the oxidative pentose phosphate pathway in developing *Brassica napus* embryos [29].

Here a relatively simple method for evaluating isotope distribution, based on GC-MS alone, in which determination of the mean fractional isotopic enrichment of the major sugar, organic acid and amino acid constituents of plant cells following incubation in ^{13}C-labeled substrates is presented. For the calculation of the label redistribution, evaluated by using a GC-MS, the proportion of the ^{13}C in specific mass fragments of the metabolites needs to be determined. Our approach is based on the premise that with the ^{13}C accumulation in a metabolite pool, mass fragments containing ^{13}C should have a mass shift of +1 *m/z* for each ^{13}C present in the molecule. To calculate the ^{13}C- enrichment as a percentage in the total carbon atoms of a metabolite pool, an automatic correction software tool can be readily used [12]. The estimation of the ^{13}C label redistribution is performed according to assumptions discussed previously [8].

Briefly the process can be summarized as follows:

1. Feeding of plant tissue in presence of heavy isotope (e.g., ^{13}C-unilabeled glucose) and simultaneously incubation with the ^{12}C-substrate.
2. Sampling the plant tissue, homogenizing, and exact weighing of sample aliquots.
3. Extracting metabolites concomitant with enzyme inactivation and the addition of internal standards and/or authentic chemical standards for peak identification or assessment of extraction efficiency.
4. Drying the polar extract.
5. Derivatizing the polar extract by methoxyamination followed by silylation, and adding retention time index standards.
6. Analyzing the derivatized samples by GC-TOF-MS for the determination of ^{13}C distribution in all metabolites measurable.
7. Calculation of the mean carbon fractional enrichment by using an automatic correction software tool.

2 Materials

2.1 ^{13}C Feeding, Sampling, and Extraction

1. Argon.
2. Centrifuge (capable of 3,700 × *g*), (e.g., Allegra® x-15R, Beckman Coulter, Krefeld, Germany).
3. Methanol and chloroform gradient grade for liquid chromatography (*see* **Note 1**).
4. MES (2-*N*-morpholino ethane sulfonic acid) buffer: 10 mM MES-KOH, pH 6.5.
5. Unlabeled and labeled substrate (e.g., glucose).
6. MilliQ water.
7. Erlenmeyer flask.
8. Oscillating ball mill MM200 (e.g., Retsch GmbH and Co.KG, Haan, Germany) or alternatively a pestle and mortar.
9. Ribitol, purity $\geq$ 99.0 %; 0.2 mg/mL in dH_2O.
10. Speed vacuum concentrator (e.g., SPD111V-230, Thermo-Electron Corporation, Langenselbold, Germany).
11. Schott glass AR-GLASS© culture tubes (soda-lime) (DURAN GMbH, Mainz, Germany).
12. Thermoblock (capable of heating to up to 70 °C).
13. Liquid nitrogen supply.
14. Vortex.

15. Scalpel blades, aluminum foil, 6-well plates, spatula, balance.
16. Microcentrifuge tubes (2 mL).

2.2 Derivatization

1. Methoxyamine hydrochloride, purity 98 % (e.g., Sigma, St. Louis, USA). Store at room temperature under dry atmosphere.
2. *N*-methyl-*N*-(trimethylsilyl)trifluor*o*acetamide (MSTFA) (Macherey and Nagel, Düren, Germany). MSTFA should be stored in opaque glass bottles under nitrogen. Contact with water generates hydrogen fluoride gas which is highly toxic. Store at 4 °C (*see* **Note 1**).
3. Pyridine, analytic grade (Merck, Darmstadt, Germany). Store at room temperature (*see* **Note 1**).
4. Retention time index standard mixture: fatty acids methyl esters (FAMES). All must be of standard grade for GC: Esters included are methylcaprylate, methyl pelargonate, methylcaprate, methyllaurate, methylmyristate, methylpalmitate, methylstearate, methyleicosanoate, methyldocosanoate, lignoceric acid methylester, methylhexacosanoate, methyloctacosanoate, and triacontanoic acid methylester. (All available via for example Sigma). The esters are dissolved in $CHCl_3$ at a final concentration of 0.8 mL/mL for liquid; 0.4 mg/mL for solid standards. Mix all well, aliquot into glass vials, and store at −20 °C.
5. Screw Top Tapered Vial-Clear Gold Grade (CHROMACOL LTD, Thermo Fisher Scientific Inc, Herts, UK).
6. Shaker (950 rpm).

2.3 GC-TOF-MS

1. Autosampler system (PAL Agilent, Santa Clara, USA).
2. Capillary column MDN-35, 30 m × 0.32 mm, 0.25 mm film thickness (SUPELCO, USA or equivalent).
3. Conical single taper split/splitless liner (Agilent, Böblingen, Germany).
4. Gas chromatograph (Agilent 6890 N), split and splitless injector with electronic pressure control up to 150 psi (Agilent, Böblingen, Germany).
5. Helium 5.0 carrier gas.
6. Pegasus III Tof mass analyzer from LECO and corresponding software (LECO, St Joseph, USA) (or equivalent).
7. *CORRECTOR* software tool (http://www-en.mpimp-golm.mpg.de/03-research/researchGroups/01-dept1/Root_Metabolism/smp/CORRECTOR/index.html).

3 Methods

3.1 Feeding and Sample Preparation

1. Collect the plant material and prepare it accordingly (e.g., leaf disks can be just removed from the plant; potato tubers needs to be peeled and disks removed; tomato fruit needs to be cut into two as well as the peel need to be removed and the pericarp chopped into small pieces). After washing thoroughly with 10 mM MES-KOH (pH 6.5) to remove damaged cells transfer the plant material to Erlenmeyer flasks containing 10 mM MES-KOH (pH 6.5) and at least 20 mm of either labeled substrate or unlabeled substrate (Fig. 1). Incubate this material for up to 12 h. After the appropriate incubation time period (*see* **Note 2**) plant material need to be washed again thoroughly with 10 mM MES-KOH (pH 6.5) to remove excess of substrate adhered to cell surface and frozen in liquid nitrogen (*see* **Note 3**).
2. Transfer the material to precooled microtubes, 6-well-plates or wrap samples in aluminum foil. Freeze immediately in liquid nitrogen (*see* **Note 3**).
3. Precool two steel cylinders and metal balls in liquid nitrogen.
4. Quickly take out two samples and place them into independent steel cylinders together with a metal ball and cover the cylinders.
5. Fix cylinders in the mixer mill and mill at 25 Hz/s for 2 min.
6. Quickly take out the cylinders and place back into liquid nitrogen.

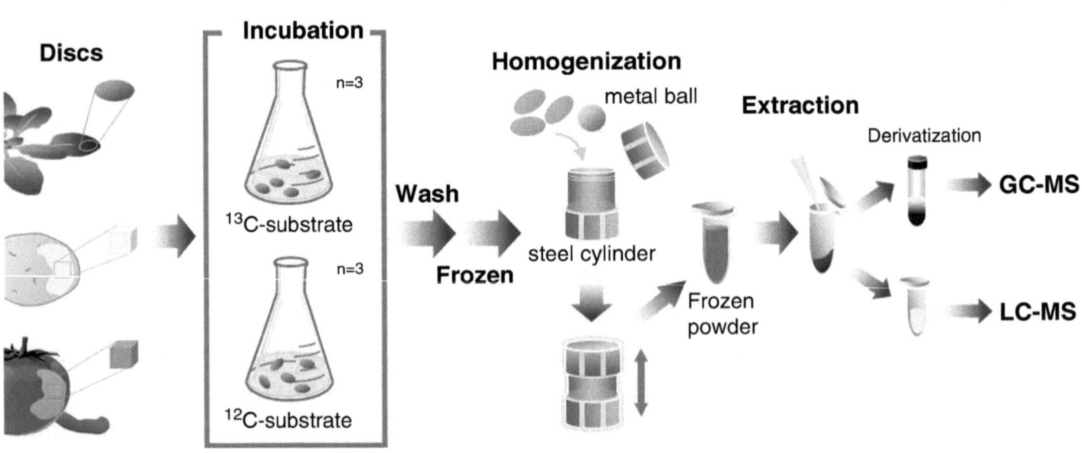

Fig. 1 General scheme of the experimental strategy for ^{13}C feeding experiment. Briefly, after adequate incubation with ^{13}C-labeled substrate, the samples are washed, harvested and can be stored at −80 °C. The samples are then homogenized, and extraction and derivatization for GCMS takes place. After the proper data Acquisition by GC-MS the isotope redistribution and ^{13}C-enrichment are calculated

7. Transfer the fine powder into a precooled tube and keep in liquid nitrogen.
8. Repeat **steps 2–6** until all samples have been homogenized (this step can also be done manually with pestle and mortar) (*see* **Note 3**).
9. Weigh out ~150 mg fine powder of each sample into a precooled 2-mL microcentrifuge tube and keep in liquid nitrogen or store at −80 °C until use (*see* **Note 3**).

3.2 Extraction

1. Remove the homogenized samples and add 1.4 mL 100 % methanol (precooled to −20 °C) to each and vortex for 10 s (*see* **Note 4**). Also prepare one tube without sample as a control (*see* **Note 5**).
2. Add 60 mL ribitol (0.2 mg/mL in dH_2O) as an internal quantitative standard in each tube and vortex for 10 s.
3. Incubate for 15 min at 70 °C in a thermoblock (after 1–2 min incubation open the tubes for a brief moment to let extra pressure go out).
4. Centrifuge for 15 min at 11,000 × *g*.
5. Transfer the supernatant to a Schott glass vial.
6. Add a further 0.75 mL 100 % chloroform (precooled to −20 °C) into the 2-mL microcentrifuge tube to wash it out.
7. Add 1.5 mL H_2O and vortex for 10 s.
8. Centrifuge for 15 min at 2,200 × *g*.
9. Aliquot 150 μL from the upper phase into a new 2-mL microcentrifuge tubes (*see* **Note 6**). The pellet can now be used for starch, protein, and/or cell wall determination or be discarded.
10. As a backup (in case you lose a sample), transfer a second aliquot to another new 2-mL tube.
11. Dry absolutely in a speed vacuum concentrator without heating for between 3 and 12 h.
12. For storage, fill the tubes with argon gas before closing. The tubes can then be stored at −80 °C for up to 3 months (*see* **Note 7**).

3.3 Derivatization

1. Take out the dried extracts from freezer and dry them absolutely in a speed vacuum concentrator for 30 min (*see* **Note 8**).
2. Prepare fresh methoxyamine solution by dissolving methoxyamine hydrochloride at 30 mg/mL in pure pyridine. Work in a fume hood (*see* **Note 1**).
3. Add 60 mL methoxyamine solution as prepared in step 2 to each sample and quickly close the tube.

4. Shake for 2 h at 37 °C at 950 rpm.
5. Spin down shortly to collect all drops on the walls and lids of the microcentrifuge tubes.
6. Prepare MSTFA reagent with FAMES (1 mL of MSTFA with 30 μL of FAMES) (*see* **Note 9**).
7. Add 70 μL of MSTFA reagent prepared in step 6 to each sample tube and quickly close the tube.
8. Shake for 30 min at 37 °C at 950 rpm.
9. Spin down shortly to collect all drops on the walls and lids of the microcentrifuge tubes.
10. Transfer reaction solutions into glass vials suitable for the GC-TOF-MS autosampler and quickly close the vials (*see* **Notes 10** and **11**).

3.4 Data Acquisition by GC-MS

1. Inject 1 μL of sample in splitless or split mode, depending on the metabolite concentration (*see* **Note 6**), with the helium carrier gas flow rate set to 2 mL/min by using the autosampler. The flow rate is kept constant with electronic pressure control enabled. The injection temperature is set to 230 °C. Injection programs must include syringe washing steps before and after each injection.
2. Perform chromatography using a 30-m MDN-35 capillary column. The temperature program should be isothermal for 2 min at 80 °C, followed by a 15 °C per min ramp to 330 °C, and holding at this temperature for 6 min. Cooling should be as rapid as the instrument specifications allow. Set the transfer line temperature to 250 °C and match ion source conditions.
3. Set the ion source to maximum instrument specifications, 250 °C. The recorded mass range should be m/z 70 to m/z 600 at 20 scan/s. Proceed the remaining monitored chromatography time with a 170-s solvent delay with filaments turned off. Manual mass defect should be set to 0, filament bias current should be −70 V, and detector voltage should be ~1,700–1,850 V. Automatically tune the instrument according to the manufacturer's instructions.
4. Transfer raw GC-MS profile chromatograms to a powerful server and regularly back up them.
5. Proceed with data (pre)processing and analysis as previously recommended [35, 36].

3.5 Isotope Redistribution and ^{13}C-Enrichment

1. Generate a peak intensity matrix containing all available mass isotopomers of characteristic mass fragments that represented the primary metabolites under investigation by TagFinder [37].

2. This matrix is processed using the *CORRECTOR* software tool (http://www-en.mpimp-golm.mpg.de/03-research/researchGroups/01-deptl/Root_Metabolism/smp/CORRECTOR/index.html) (*see* **Note 12**).
3. By using this batch processing tool, it is possible to calculated the sum of mass isotopomer intensities and the ^{13}C enrichments of mass fragments that had been annotated previously [13] using previously described methods [38, 39] (*see* **Note 13**).
4. After correction, quantification of the metabolites is performed based on calibration curves obtained with authentic standards whose matrices are also corrected. Relative isotopomer abundance (m_{n}) for each metabolite in which $n^{13}C$ atoms are incorporated is calculated in mass fragments [13, 38] by the following:

$$m_{\mathrm{n}}(\%) = \frac{M_n}{\sum_{j=0}^{i} M_{\mathrm{j}}} \times 100$$

where M_n represents the isotopomer abundance for each metabolite. The ^{13}C enrichment of the metabolite possessing i carbon atoms is calculated by the following:

$$^{13}\mathrm{C}\ \text{enrichment}\ (\%) = \sum_{n=1}^{i} \frac{i \times m_{\mathrm{n}}}{i}$$

4 Notes

1. Reagents are extremely toxic and should be handled in a fume hood with gloves.
2. The majority of metabolites appear to approach isotopic steady-state after 5 h incubation [8], however depending on the tissue used and metabolite desired this can be further complicated and caution must be taken.
3. Once the sample has been frozen in liquid nitrogen it must not thaw out even slightly before extraction. Make sure therefore to keep samples at constant freezing temperature (in liquid nitrogen) to avoid degradation of metabolites and precool all components to be used (spatulas, vials, etc.) in liquid nitrogen before they come into contact with the sample.
4. Enzymatic activity stops on adding methanol.
5. It is necessary to include the control tube without metabolite extract in order to identify any contaminants. It is important that all chemicals and containers need to be of the highest

available purity. Please consider that autoclaved material, although sterile, may nevertheless be chemically contaminated. It is important to consider the use of one already known sample as control. It can help to decide whether any error is sample related or machine related.

6. Some samples (e.g., tomato fruits) accumulate high amounts of sugars such as fructose, glucose, sucrose, as well as some organic acids such as citric acid and malic acid. 50 μL of extract in a splitless mode GC-MS run gives overloaded peaks for these compounds. Therefore, in order to measure accurately both high and low abundant metabolites, two approaches can be used. Either two modes of run, one splitless (for low abundant metabolites) and one split (for high abundant metabolites) using 50 μL of extract are performed, or two different extracts of 50 μL (for low abundant metabolites) and 5 μL (for high abundant metabolites) in splitless mode runs are recommended [40].
7. Argon-filled sample tubes prevent the extract from oxidization and degradation by reactions through components of atmospheric air. Extracts can be stored at −80 °C for up to 3 months. A longer storage time has not yet been investigated.
8. The most critical point is to avoid any water or moisture during derivatization. The silylating step is highly especially sensitive and even minor contamination with water will lead to inconsistent results.
9. For amount of FAMES, if two modes of run (splitless and split) are applied, 50 μL of FAMES per mL MSTFA is required. If only splitless mode run is used, a reduced amount of FAMES to 20–30 μL/mL MSTFA is possible.
10. The rest of derivatized samples can be stored in glass vials (in case something goes wrong with injection or measurement), but always in the dark at room temperature, for up to 2 days. Avoid storage in a cold room.
11. Samples must be injected in statistically valid randomized order to minimize the influence of experiment handling.
12. The mass-spectroscopic data are corrected for natural abundance of the isotopes [38]. Therefore, the metabolic pools are assumed to be completely unlabeled at the beginning of the experiment.
13. Since MS detects ionized compounds separated by their mass to charge ratio (m/z), the m/z of ^{13}C-labeled compounds is increasing by an amount which equals the number of

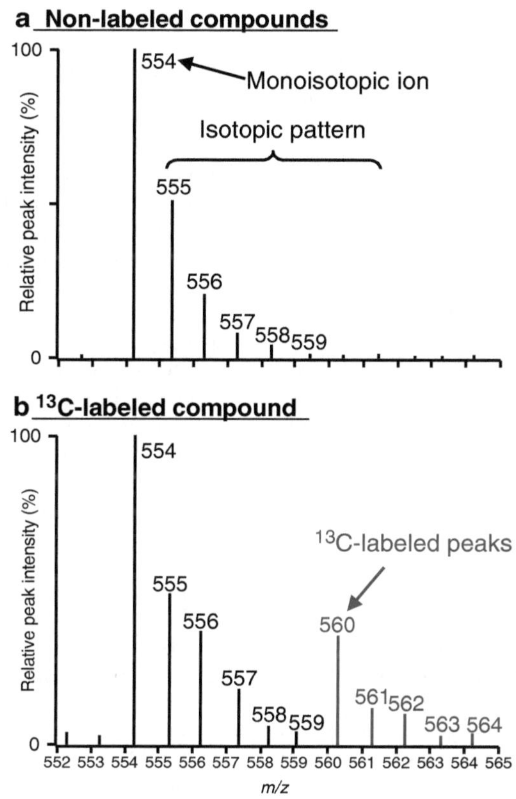

Fig. 2 Exemplary mass distribution following ^{13}C feeding. The normal isotopic pattern is observed in A while after incubation with ^{13}C substrate (B) the mass is increasing by an amount equal to the number of stable isotopes incorporated

incorporated stable isotopes (*see* Fig. 2). Therefore, by determining the ratio of intensity of the monoisotopic ion and its isotopic ions, the ratio of stable isotope labeling can be quantified [41].

Acknowledgements

Financial support from the Max-Planck-Society (to WLA and ARF), the Deutsche Forschungsgemeinschaft (grant no. DFG-SFB429 to ARF), and the National Council for Scientific and Technological Development CNPq-Brazil (grant number 472787/2011-0 to WLA) is gratefully acknowledged.

Conflict of interest: The authors declare that they have no conflict of interest.

References

1. Morgan MJ, Lehmann M, Schwarzlander M et al (2008) Decrease in manganese superoxide dismutase leads to reduced root growth and affects tricarboxylic acid cycle flux and mitochondrial redox homeostasis. Plant Physiol 147:101–114
2. Giegé P, Heazlewood JL, Roessner-Tunali U et al (2003) Enzymes of glycolysis are functionally associated with the mitochondrion in Arabidopsis cells. Plant Cell 15:2140–2151
3. Graham JWA, Williams TCR, Morgan M et al (2007) Glycolytic enzymes associate dynamically with mitochondria in response to respiratory demand and support substrate channeling. Plant Cell 19:3723–3738
4. Nunes-Nesi A, Carrari F, Lytovchenko A et al (2005) Enhanced photosynthetic performance and growth as a consequence of decreasing mitochondrial malate dehydrogenase activity in transgenic tomato plants. Plant Physiol 137:611–622
5. Araújo WL, Nunes-Nesi A, Osorio S et al (2011) Antisense inhibition of the iron-sulphur subunit of succinate dehydrogenase enhances photosynthesis and growth in tomato via an organic acid-mediated effect on stomatal aperture. Plant Cell 23:600–627
6. Araújo WL, Tohge T, Osorio S et al (2012) Antisense inhibition of the 2-oxoglutarate dehydrogenase complex in tomato demonstrates its importance for plant respiration and during leaf senescence and fruit maturation. Plant Cell 24:2328–2351
7. Schwender J, Shachar-Hill Y, Ohlrogge JB (2006) Mitochondrial metabolism in developing embryos of *Brassica napus*. J Biol Chem 281:34040–34047
8. Roessner-Tunali U, Liu J, Leisse A et al (2004) Kinetics of labelling of organic and amino acids in potato tubers by gas chromatography–mass spectrometry following incubation in 13C labelled isotopes. Plant J 39:668–679
9. Yuan J, Bennett BD, Rabinowitz JD (2008) Kinetic flux profiling for quantitation of cellular metabolic fluxes. Nat Protocols 3: 1328–1340
10. Haverkorn van Rijsewijk BRB, Nanchen A, Nallet S et al (2011) Large-scale 13C-flux analysis reveals distinct transcriptional control of respiratory and fermentative metabolism in Escherichia coli. Mol Syst Biol 7
11. Birkemeyer C, Luedemann A, Wagner C et al (2005) Metabolome analysis: the potential of in vivo labeling with stable isotopes for metabolite profiling. Trends Biotechnol 23:28–33
12. Huege J, Goetze J, Schwarz D et al (2011) Modulation of the major paths of carbon in photorespiratory mutants of *Synechocystis*. Plos One 6:e16278
13. Huege J, Sulpice R, Gibon Y et al (2007) GC-EI-TOF-MS analysis of in vivo carbon-partitioning into soluble metabolite pools of higher plants by monitoring isotope dilution after $^{13}CO_2$ labelling. Phytochemistry 68: 2258–2272
14. Araújo WL, Nunes-Nesi A, Trenkamp S et al (2008) Inhibition of 2-oxoglutarate dehydrogenase in potato tuber suggests the enzyme is limiting for respiration and confirms its importance in nitrogen assimilation. Plant Physiol 148:1782–1796
15. Studart-Guimarães C, Fait A, Nunes-Nesi A et al (2007) Reduced expression of succinyl-coenzyme A ligase can be compensated for by up-regulation of the γ-aminobutyrate shunt in illuminated tomato leaves. Plant Physiol 145:626–639
16. Dal CV, Tieman DM, Tohge T et al (2011) Identification of genes in the phenylalanine metabolic pathway by ectopic expression of a MYB transcription factor in tomato fruit. Plant Cell 23:2738–2753
17. Araújo WL, Ishizaki K, Nunes-Nesi A et al (2010) Identification of the 2-hydroxyglutarate and isovaleryl-CoA dehydrogenases as alternative electron donors linking lysine catabolism to the electron transport chain of Arabidopsis mitochondria. Plant Cell 22:1549–1563
18. Kleessen S, Araújo WL, Fernie AR et al (2012) Model-based confirmation of alternative substrates of mitochondrial electron transport chain. J Biol Chem 287:11122–11131
19. Ratcliffe RG, Shachar-Hill Y (2006) Measuring multiple fluxes through plant metabolic networks. Plant J 45:490–511
20. Libourel IGL, Shachar-Hill Y (2008) Metabolic flux analysis in plants: from intelligent design to rational engineering. Annu Rev Plant Physiol Plant Mol Biol 59:625–650
21. Schwender J, Ohlrogge J, Shachar-Hill Y (2004) Understanding flux in plant metabolic networks. Curr Opin Plant Biol 7:309–317
22. Schwender J (2008) Metabolic flux analysis as a tool in metabolic engineering of plants. Curr Opin Biotechnol 19:131–137
23. O'Grady J, Schwender J, Shachar-Hill Y et al (2012) Metabolic cartography: experimental quantification of metabolic fluxes from isotopic labelling studies. J Exp Bot 63:2293–2308

24. Allen DK, Ohlrogge JB, Shachar-Hill Y (2009) The role of light in soybean seed filling metabolism. Plant J 58:220–234
25. Blank L, Kuepfer L, Sauer U (2005) Large-scale ^{13}C-flux analysis reveals mechanistic principles of metabolic network robustness to null mutations in yeast. Genome Biol 6:R49
26. Fischer E, Sauer U (2005) Large-scale in vivo flux analysis shows rigidity and suboptimal performance of Bacillus subtilis metabolism. Nat Genet 37:636–640
27. Lonien J, Schwender J (2009) Analysis of metabolic flux phenotypes for two Arabidopsis mutants with severe impairment in seed storage lipid synthesis. Plant Physiol 151:1617–1634
28. Fernie AR, Roscher A, Ratcliffe RG et al (2001) Fructose 2,6-bisphosphate activates pyrophosphate: fructose-6-phosphate 1-phosphotransferase and increases triose phosphate to hexose phosphate cycling in heterotrophic cells. Planta 212:250–263
29. Schwender J, Ohlrogge JB, Shachar-Hill Y (2003) A flux model of glycolysis and the oxidative pentosephosphate pathway in developing *Brassica napus* embryos. J Biol Chem 278:29442–29453
30. Dieuaide-Noubhani M, Raffard G, Canioni P et al (1995) Quantification of compartmented metabolic fluxes in maize root tips using isotope distribution from ^{13}C- or ^{14}C-labeled glucose. J Biol Chem 270:13147–13159
31. Roscher A, Kruger NJ, Ratcliffe RG (2000) Strategies for metabolic flux analysis in plants using isotope labelling. J Biotechnol 77:81–102
32. Glawischnig E, Gierl A, Tomas A et al (2001) Retrobiosynthetic nuclear magnetic resonance analysis of amino acid biosynthesis and intermediary metabolism. Metabolic flux in developing maize kernels. Plant Physiol 125: 1178–1186
33. Rontein D, Dieuaide-Noubhani M, Dufourc EJ et al (2002) The metabolic architecture of plant cells - Stability of central metabolism and flexibility of anabolic pathways during the growth cycle of tomato cells. J Biol Chem 277:43948–43960
34. Edwards S, Nguyen BT, Do B et al (1998) Contribution of malic enzyme, pyruvate kinase, phosphoenolpyruvate carboxylase, and the Krebs cycle to respiration and biosynthesis and to intracellular pH regulation during hypoxia in maize root tips observed by nuclear magnetic resonance imaging and gas chromatography mass spectrometry. Plant Physiol 116:1073–1081
35. Luedemann A, Malotky L, Erban A et al (2012) TagFinder: Preprocessing software for the fingerprinting and the profiling of gas chromatography–mass spectrometry based metabolome analyses. Methods Mol Biol 860: 255–286
36. Lommen A (2012) Data (Pre-)processing of Nominal and Accurate Mass LC-MS or GC-MS Data Using MetAlign. Methods Mol Biol 860:229–253
37. Luedemann A, Strassburg K, Erban A et al (2008) TagFinder for the quantitative analysis of gas chromatography–mass spectrometry (GC-MS)-based metabolite profiling experiments. Bioinformatics 24:732–737
38. Van Winden WA, Wittmann C, Heinzle E et al (2002) Correcting mass isotopomer distributions for naturally occurring isotopes. Biotechnol Bioeng 80:477–479
39. Wittmann C, Heinzle E (1999) Mass spectrometry for metabolic flux analysis. Biotechnol Bioeng 62:739–750
40. Osorio S, Do PT, Fernie AR (2012) Profiling primary metabolites of tomato fruit with gas chromatography/mass spectrometry. Methods Mol Biol 860:101–109
41. Hasunuma T, Harada K, Miyazawa S-I et al (2010) Metabolic turnover analysis by a combination of in vivo ^{13}C-labelling from $^{13}CO_2$ and metabolic profiling with CE-MS/MS reveals rate-limiting steps of the C3 photosynthetic pathway in *Nicotiana tabacum* leaves. J Exp Bot 61:1041–1051

Chapter 8

Quantifying ^{13}C-labeling in Free Sugars and Starch by GC-MS

Mohamed Koubaa, Brigitte Thomasset, and Albrecht Roscher

Abstract

We describe an approach to extract ^{13}C-labeled sugars (glucose, fructose, maltose, sucrose, *myo*-inositol as well as glucose from starch) from plant tissues and to analyze their isotopomer distribution by gas chromatography–mass spectrometry (GC-MS). Sugars are derivatized with *N,O*-bis(trimethylsilyl) trifluoroacetamide (BSTFA) into their $Si(CH_3)_3$ derivatives. Electronic and chemical ionizations are used to obtain suitable fragments for metabolic flux analysis (MFA). Unique fragments are identified by computer simulation and experimental verification with labeled standards. Linear equations for separating information from glucosyl and fructosyl moieties of sucrose are presented. Finally, mass distributions are corrected for natural isotope abundance using a home-written program. The method is illustrated by sugar isotopomer analysis of ^{13}C-labeled rapeseed embryos.

Key words Metabolic flux analysis, Isotopomers, GC-MS, ^{13}C-labeling, Sugars, *Brassica napus*

1 Introduction

^{13}C metabolic flux analysis (MFA) needs ^{13}C isotopomer distribution data as an input. These data should stem from metabolites that are well distributed over the metabolic network and are at isotopic steady-state. The most widely used metabolites are amino acids released from protein hydrolysis (Chapter 7 of this volume). However, in plant tissues, protein turnover is generally slow for reaching isotopic steady-state, and therefore data has to be corrected for the existing unlabeled biomass. Free amino and organic acids can be alternatively analyzed (Chapter 7 of this volume) to avoid this dilution [1]. Nevertheless, coverage of the strongly compartmented plant metabolic network is not optimal, and analysis of sugars [2] is an ideal complement. Indeed, most saccharides derive from hexose phosphates, which are not direct amino acid precursors, and synthesis of some saccharides can be assigned to specific compartments, e.g., sucrose to the cytosol and starch to the plastid.

Martine Dieuaide-Noubhani and Ana Paula Alonso (eds.), *Plant Metabolic Flux Analysis: Methods and Protocols*, Methods in Molecular Biology, vol. 1090, DOI 10.1007/978-1-62703-688-7_8, © Springer Science+Business Media New York 2014

Sugars as well as other water-soluble metabolites can be analyzed by nuclear magnetic resonance (NMR), which gives information on positional labeling and on carbon–carbon bond labeling. However, NMR is limited by its sensitivity (millimolar range). An alternative and/or complementary technique is gas chromatography–mass spectrometry (GC-MS) that yields information on mass isotopomer distributions of whole molecules or fragments [3, 4]. However, most naturally occurring metabolites including sugars are not sufficiently volatile and their GC separation requires a derivatization step [5–7].

The most common derivatization methods for GC-MS are silylation, acylation, esterification, and alkylation [8]. Silylation technique involves the replacement of an acidic hydrogen with an alkylsilyl group, for example –Si $(CH_3)_3$ to form tri-methyl silyl (TMS) derivatives [5–7]. The silylation of sugars and their derivatives (sugar alcohols, amino sugars, and others) is routinely employed for GC-MS analysis [5, 6, 9]. The most commonly used silyl derivatizing reagents are BSTFA (*N,O*-bis(trimethylsilyl) trifluoroacetamide), BSA (*N,O*-bis(trimethylsilyl)acetamide), TMCS (Trimethyl-chlorosilane) and TMSIM (Trimethylsilylimidazole) [7]. A relative drawback of silylation compared to methylation or acetylation is the necessity to correct isotopomer abundances for ^{29}Si and ^{30}Si as well as C, O, and H heavy isotopes.

The protocol described in this chapter consists of the following steps: First, sugars (mono-, di-, and polysaccharides) are extracted from the plant tissues or cells by hot ethanol. If starch needs to be analyzed, a hydrolysis into its glucose units is necessary. The extracted free sugars and/or glucose units from starch are then derivatized with BSTFA. The obtained TMS derivatives are analyzed by GC-MS in electron ionization (EI) and chemical ionization (CI) modes. CI produces ions that are less fragmented, and therefore chances are better to observe unique non overlapping fragments whose isotopomer distribution can be quantified. Indeed, chromatographic peaks and mass fragments have first to be identified, and the absence of overlapping mass peaks has to be verified. The suitability of a fragment for isotopomer analysis is further tested by verifying for unlabeled material that the observed isotopomer distribution corresponds to the theoretically expected values. For "good" fragments, the raw isotopomer distribution for labeled material is then corrected for natural heavy isotope abundance of atoms other than the targeted carbon atoms. A last useful verification for the robustness of the data is to compute the fractional labeling and to compare it to the fractional labeling of the fed carbon source. This procedure should ensure the acquisition of reliable data for MFA.

This protocol was used to determine isotopomer distribution in sugars from rapeseed (*Brassica napus*) embryos incubated with [U-$^{13}C_6$]glucose. Free glucose, glucose from starch, fructose, inositol, sucrose, and maltose were amenable to analysis [2].

2 Materials

1. Buffer for starch hydrolysis: 125 mM sodium acetate adjusted to pH 4.8.
2. Enzyme solutions: 200 U/ml α-amylase and 200 U/ml amyloglucosidase prepared in water and stored at −20 °C (*see* **Note 1**).
3. Derivatization solvent: *N,N*-dimethyl formamide containing 0.1 % pyridine prepared in a screw glass tube (*see* **Note 2**).
4. A GC-MS instrument equipped with a suitable general-purpose nonpolar column and capable of using electron ionization or chemical ionization with methane. A Thermo-Fisher TSQ Quantum GC instrument is used. It consists of a Triplus autosampler, a trace GC Ultra for metabolite separation equipped with an Equity®-5 Capillary GC Column (Sigma-Aldrich) (5 % phenyl-methyl-siloxane diphenylpolysiloxane, 30 m × 0.25 mm) and a TSQ Quantum for mass spectrometry. MS control and spectral processing are carried out using Xcalibur software, version 2.0.7.
5. Fragmentation simulation software: Mass Frontier 5.0 (High-Chem Ltd.).
6. Program for correction of mass distributions for natural abundance: home-written routine under Scilab (http://www.scilab.org/en) that is available upon request.

3 Methods

3.1 Sugar Extraction from Plant Tissue

1. Weigh 100 mg of fresh tissue previously labeled with ^{13}C (*see* **Note 3**).
2. Crush in liquid nitrogen twice using a precooled mortar and pestle.
3. Add 1 ml ethanol (80 %) and continue crushing until the visible debris and cell wall have disappeared.
4. Using a 1 ml micropipette, transfer the mixture into a 15-ml Falcon tube.
5. Wash the mortar with 4 ml ethanol (80 %) and transfer to the Falcon tube using the same tip in order to wash it too.
6. Vortex the combined fractions vigorously for at least 1 min.
7. Place the tube for 10 min in a water bath set at 90 °C and then centrifuge it for 10 min at 1,600 × *g*.
8. Transfer the supernatant to a new 15-ml Falcon tube.
9. Wash the pellet with 5 ml ethanol (80 %), vortex, and centrifuge as previously.

10. Combine the supernatant with the one from **step 8** and evaporate ethanol under a stream of nitrogen at 80 °C. Freeze-dry and proceed to the soluble sugar analysis.

3.2 Starch Hydrolysis

1. Homogenize the starch-containing pellet in 560 μl acetate buffer (125 mM, pH 4.8).
2. Transfer the resuspended pellet to a 2-ml screw tube.
3. Warm it up for 90 min at 80 °C in a water-bath.
4. Transfer the tube to a dry bath for 20 min at 120 °C.
5. Cool the tube at room temperature.
6. Add 200 μl amyloglucosidase (40 U) and 200 μl α-amylase (40 U) to the homogenized pellet.
7. Complete to 1 ml with water to obtain a final buffer acetate concentration of 70 mM.
8. Mix the tube gently (do not vortex) and incubate it overnight at 37 °C in a water bath (*see* **Note 4**).
9. After incubation overnight, spin at 1,600 × *g* for 1 min at room temperature to recover the supernatant containing glucose units from starch.
10. Freeze-dry the supernatant overnight.

3.3 Derivatization Reaction

1. Under a fume hood add 150 μl and 400 μl of the derivatization solvent to the free sugars and the starch hydrolysis tubes, respectively.
2. Add 100 μl BSTFA to each tube, and incubate at 80 °C for 30 min. The derivatization time should be sufficient to obtain sugar derivatives from all extracted sugars.

3.4 GC-MS Analysis

$Si(CH_3)_3$ sugar derivatives are analyzed as described by Koubaa et al. [2] (*see* **Note 5**).

1. Use helium carrier gas with a linear velocity flow control mode of 1.5 ml/min.
2. Set the injection temperature to 300 °C and the injection mode to split with a split ratio of 100.
3. Set the initial oven temperature to 120 °C.
4. Hold it for 5 min to remove the most volatile compounds.
5. Increase the temperature to 270 °C at 4 °C/min and finally to 320 °C at 20 °C/min to remove the less volatile compounds.
6. Set both the ion source and the interface temperature to 320 °C with electron impact ionization (EI) or chemical ionization (CI) mode. Positive ionization is used in both cases.
7. Use the full scan detection mode between *m/z* 30 and 950 with a dwell time of 0.2 s.

Table 1
The most prominent used fragments of $Si(CH_3)_3$-derivatized sugars in rapeseed embryos

Compound(s)	Retention time (min)	Ionization mode	Fragment(s)	#C
Fructose	17.1 and 17.3	CI	525	6
Glucose & Starch	19.2 and 21.7	CI	Mean 361, 435, 525	6
Myo-Inositol	23.7	CI	507	6
Glc-sucrose	35.4	CI	Calculated from 361 and 451	6
Fru-sucrose		CI	Calculated from 361 and 451	6
Maltose	36.1 and 36.9	CI	361	6

Note. #C denotes the number of carbons in the fragment. Glc-sucrose and Fru-sucrose denote the glucosyl and fructosyl moieties of sucrose, respectively

3.5 Peak and Fragment Identification from Unlabeled Extracts and Standards

1. Identify chromatographic peaks based on retention times (*see* Table 1) and compare the fragmentation patterns with the literature or databases. If necessary, confirm assignments using sugar standards.
2. Identify fragments in order to know their elemental composition and the number of carbons from the sugar that can potentially be labeled (*see* **Note 6**).
3. Verify that there is no other peak even small in the region from $M-2$ to $M + C + 1$ where M is the fragment's mass and C the number of carbons retained from the sugar skeleton (*see* **Note 7**).
4. It has been shown that for MFA, sugar CI mass spectra are generally more informative than EI spectra [2]. Table 1 shows all fragments identified in free sugars and starch extracted from rapeseed embryos that can be used for MFA.
5. Note that the sucrose moieties do not correspond directly to a fragment but to linear combinations of two fragments (*see* **Note 8**). Figure 1 shows the chromatogram and the mass spectrum obtained in CI mode of an unlabeled sucrose standard. The retention time, 35.4 min, can be used for chromatographic assignment of extracts (**step 1** in this section).

3.6 Mass Distribution Correction

Mass distribution of the identified fragments must be corrected for natural isotope abundance of all atoms in the molecule (*see* Table 2) except the carbon skeleton of interest. This correction can be done by a routine using equations and algorithms proposed in [10, 11]. The corrected mass distributions can then be used as input data for MFA. We usually set the corresponding error margins for MFA to 0.01 unless the verification in **Note 7** gave a higher value. In this case, use this one as error margins.

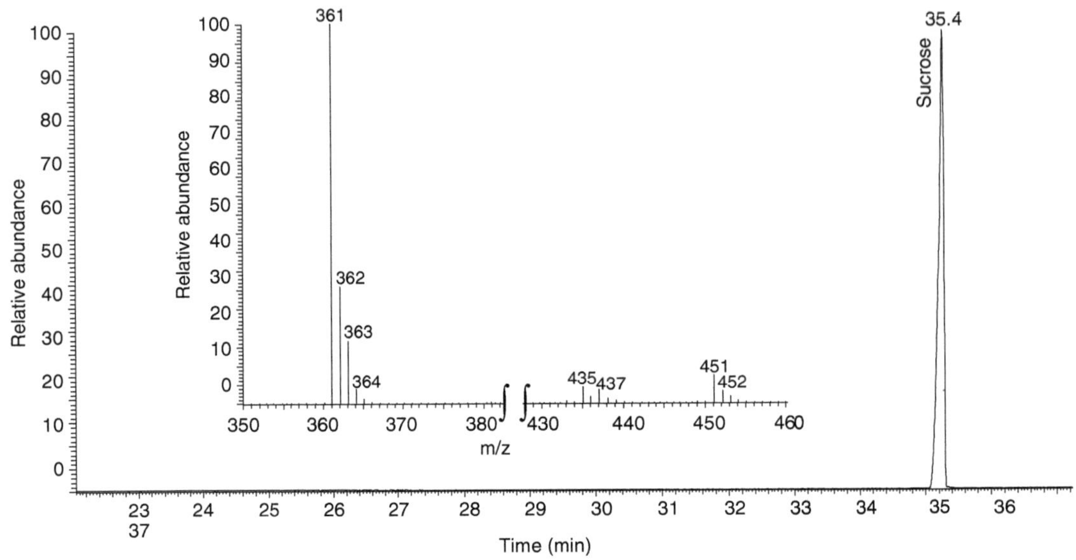

Fig. 1 Gas chromatogram and mass spectrum of unlabeled sucrose standard. 1 mg of sucrose was derivatized (Silylation) with 400 μl of derivatization solvent and 100 μl BSTFA. Silylated sucrose was analyzed by GC-MS using CI mode with a methane flow set at 2 ml/min

Table 2
Isotopomer abundance in unlabeled $Si(CH_3)_3$ sugar derivatives [12]

Atoms	M abundance	M + 1 abundance	M + 2 abundance
^{12}C	0.9893	0.0107	
^{16}O	0.99757	0.00038	0.00205
^{1}H	0.999885	0.000115	
^{28}Si	0.92223	0.04685	0.03092

3.7 Verification of Fractional Labeling

A useful test for data quality is to calculate the fractional labeling f for each fragment. It is obtained by summing the mass isotopomer abundances multiplied with the number of ^{13}C in the isotopomer, $f = (M_1 + 2\ M_2 + 3\ M_3 + \ldots + CM_C)/C$ where M_i is the mass isotopomer abundance with mass $M_0 + i$ and C the number of carbon atoms in the fragment (*see* **Note 9**).

4 Notes

1. Prepare 1,200 U/ml amyloglucosidase stock solution. Preparation of a 200 U/ml solution would require only 3.33 mg for 1 ml, which is difficult to weight with precision.

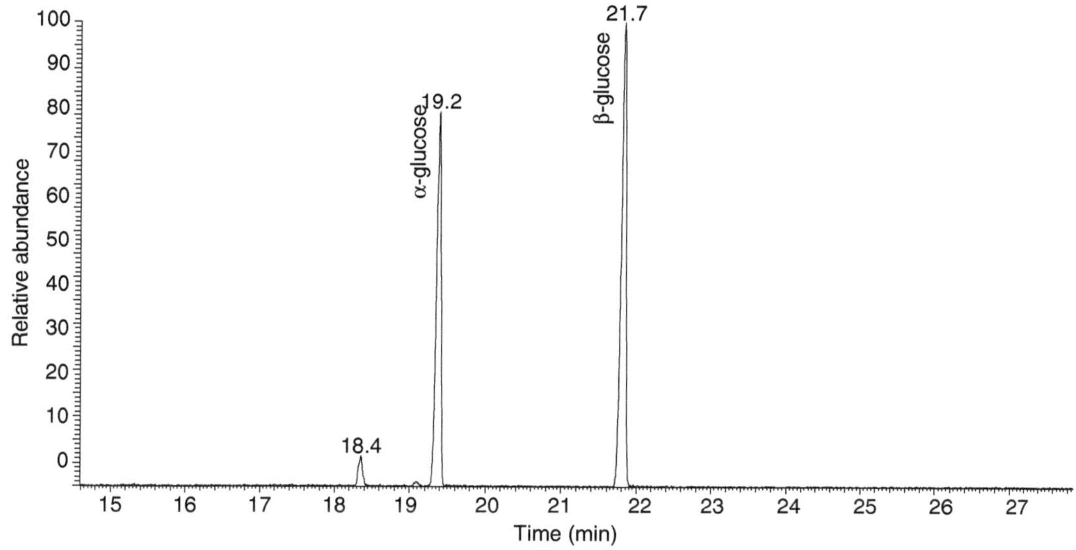

Fig. 2 Chromatogram obtained by GC-MS after starch hydrolysis and $Si(CH_3)_3$ derivatization of the generated glucose units. Analysis was performed using CI mode with a methane flow at 2 ml/min

2. Add 4 μl pyridine to 3,996 μl *N,N*-dimethyl formamide under a fume hood and mix vigorously for 10 s. Screw glass tubes should always be used for organic solvents, especially when samples are analyzed by GC-MS.
3. For steady-state metabolic flux analysis, we do not need to quantify metabolite contents, so fresh tissue could be used instead of dry tissue. Sugars could also be extracted after freeze-drying. Use at least 20 mg of dry weight tissue in order to obtain accurate GC-MS results.
4. The hydrolysis can be done with less enzyme units depending on the tissue starch content. Using 40 units of each enzyme improves the hydrolysis yield. Less enzyme units could be used with a longer incubation time (e.g., 48 h incubation). After hydrolysis, glucose units are analyzed by GC-MS as described in Subheadings 3.3 and 3.4. The corresponding chromatogram is represented in Fig. 2.
5. GC-MS conditions are given for the instrument specified in Subheading 2. They might need to be optimized for your instrument setup if peak separation is insufficient.
6. Manual identification will look for loss of one or several typical groups like CH_3 ($\Delta M = 15$), $Si(CH_3)_3$ ($\Delta M = 73$), O-Si $(CH_3)_3$ ($\Delta M = 89$), sometimes in combination with loss of an H ($\Delta M = 1$). As long as the sugar carbon skeleton remains intact, the exact site of lost groups does not need to be known. Further fragment identification uses simulation software like Mass Frontier. Interesting fragments for MFA should be

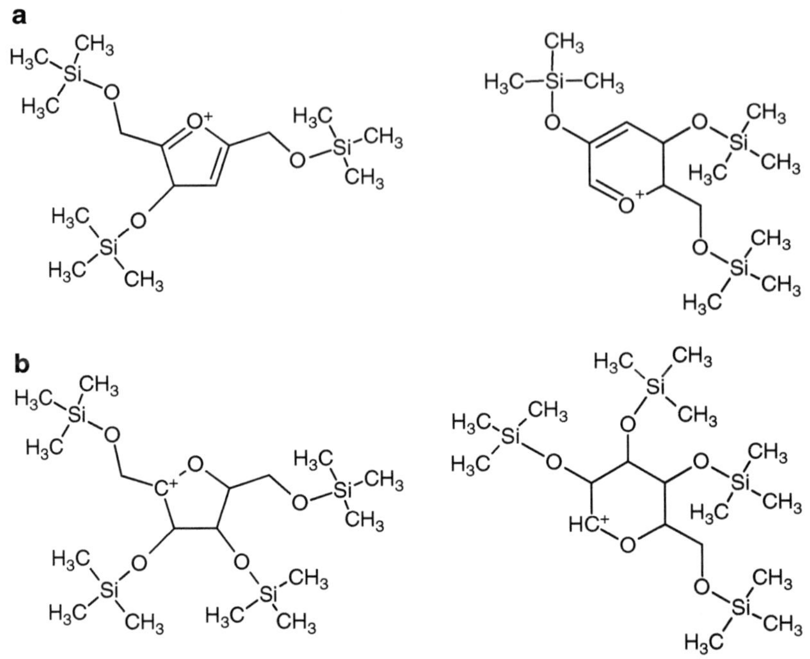

Fig. 3 Sucrose fragment identification by Mass Frontier simulation. Fragments 361 (**a**) and 451 (**b**) were determined for sucrose derivative. Molecules on the right and left sides derive from the glucosyl and fructosyl moieties, respectively

confirmed experimentally using specifically ^{13}C-labeled standards (first fully labeled to determine the number of sugar carbons; then if carbons are lost from the sugar skeleton use positional labeled standards to determine which carbons were retained).

7. Visual inspection of the unlabeled standard spectra allows to detect some cases but it is best practice to verify the absence of further peaks by correcting the unlabeled standard spectrum for all heavy isotopes including the sugar carbon skeleton (*see* Subheading 3.6). The result should not be significantly different (less than 1 %) from 1 for mass *M* and 0 for all other masses.
8. As sucrose is the most abundant free sugar in plant tissues [2], we will illustrate here all steps with sucrose as an example. We also point a difficulty arising for oligosaccharides containing different monomers of the same mass. The mass spectrum in Fig. 1 shows the major peaks in the range of *m/z* 100 to 950 which are 361 and 451 and two others at 435 and 437. The simulation by Mass Frontier suggests that both peaks 361 and 451 contain six sugar carbon atoms but as a mixture of molecules stemming from the fructosyl and the glucosyl moieties of sucrose (*see* Fig. 3). Following Subheading 3.5, **step 3**, peaks 435 and 437 (shifted to 442 in both spectra thus representing

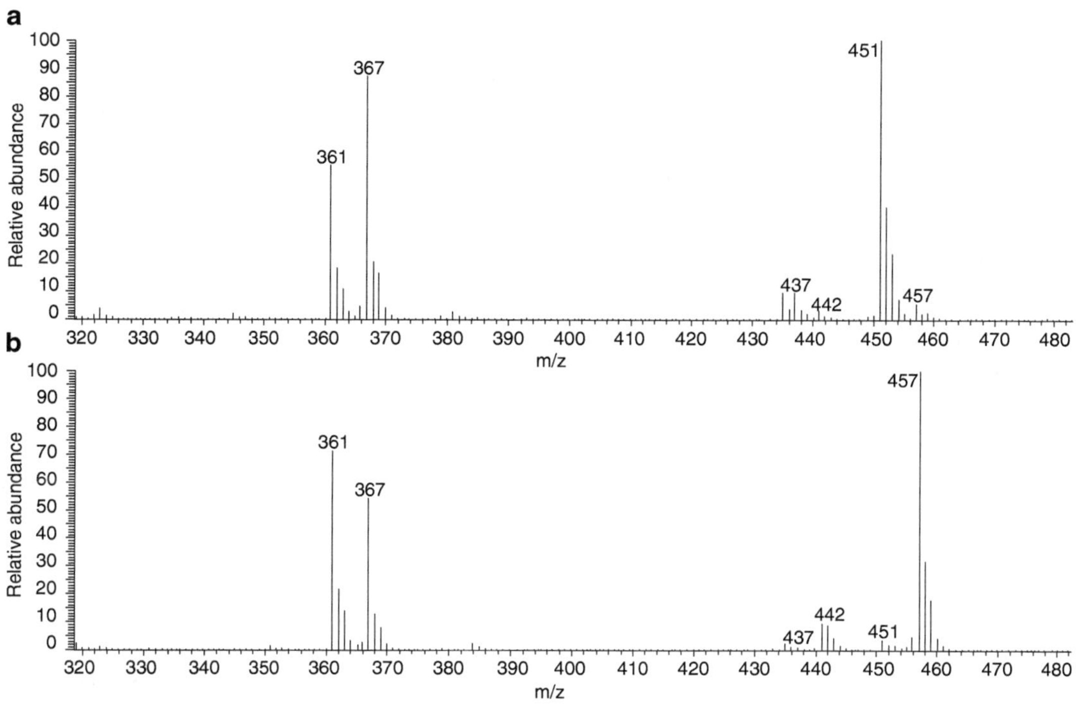

Fig. 4 Mass spectra of labeled sucrose fragmented in CI mode. (**a**) and (**b**) correspond to [*glc*-$^{13}C_6$]sucrose and [*fru*-$^{13}C_6$]sucrose, respectively. Analyses were performed using CI positive mode with a methane flow set up to 2 ml/min

5-carbon fragments from glucosyl and fructosyl moieties) are classified as unexploitable because they are overlapping. For peaks 361 and 451, natural abundance correction confirms their purity. The mass spectra of [*glc*-$^{13}C_6$]sucrose and [*fru*-$^{13}C_6$]sucrose in Fig. 4 confirm the presence of six carbon atoms both in fragment 361 (shifted to 367) and in fragment 451 (shifted to 457). Furthermore the contribution of both glucosyl and fructosyl moieties to both peaks is confirmed (*see* Subheading 3.5, **step 2**). Quantification after correction for natural abundance (*see* Subheading 3.6) allows to calculate the ^{13}C isotopomer distributions of both sucrose moieties as a linear combination of the isotopomer distributions of peaks 361 and 451: *glucosyl* = 1.73* *(361)* – 0.73* *(451)* and *fructosyl* = 1.09* *(451)* – 0.09* *(361)* [2]. This linear combination should be verified and corrected for each GC-MS instrument. Based on the analysis of unlabeled and labeled sucrose and the relative abundance of each peak, the combination is determined after natural abundance correction.

9. For metabolites at metabolic and isotopic steady state, the fractional labeling should correspond to the fractional labeling of the fed substrate. It will be lower for pools that are not at

isotopic steady state due to slow turnover, e.g., starch. For the starch, the unlabeled fraction has to be integrated into the metabolic model. For pools that should be at isotopic steady state and are not, both the analytical procedure and the underlying metabolic model have to be verified to find the reason for the observed discrepancy.

References

1. Koubaa M, Cocuron JC, Thomasset B, Alonso AP (2013) Highlighting the tricarboxylic acid cycle: liquid and gas chromatography–mass spectrometry analyses of ^{13}C-labeled organic acids. Anal Biochem 436:151–159
2. Koubaa M, Mgaieth S, Thomasset B, Roscher A (2012) Gas chromatography–mass spectrometry analysis of ^{13}C labeling in sugars for metabolic flux analysis. Anal Biochem 425:183–188
3. Blau K, Halket JM (1993) Handbook of derivatives for chromatography. Wiley, Chichester, UK
4. Villas-Bôas SG, Smart KF, Sivakumaran S, Lane GA (2011) Alkylation or silylation for analysis of amino and non-amino organic acids by GC-MS? Metabolites 1:3–20
5. Villas-Bôas SG, Mas S, Akesson M, Smedsgaard J, Nielsen J (2005) Mass spectrometry in metabolome analysis. Mass Spectrom Rev 24:613–646
6. Villas-Bôas SG, Koulman A, Lane GA (2007) Analytical methods from the perspective of method standardization. In: Jewett MC, Nielsen J (eds) Topics in current genetics - metabolomics. Elsevier-Verlag, Berlin, Germany, p 11
7. Sparkman OD, Penton Z, Kitson FG (2011) Gas chromatography and mass spectrometry: a practical guide, 2nd edn. Academic, USA, pp 185–187
8. Thurman EM, Mills MS (1998) Solid phase extraction: principles and practice. Wiley Europe, Hoboken, USA, p 372
9. Little JL (1999) Artifacts in trimethylsilyl derivatization reactions and ways to avoid them. J Chromatogr A 844:1–22
10. Lee WN, Byerley LO, Bergner EA, Edmond J (1991) Mass isotopomer analysis: theoretical and practical considerations. Biol Mass Spectrom 20:451–458
11. Nanchen A, Fuhrer T, Sauer U (2007) Determination of metabolic flux ratios from ^{13}C experiments and gas chromatography–mass spectrometry data: protocol and principles. In: Weckwerth W (ed) Metabolomics: methods and protocols. Humana, Totowa, NJ, pp 177–197
12. Böhlke JK, De Laeter JR, De Bièvre P et al (2005) Isotopic compositions of the elements, 2001. J Phys Chem Ref Data 34:57–67

Chapter 9

Liquid Chromatography Tandem Mass Spectrometry for Measuring ^{13}C-labeling in Intermediates of the Glycolysis and Pentose Phosphate Pathway

Jean-Christophe Cocuron and Ana Paula Alonso

Abstract

This chapter describes a procedure to analyze ^{13}C-labeled phosphorylated compounds by liquid chromatography tandem mass spectrometry. Phosphorylated compounds, intermediaries of the glycolysis and pentose phosphate pathway, are separated by anion exchange chromatography and their isotopic labeling is determined by mass spectrometry. A sensitivity in the fmole range is achieved using scheduled multiple reaction monitoring mode.

Key words Anion exchange chromatography, LC-MS/MS, Phosphorylated compounds, Multiple reaction monitoring, ^{13}C-labeling, Metabolic flux analysis

Abbreviations

2/3-PGA	2/3-Phosphoglycerate
6PG	6-Phosphogluconic acid
disac-P	Disaccharide-phosphate
DW	Dry weight
fruc 1,6-bisP	Fructose 1,6-bisphosphate
fruc 6-P	Fructose 6-phosphate
gal 1-P	Galactose 1-phosphate
glc 1-P	Glucose 1-phosphate
glc 6-P	Glucose 6-phosphate
glycerol 3-P	Glycerol 3-phosphate
hexose-Ps	Hexose phosphates
KOH	Potassium hydroxide
LC	Liquid chromatography
man 1-P	Mannose 1-phosphate
man 6-P	Mannose 6-phosphate
MRM	Multiple reaction monitoring
MS	Mass spectrometry

Martine Dieuaide-Noubhani and Ana Paula Alonso (eds.), *Plant Metabolic Flux Analysis: Methods and Protocols*, Methods in Molecular Biology, vol. 1090, DOI 10.1007/978-1-62703-688-7_9, © Springer Science+Business Media New York 2014

P5Ps	Pentose 5-phosphates
PEP	Phosphoenolpyruvate
rib 1,5-bisP	Ribulose 1,5-bisphosphate
S6P	Sucrose 6-phosphate
S7P	Sedoheptulose 7-phosphate
T6P	Trehalose 6-phosphate

1 Introduction

Intracellular metabolic rates (fluxes) provide critical insights into the metabolic state of an organ or cell. ^{13}C-metabolic flux analysis (^{13}C-MFA) is a powerful tool to analyze the flow of carbon through metabolic networks. For this purpose, a plant organ is incubated with a combination of ^{13}C-labeled substrates (that mimics its growth *in planta*) until the isotopic steady state is reached. Labeled substrates are metabolized by cells, resulting in the incorporation of ^{13}C-atoms into intracellular metabolites and biomass products. ^{13}C-labeling is usually measured in hydrolyzed macromolecules (proteins, starch, and triacylglycerols) [1–3] as they are more abundant than intermediary metabolites. These ^{13}C-measurements, directly reflecting the labeling of macromolecules precursors (amino acids, plastidic glucose 6-phosphate, acetyl CoA) are used within the context of a retrobiosynthetic approach to map carbon fluxes through central metabolism. Indeed, metabolic fluxes are calculated from these labeling measurements using a model-based approach that maximizes the best fit between the measured and model-predicted labeling distributions [4–8]. This approach has been successfully applied to characterize the in vivo metabolic fluxes in plants for: (1) understanding carbon partitioning in maize root tips [1, 9, 10], tomato cells [11], tobacco cells [12], and Arabidopsis cells [13]; (2) unraveling new pathways/reactions in plants, such as the "RuBisCo bypass" in *Brassica napus* embryos [14] and new "futile" cycles [1, 15]; (3) understanding the storage metabolism in seeds, such as maize kernels [16, 17], and soybean [2, 18], *Brassica napus* [19, 20], sunflower [21], and Arabidopsis embryos [22]; (4) determining the impact of a mutation on carbon partitioning in maize root tips [9] and Arabidopsis embryos [22]. However, ^{13}C-MFA and retrobiosynthetic approaches are limited to plant organs/cells that can reach metabolic and isotopic steady and pathways resulting in carbon rearrangement. For systems that do not meet those requirements, such as leaves or linear biosynthetic pathways, the direct labeling in intermediary metabolites is required.

Several metabolic pathways, such as glycolysis and oxidative pentose-phosphate pathway (OPPP), exclusively involve phosphorylated compounds. Determining the levels and labeling distribution in these metabolites is therefore important to access carbon fluxes in central metabolism. Several studies have reported the separation of phosphorylated metabolites using liquid chromatography coupled with tandem mass spectrometry [23–25]. Anion exchange chromatography was shown to be effective at separating various intermediates involved in central metabolism [26–30], such as but not limited to glycolysis, Calvin cycle, plant cell wall biosynthesis, and tricarboxylic acid cycle, especially when coupled with the high specificity and sensitivity of a triple quadrupole mass spectrometer. Moreover, this technique can be used to determine the mass isotopomer distribution of labeled compounds [31], specifically in phosphorylated metabolites [26, 32, 33].

This protocol describes the use of LC-MS/MS for measuring ^{13}C-labeling in intermediates of the glycolysis and pentose-phosphate pathway. Metabolites are extracted from ^{13}C-labeled plant organs using boiling water. Phosphorylated compounds are separated by an anion exchange column, and then detected by a triple quadrupole mass spectrometer using multiple reaction monitoring (MRM) mode: (1) the first quadrupole selects for particular m/z ratios of the ionized precursor compounds; (2) the second one fragments the precursor into product ions; and (3) the third quadrupole selects for a specific product ion which is here the phosphate group. For a phosphorylated metabolite containing n carbon atoms, $n + 1$ transitions are followed. Mass isotopomer distribution is determined for each phosphorylated compound extracted from ^{13}C-labeled plant organ; these labeling data can then be used for MFA.

2 Materials

2.1 Liquid Chromatography Supplies

Chromatography supplies were purchased from Dionex:

1. IonPac® ATC-3 Anion Trap Column (4 × 35 mm).
2. Guard column AG11 (50 × 2 mm).
3. IonPac AS11 column (250 × 2 mm).
4. ASRS 300 suppressor.
5. SRS controller.

2.2 Liquid Chromatography Solutions

1. Degas fresh ultrapure water (18.2 MΩ) in a 2 L flask containing a stir bar for 30 min using a pump system.
2. Make a 2 M KOH solution in degassed ultrapure water. Preweigh a 100 mL graduated cylinder and add KOH from a 45 % KOH liquid solution (cat # SP236500, ThermoFisher) to have 2 M final. Complete to 100 mL and flush with helium. This KOH solution is used to condition the IonPac® ATC-3 Anion Trap Column (*see* **Note 1**).

3. Prepare 0.5 mM (solvent A) and 75 mM (solvent B) potassium hydroxide solutions from a 45 % KOH liquid solution (cat # SP236500, ThermoFisher). First, weigh KOH in a weighing boat to have a final concentration of 75 mM in a desired volume. Transfer KOH into a graduated cylinder and add degassed ultrapure water to have a final concentration of 75 mM KOH (solvent B). Cover top of graduated cylinder with parafilm and invert two to three times. Transfer the solution into a 2 L plastic bottle (cat # 45903, Dionex). Do not use borosilicate bottle to avoid leaching of borate that will interfere with the separation of phosphorylated compounds. Then, take a desired volume from the 75 mM KOH solution to make the 0.5 mM (solvent A) KOH solution. Finally, flush both solutions of KOH with helium as well as the 45 % KOH stock solution.
4. Solvent for needle wash: 50 % (v/v) LC-MS grade methanol in ultrapure water.

2.3 Sample Preparation

1. Phosphorylated compounds were purchased from Sigma and stored accordingly to the manufacturer instructions.
2. Plant tissue labeled with ^{13}C substrates were weighed (~10 mg dry weight) using an analytical scale.

3 Method

3.1 Extraction of Phosphorylated Compounds

1. Place a 50 mL conical tube of ultrapure water in a water bath set at 100 °C.
2. Add one tungsten bead (5 mm) per 2 mL screw-cap tube containing the 10 mg dry weight of plant tissue.
3. Shake in a mill mixer for 3 min at a constant frequency of 30 Hz. Remove tubes and insure that the tissue is ground into a powder.
4. Add 1 mL of boiling ultrapure water to the 2 mL screw-cap tubes and incubate at 100 °C for 5 min in a water bath (*see* **Note 2**).
5. Take out, vortex the samples and put them back in the boiling water bath for five additional minutes.
6. Place quickly the tubes on ice. Wait 3–5 min until samples cool down.
7. Centrifuge the tubes at 20,000 × *g* at 4 °C for 5 min.
8. Collect and filter each supernatant through a 5 mL syringe with a 0.22 μm filter into a 15 mL tube kept on ice. Syringes and filters are conserved.
9. Wash tubes containing disrupted plant tissue with 1 mL of 4 °C ultrapure water, vortex to resuspend material, and centrifuge at 20,000 × *g* at 4 °C for 5 min.

10. Collect and filter each supernatant as explained above, using previous syringes (so syringes and filters get rinsed too).
11. Rinse syringes and filters with 1 mL of cold ultrapure water.
12. Spin down 15 mL tubes, poke holes on each cap.
13. Freeze-dry extracts using lyophilizer.
14. Dried samples are resuspended in 500 μL of cold ultrapure water and applied on a 0.5 mL Amicon Ultra filter (Millipore) with a 3 kDa cut off.
15. Tubes are centrifuged at 20,000 × *g* at 4 °C for 30 min.
16. The flow-through, containing ^{13}C labeled phosphorylated metabolites, is ready for LC-MS/MS analyses. Samples are diluted (5:1) in water.

3.2 Optimization of MS Parameters for Each Compound

Tandem mass spectrometry analysis is performed with a hybrid Triple Quadrupole/Ion trap mass spectrometer QTRAP 5500 (AB Sciex). Mass spectra are acquired with Analyst 1.6 software using Turbo Spray ionization in negative ion mode.

Automatic optimization of MS parameters (*see* Table 1) associated to the different compounds will give the best sensitivity for MS/MS analysis.

Table 1
Specific metabolite-dependent MS parameters used for LC-MS/MS

Analyte	Parent ion formula	Product ion formula	Parent/product transition	DP[a] (V)	EP[b] (V)	CE[c] (V)	CXP[d] (V)
Disac-P	$C_{12}H_{22}O_{14}P^-$	$[PO_3]^-$	421/79	−95	−10	−130	−15
E4P	$C_4H_8O_{14}P^-$	$[PO_3]^-$	199/79	−70	−10	−84	−11
Fruc 1,6-bisP	$C_6H_{13}O_{12}P_2^-$	$[PO_3]^-$	339/79	−50	−10	−88	−13
Glycerol-Ps	$C_3H_8O_6P^-$	$[PO_3]^-$	171/79	−60	−10	−42	−9
Hexose-Ps	$C_6H_{12}O_9P^-$	$[PO_3]^-$	259/79	−40	−10	−90	−11
P5Ps/Rib 1,5-bisP	$C_5H_8O_8P^-$	$[PO_3]^-$	229/79	−45	−10	−74	−37
PEP	$C_3H_4O_6P^-$	$[PO_3]^-$	167/79	−40	−10	−52	−9
6PG	$C_6H_{12}O_{10}P^-$	$[PO_3]^-$	275/79	−45	−10	−108	−15
2/3-PGA	$C_3H_6O_7P^-$	$[PO_3]^-$	185/79	−35	−10	−63	−12
S7P	$C_7H_{14}O_{10}P^-$	$[PO_3]^-$	291/79	−80	−10	−100	−9

[a]Declustering potential
[b]Entrance potential
[c]Collision energy
[d]Collision exit potential

1. Prepare each metabolite individually in water at a final concentration of 10 μM.
2. Load metabolite in a 1 mL Hamilton syringe and inject by direct infusion at a flow rate of 7 μL/min.
3. Insure that compound is detected (*see* **Note 3**).
4. Proceed to automatic compound optimization (*see* **Note 4**) and save file. This file contains optimized declustering potential (DP), entrance potential (EP), collision energy (CE), and collision exit potential (CXP) for each metabolite. Note that automatic compound optimization can generate five-product ions by metabolite that will be optimized for the parameters cited above.

3.3 Flow Injection Analysis

To optimize gas/source parameters to have the best sensitivity for MS/MS analysis.

1. Prepare a vial with a mixture of phosphorylated compounds (avoid isomers) in water at a final concentration of 10 μM.
2. Select compound optimization and then flow injection analysis (FIA) experiment for MS/MS analysis. 2 μL injection for FIA experiments is suitable to test different value for curtain gas, collision gas (CAD), ion spray voltage (IS), heater temperature (TEM), nebulizer gas (GS1), and heater gas (GS2) (*see* **Note 5**). Water is used as mobile phase at a 350 μL/min flow rate.
3. Optimized curtain gas, CAD, IS, TEM, GS1, and GS2 are saved for the mixture of phosphorylated compounds (*see* Table 2).

3.4 Chromatographic Conditions

1. LC is performed with an Agilent *U*ltra *H*igh *P*ressure *L*iquid *C*hromatography (UHPLC) 1290.
2. Column: Anion exchange IonPac® AS11 column (250 × 2 mm). Column is kept at 23 °C. Guard column AG11 (50 × 2 mm). IonPac® ATC-3 Anion Trap Column (4 × 35 mm) is used to remove contaminant ions from KOH solvents.

Table 2
Optimal gas/source parameters for scheduled MRM

Source/gas parameters	Value
Curtain gas (CUR)	35
Collision gas (CAD)	Medium
Ion Spray voltage (IS)	–4,500
Temperature (TEM)	550
Ion source gas 1 (GS1)	60
Ion source gas 2 (GS2)	40

Table 3
Liquid chromatography gradient used for phosphorylated compounds

Time (min)	Flow rate (μL/min)	A (%)	B (%)
0.0	350	100.0	0.0
2.0	350	100.0	0.0
13.0	350	95.2	4.8
20.0	350	95.2	4.8
30.0	350	87.3	12.7
45.0	350	73.8	26.2
65.0	350	33.6	66.4
70.0	350	0.0	100.0
75.0	350	0.0	100.0
75.1	350	100.0	0.0
80.0	350	100.0	0.0

3. Gradient for LC-MS/MS analysis is shown in Table 3.
4. Sample injection: 5 μL of standard or biological sample are injected with a 30 s needle wash.
5. Salt suppression (*see* **Note 6**): KOH is removed using a post-column anion self-regenerating suppressor ASRS 300 (Dionex). It is operated in the Autosuppression External Water Mode for high sensitivity analysis. The current and the reagent flow rate are set respectively at 50 mA and 2 mL/min.

3.5 Determination of Retention Time

1. Specific metabolite-dependent MS parameters (*see* Table 1) are used to determine retention time of phosphorylated standards on a full Multiple Reaction Monitoring (MRM) run.
2. Retention time of phosphorylated compounds is shown in Fig. 1. Each metabolite is associated to a specific parent/product ion pair (aka transition).

3.6 Scheduled MRM for ^{13}C-labeling Determination

1. Retention time of phosphorylated standards, previously determined, is used to define the different windows for scheduled MRM experiment (*see* Table 4). A 180 s window is applied for each ^{13}C-labeled phosphorylated compound.
2. Scheduled MRM for ^{13}C-S7P are shown in Fig. 2.

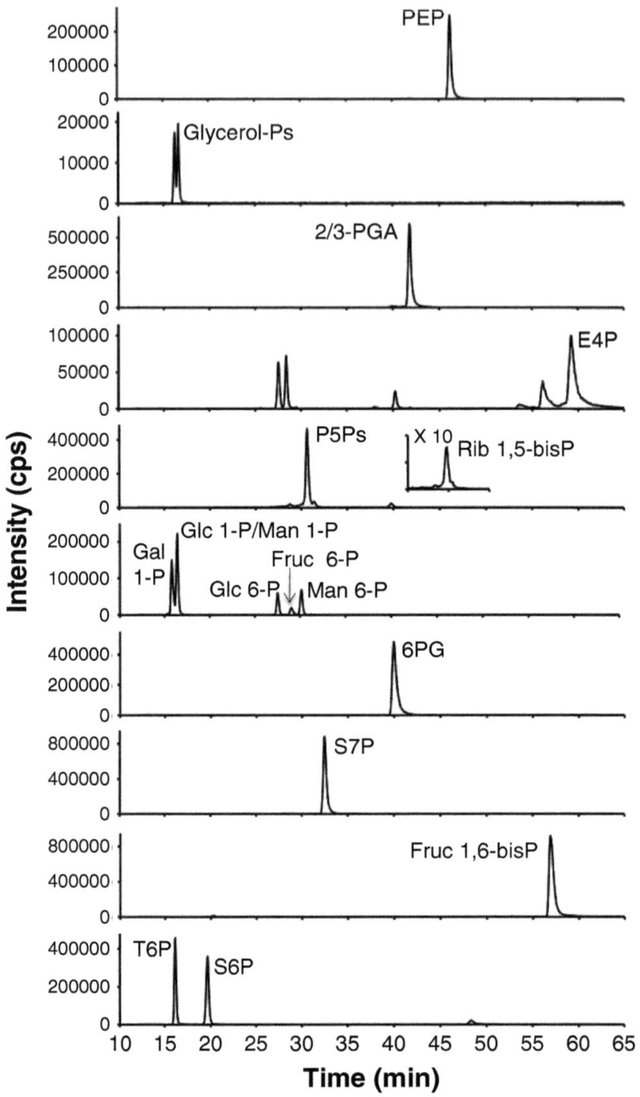

Fig. 1 Analyses of a mixture of phosphorylated metabolites by LC-MS/MS using MRM. Phosphorylated metabolite standards, approximately 10 μM each, are monitored by LC-MS/MS at the transition $[M\text{-}H]^-/79$ using Table 1 MS parameters and Table 4 gradient. Total ion count is reported for each phosphorylated metabolite: *PEP* phosphoenolpyruvate, *Glycerol-Ps* glycerol-phosphates, *2/3-PGA* 2/3-phosphoglyceric acids, *E4P* erythrose 4-phosphate, *P5Ps* pentose 5-phosphates, *Rib 1,5-bisP* ribulose 1,5-bisphosphate, *Gal 1-P* galactose 1-phosphate, *Glc 1-P/Man-1P* glucose 1-phosphate/mannose 1-phosphate, *Glc 6-P* glucose 6-phosphate, *Fruc 6-P* fructose 6-phosphate, *Man 6-P* mannose 6-phosphate, *6PG* 6-phosphogluconic acid, *S7P* sedoheptulose 7-phosphate, *T6P* trehalose 6-phosphate, *S6P* sucrose 6-phosphate

4 Notes

1. The IonPac® ATC-3 Anion Trap Column needs to be conditioned following Document No. 031835–01 (Dionex) to remove trace anions from KOH gradient. The use of the

Table 4
Transitions and retention time used in scheduled MRM

Analyte	Retention time (min)	Mass isotopomers	Parent/product transition
PEP	46.16	m_0 to m_{+3}	167/79 to 170/79
Glycerol-Ps	16.23/16.66[a]	m_0 to m_{+3}	171/79 to 174/79
2/3PGA	41.75	m_0 to m_{+3}	185/79 to 188/79
E4P	59.30	m_0 to m_{+4}	199/79 to 203/79
P5Ps	30.63	m_0 to m_{+5}	229/79 to 234/79
Rib 1,5-bisP[b]	39.82	m_0 to m_{+5}	229/79 to 234/79
Gal 1-P	15.80	m_0 to m_{+6}	259/79 to 265/79
Glc 1-P/Man 1-P	16.45	m_0 to m_{+6}	259/79 to 265/79
Glc 6-P	27.56	m_0 to m_{+6}	259/79 to 265/79
Fruc 6-P	29.03	m_0 to m_{+6}	259/79 to 265/79
Man 6-P	30.17	m_0 to m_{+6}	259/79 to 265/79
6PG	39.94	m_0 to m_{+6}	275/79 to 281/79
S7P	32.49	m_0 to m_{+7}	291/79 to 298/79
Fruc 1,6-bisP	56.91	m_0 to m_{+6}	339/79 to 345/79
S6P	19.71	m_0 to m_{+12}	421/79 to 433/79
T6P	16.16	m_0 to m_{+12}	421/79 to 433/79

[a]Two peaks for the glycerol-phosphate standard meaning that two glycerol-phosphate isomers are present. The two peaks are usually integrated into one

[b]The transition of pentose phosphates (P5Ps) is used to follow the ribulose 1,5-bisphosphate which looses a phosphate in the source

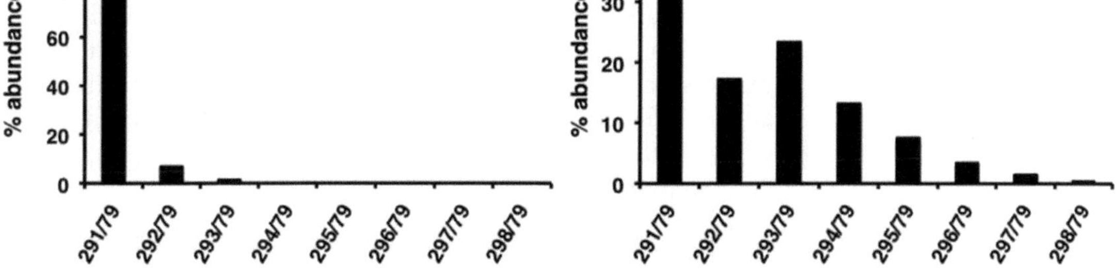

Fig. 2 Scheduled MRM application for unlabeled and ^{13}C-labeled S7P. Corn embryos were grown with unlabeled or ^{13}C-labeled substrate sources following the procedure described previously [26]. Phosphorylated metabolites were extracted as previously described and analyzed by LC-MS/MS. The selected product ion was $[PO_3]^-$ ($m/z = 79$) for S7P. For the S7P containing seven atoms of carbon we follow the transitions 291/79 through 298/79 (*see* Table 4). Mass isotopomer distribution of S7P after 7 days of corn embryo culturing with respectively unlabeled (**a**) or ^{13}C labeled (**b**) substrates

Table 5
Parameters for FIA experiments

Source/gas parameters	Value
Curtain gas (CUR)	30; 35; 40
Collision gas (CAD)	Low; medium; high
Ion Spray voltage (IS)	−4,500; −4,000; −3,500; −3,000; −2,500; −2,000
Temperature (TEM)	300; 350; 400; 450; 500; 550; 600; 650
Ion source gas 1 (GS1)	35; 40; 45; 50; 55; 60; 65; 70
Ion source gas 2 (GS2)	35; 40; 45; 50; 55; 60; 65; 70

ATC-3 column permits to decrease the background and/or eliminate contaminant peaks. Therefore, the signal to noise of each metabolite is increased as well as the sensitivity.

2. Add 1 mL of boiling ultrapure water to one tube at a time and quickly transfer the tube in the water bath set at 100 °C. Proceed one tube at a time: six to eight samples per extraction are recommended.
3. A Q1 scan survey is performed to see whether or not the metabolite injected through direct infusion is detected. The mass range, EP, and DP are set respectively from 100 to 500 amu, −10 V, and −50 V. In the case that the compound is not detected, the DP is optimized from −250 to 0 V. The DP value giving the best signal is entered in the new Q1 scan survey which allows the detection of the metabolite.
4. Select compound optimization as well as MS/MS experiment. Enter parameters in the following order: negative polarity, unit resolution, product ion of at least 14 amu, compound name, monoisotopic mass, charge of 1. Start compound optimization and save file corresponding to each individual metabolite.
5. For FIA experiments a set of values is entered for each parameter (*see* Table 5)
6. Before connecting the liquid chromatography system to the mass spectrometer, a conductometer is used to check whether or not salts are removed according to Dionex instructions. Indeed, the conductivity should not exceed 4–7 μS.

Acknowledgements

We are grateful to The Ohio State University Targeted Metabolomics Laboratory (metabolomics.osu.edu) for access to the LC-MS/MS equipment funded by the Translational Plant Sciences Targeted Investment in Excellence (TIE).

References

1. Dieuaide-Noubhani M, Raffard G, Canioni P, Pradet A, Raymond P (1995) Quantification of compartmented metabolic fluxes in maize root-tips using isotope distribution from C-13-labeled or C-14-labeled glucose. J Biol Chem 270:13147–13159
2. Sriram G, Fulton DB, Iyer VV et al (2004) Quantification of compartmented metabolic fluxes in developing soybean embryos by employing biosynthetically directed fractional (13)C labeling, two-dimensional [(13)C, (1) H] nuclear magnetic resonance, and comprehensive isotopomer balancing. Plant Physiol 136:3043–3057
3. Allen DK, Shachar-Hill Y, Ohlrogge JB (2007) Compartment-specific labeling information in C-13 metabolic flux analysis of plants. Phytochemistry 68:2197–2210
4. Mollney M, Wiechert W, Kownatzki D, de Graaf AA (1999) Bidirectional reaction steps in metabolic networks: IV. Optimal design of isotopomer labeling experiments. Biotechnol Bioeng 66:86–103
5. Wiechert W, deGraaf AA (1997) Bidirectional reaction steps in metabolic networks.1. Modeling and simulation of carbon isotope labeling experiments. Biotechnol Bioeng 55:101–117
6. Wiechert W, Mollney M, Isermann N, Wurzel W, de Graaf AA (1999) Bidirectional reaction steps in metabolic networks: III. Explicit solution and analysis of isotopomer labeling systems. Biotechnol Bioeng 66:69–85
7. Wiechert W, Mollney M, Petersen S, de Graaf AA (2001) A universal framework for C-13 metabolic flux analysis. Metab Eng 3:265–283
8. Wiechert W, Siefke C, deGraaf AA, Marx A (1997) Bidirectional reaction steps in metabolic networks.2. Flux estimation and statistical analysis. Biotechnol Bioeng 55:118–135
9. Alonso AP, Raymond P, Hernould M et al (2007) A metabolic flux analysis to study the role of sucrose synthase in the regulation of the carbon partitioning in central metabolism in maize root tips. Metab Eng 9:419–432
10. Alonso AP, Raymond P, Rolin D, Dieuaide-Noubhani M (2007) Substrate cycles in the central metabolism of maize root tips under hypoxia. Phytochemistry 68:2222–2231
11. Rontein D, Dieuaide-Noubhani M, Dufourc EJ, Raymond P, Rolin D (2002) The metabolic architecture of plant cells—Stability of central metabolism and flexibility of anabolic pathways during the growth cycle of tomato cells. J Biol Chem 277:43948–43960
12. Fernie AR, Roscher A, Ratcliffe RG, Kruger NJ (2001) Fructose 2,6-bisphosphate activates pyrophosphate: fructose-6-phosphate 1-phosphotransferase and increases triose phosphate to hexose phosphate cycling in heterotrophic cells. Planta 212:250–263
13. Williams TC, Miguet L, Masakapalli SK et al (2008) Metabolic network fluxes in heterotrophic Arabidopsis cells: stability of the flux distribution under different oxygenation conditions. Plant Physiol 148:704–718
14. Schwender J, Goffman F, Ohlrogge JB, Shachar-Hill Y (2004) Rubisco without the Calvin cycle improves the carbon efficiency of developing green seeds. Nature 432:779–782
15. Alonso AP, Vigeolas H, Raymond P, Rolin D, Dieuaide-Noubhani M (2005) A new substrate cycle in plants. evidence for a high glucose-phosphate-to-glucose turnover from in vivo steady-state and pulse-labeling experiments with [(13)C] glucose and [(14)C] glucose. Plant Physiol 138:2220–2232
16. Alonso AP, Dale VL, Shachar-Hill Y (2010) Understanding fatty acid synthesis in developing maize embryos using metabolic flux analysis. Metab Eng 12:488–497
17. Alonso AP, Val DL, Shachar-Hill Y (2011) Central metabolic fluxes in the endosperm of developing maize seeds and their implications for metabolic engineering. Metab Eng 13:96–107
18. Allen DK, Ohlrogge JB, Shachar-Hill Y (2009) The role of light in soybean seed filling metabolism. Plant J 58:220–234

19. Schwender J, Ohlrogge JB, Shachar-Hill Y (2003) A flux model of glycolysis and the oxidative pentosephosphate pathway in developing *Brassica napus* embryos. J Biol Chem 278:29442–29453
20. Schwender J, Shachar-Hill Y, Ohlrogge JB (2006) Mitochondrial metabolism in developing embryos of *Brassica napus*. J Biol Chem 281:34040–34047
21. Alonso AP, Goffman FD, Ohlrogge JB, Shachar-Hill Y (2007) Carbon conversion efficiency and central metabolic fluxes in developing sunflower (*Helianthus annuus* L.) embryos. Plant J 52:296–308
22. Lonien J, Schwender J (2009) Analysis of metabolic flux phenotypes for two Arabidopsis mutants with severe impairment in seed storage lipid synthesis. Plant Physiol 151:1617–1634
23. Bajad SU, Lu W, Kimball EH et al (2006) Separation and quantitation of water soluble cellular metabolites by hydrophilic interaction chromatography-tandem mass spectrometry. J Chromatogr A 1125:76–88
24. Huck JHJ, Struys EA, Verhoeven NM, Jakobs C, Van der Knaap MS (2003) Profiling of pentose phosphate pathway intermediates in blood spots by tandem mass spectrometry: application to transaldolase deficiency. Clin Chem 49:1375–1380
25. Luo B, Groenke K, Takors R, Wandrey C, Oldiges M (2007) Simultaneous determination of multiple intracellular metabolites in glycolysis, pentose phosphate pathway and tricarboxylic acid cycle by liquid chromatography-mass spectrometry. J Chromatogr A 1147:153–164
26. Alonso AP, Piasecki RJ, Wang Y, LaClair RW, Shachar-Hill Y (2010) Quantifying the labeling and the levels of plant cell wall precursors using ion chromatography tandem mass spectrometry. Plant Physiol 153:915–924
27. Arrivault S, Guenther M, Ivakov A et al (2009) Use of reverse-phase liquid chromatography, linked to tandem mass spectrometry, to profile the Calvin cycle and other metabolic intermediates in Arabidopsis rosettes at different carbon dioxide concentrations. Plant J 59:826–839
28. Cruz JA, Emery C, Wust M, Kramer DM, Lange BM (2008) Metabolite profiling of Calvin cycle intermediates by HPLC-MS using mixed-mode stationary phases. Plant J 55:1047–1060
29. Koubaa M, Cocuron J-C, Thomasset B, Alonso AP (2013) Highlighting the tricarboxilic acid cycle: liquid and gas chromatography–mass spectrometry analyses of 13C-labeled organic acids. Anal Biochem 436(2):151–159
30. van Dam JC, Eman MR, Frank J et al (2002) Analysis of glycolysis intermediates in *Saccharomyces cerevisiae* using anion exchange chomatography and electrospray ionization with tandem mass spectrometric detection. Anal Chim Acta 460:209–218
31. Antoniewicz MR (2013) Tandem mass spectrometry for measuring stable-isotope labeling. Curr Opin Biotechnol 24:48–53
32. Chen XW, Alonso AP, Allen DK, Reed JL, hachar-Hill Y (2011) Synergy between (13)C-metabolic flux analysis and flux balance analysis for understanding metabolic adaption to anaerobiosis in *E. coli*. Metab Eng 13:38–48
33. Kiefer P, Nicolas C, Letisse F, Portais JC (2007) Determination of carbon labeling distribution of intracellular metabolites from single fragment ions by ion chromatography tandem mass spectrometry. Anal Biochem 360:182–188

Chapter 10

In Vivo NMR for ^{13}C metabolic Flux Analysis

Albrecht Roscher, Stéphanie Troufflard, and Abdelghani Idrissi Taghki

Abstract

The use of in vivo NMR within the framework of Metabolic Flux Analysis in plants is presented. In vivo NMR allows to visualize the active metabolic network, to determine metabolic and isotopic steady state and to measure metabolic fluxes which are not necessarily accessible by isotopic steady state (stationary) Metabolic Flux Analysis. The kinetic data can be used as input for dynamic (nonstationary) Metabolic Flux Analysis. Both 1D and 2D NMR methods are employed.

Key words In vivo NMR, Plant metabolic NMR, ^{13}C NMR, Kinetic NMR, Metabolic flux analysis

1 Introduction

In vivo NMR allows the direct observation of metabolic events as they happen without destroying or even perturbing the sample by the measurement. This is particularly true when used in conjunction with stable isotope labeling.

Indeed the first useful information from such experiments is of qualitative nature: the propagation of isotope from a labeled substrate visualizes the active metabolic pathways thus helping to build the network for metabolic flux analysis (MFA) [1]. In vivo NMR also allows to judge if metabolic and isotopic steady state are reached [2], and the in vivo experiment can then be stopped, the sample be quickly frozen for later extraction, and steady state MFA be performed with extract data.

But in vivo NMR gives also quantitative flux information which is complementary to steady state MFA, and it is the purpose of this chapter to provide the methods to obtain such information. Recording the timecourse of labeling in vivo provides input data for nonstationary (or dynamic) MFA, a concept which is increasingly used for analysis of microbial metabolism [3] but has yet to be performed on large scale plant networks (*see* Chapter 11). Also short of such a comprehensive analysis, the determination of initial slopes from ^{13}C NMR kinetic data together with knowledge of the

Martine Dieuaide-Noubhani and Ana Paula Alonso (eds.), *Plant Metabolic Flux Analysis: Methods and Protocols*, Methods in Molecular Biology, vol. 1090, DOI 10.1007/978-1-62703-688-7_10, © Springer Science+Business Media New York 2014

precursor pool enrichment allows to determine unidirectional metabolic rates which are not always available in steady state MFA, e.g., for sucrose synthesis [2]. Steady state slopes from ^{1}H or ^{13}C NMR give flux information for accumulating metabolites, e.g., lipids or soluble sugars, which is useful input data for MFA [2].

Different types of NMR spectra, both 1D and 2D [4, 5], can be recorded for these purposes, and the choice depends on the information sought for, the spectrometer hardware, especially if gradient equipped probeheads are available, and the operator's expertise. ^{1}H spectra allow to monitor sample stability and changes in major pool sizes. Incorporation of ^{13}C label is visible as satellite signals but in practice only few if any peaks in the ^{1}H 1D spectrum are sufficiently resolved to observe these. An interesting alternative is provided by ^{1}H TOCSY spectra which can be recorded in vivo [5] and have the necessary resolution to quantify ^{13}C satellites [6].

The incorporation of ^{13}C can be directly observed with useful resolution by 1D ^{13}C NMR, preferably in the power-gated version to make use of nOe enhancement and with short repetition delays (e.g., 1.5 s interpulse delay [2]). Thus signals for different time points for a given metabolite can be compared (on the assumption that the relaxation properties for each peak remain constant over time). Different peaks can be compared after normalization with respect to quantification from extracts obtained after the last in vivo time point. It is therefore not necessary to record quantitative spectra with long relaxation delays in vivo. The intensity of the ^{13}C signal reflects both accumulation of ^{13}C labeled metabolites and higher fractional enrichments by replacement of ^{12}C by ^{13}C due to metabolite turnover. The two can be separated by inspection of both ^{1}H and ^{13}C spectra (also from the extract if the in vivo ^{1}H spectra do not have sufficient resolution) or by comparing the evolution of different carbon positions from the same molecule [2].

The sensitivity of ^{13}C NMR can be increased by using INEPT or DEPT coherence transfer from ^{1}H, however at the expense of losing information on quaternary carbons. A further gain in sensitivity and also a gain in resolution which makes observation of ^{13}C-^{13}C satellites easier are afforded by 2D indirect detection techniques like HSQC or HMQC which are perfectly feasible provided that the probehead is equipped with pulsed field gradients [7].

2 Materials

2.1 Plant Material

In vivo NMR can be performed with nearly any plant material that (1) is sufficiently small to fit into the chosen NMR tube, (2) can be packed sufficiently densely to obtain good signal, and (3) does not need light. Typical samples are heterotrophic cell suspension cultures, dissected small organs like developing embryos or root tips, or larger organs cut into small pieces like roots, stems, or leaves, the latter of course in dark conditions. The necessary quantity depends on the NMR tube diameter and attainable tissue density: typically about 70–150 mg for a 5 mm tube, 250–500 mg for a 8 mm tube, 450–900 mg for a 10 mm tube, 2.5–5 g for a 20 mm tube, and 4–8 g for a 25 mm tube.

2.2 Culture Medium

The culture medium must contain a carbon source which can be replaced by a ^{13}C labeled counterpart. The presence of other organic nutrient compounds might be necessary on physiological grounds but it complicates subsequent analysis of carbon fluxes. Furthermore the medium should contain mineral macronutrients (N, P, K, S, Ca, Mg) in appropriate forms and proportions. The presence of mineral or organic micronutrients is normally not necessary since in vivo NMR experiments are rarely conducted for more than 48 h, and intracellular stocks can cover the needs. In any case, the presence of paramagnetic ions like Mn or Fe should be avoided (*see* **Note 1**). The medium should contain sufficient pH buffer to maintain the medium pH fairly stable for the experiment duration unless a pH control device is used. Inorganic buffer compounds like phosphate or borate are ideal since they do not give rise to ^{13}C NMR signals but if they do not cover the chosen pH range, organic buffers have to be used. Especially for the typical plant medium pH around 6.0, MES buffer is a good choice since it gives rise to only three ^{13}C NMR signals which do not overlap with major metabolites at pH 6. Furthermore, the exact position of these signals can be used to continuously measure the medium pH. Finally, the presence of an osmolyte might be necessary, and the choice of compound again is a compromise between physiological compatibility and the minimization of ^{13}C NMR signals in spectral regions of interest (*see* **Note 2**).

Prepare the medium without the carbon source eight to ten times the volume of the circulation system. Then add the unlabeled carbon source at its final concentration to half of this volume. Weigh the quantity of labeled carbon source necessary for one volume of the perfusion system.

2.3 Circulation or Airlift System

During the in vivo NMR experiment, the plant sample has to be oxygenated and CO_2 has to be removed. This can be achieved by different airlift or circulation system designs [5]. Airlift systems use minimal medium volumes, and are thus ideal when it is primordial to keep consumption of a valuable labeled compound to a minimum. For metabolic flux analysis, however, it is in most cases desirable to keep the sample saturated with substrate for one or two days, and this is easier with the larger volume in a circulation system. A simple system which we use with 8 and 10 mm NMR tubes is described in Fig. 1 (*see* **Note 3**). For 8 mm tubes, we use it with 30 ml medium, for 10 mm tubes with 40 ml medium.

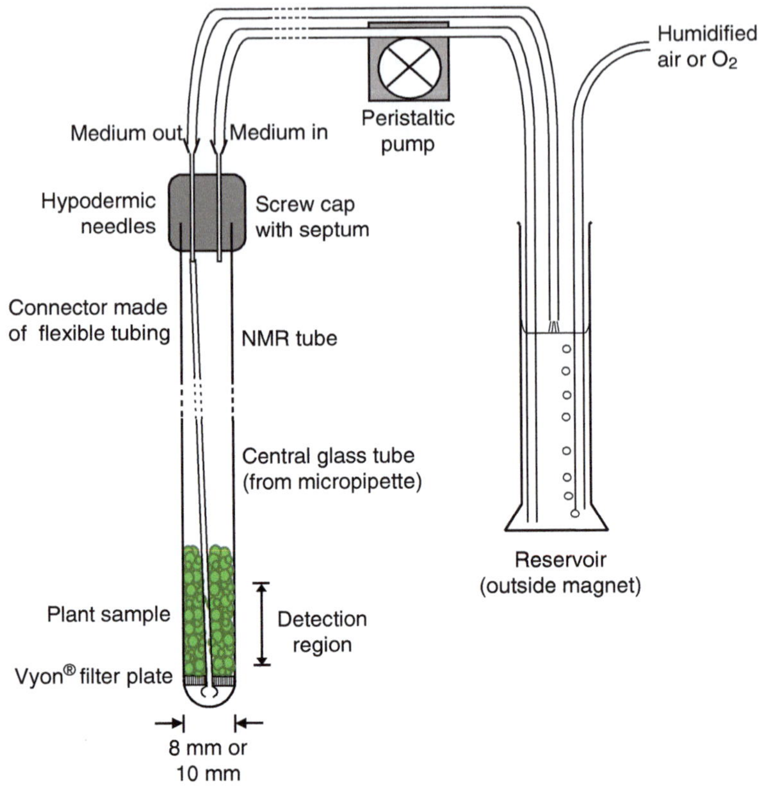

Fig. 1 A simple circulation system for in vivo NMR. The aerated or oxygenated culture medium is circulated with a peristaltic pump through Tygon tubing to the top of a screw-cap NMR tube. The septum of the screw cap is traversed by a hypodermic needle. Since the screw cap is airtight, the medium flows through the sample and the sintered polyethylene (Vyon®) filter plate to the bottom of the tube, and from there through a central glass capillary (a 100 μl glass micropipette of which one end has been melted and blown out over a Bunsen burner so as to have a thickening with which the filter plate can be pulled up), a flexible Tygon tube connection, a second hypodermic needle and Tygon tubing back into the medium reservoir, a 50 ml measuring cylinder

3 Methods

3.1 Sample Preparation and Circulation System Setup

1. Collect and dissect the necessary quantity of plant tissue or use cell cultures. Dissected tissues should be kept during preparation in well-aerated medium containing the unlabeled carbon source (*see* **Note 4**).
2. Weigh the sample.
3. Measure the necessary volume of unlabeled medium for the circulation system.

4. Fill the bottom of the NMR tube and chase any air bubbles from the bottom or the filter plate by pulling and pushing the filter (*see* **Note 5**).
5. Place the filter just below the NMR detection zone.
6. Pour the remaining medium into the reservoir and start the aeration or oxygenation system. If possible, start circulating the medium in a closed loop without the NMR tube.
7. Rinse the tissue or cells with fresh unlabeled medium and place it in the NMR tube.
8. Close the NMR tube loosely and connect the tubing to the circuit.
9. Start the pump, and when the NMR tube is filled tighten the screw cap. Verify that the system is not leaking and adjust the circulation rate (*see* **Note 6**) so as to adequately supply oxygen to the sample. Though the oxygen consumption rate depends on several parameters (cell/tissue type, temperature, partial oxygen pressure etc.), as a rule of thumb the circulation rate should be set to at least 10 ml/min (g FW). Higher rates are fine but are often limited by the associated higher pressure and therefore need more sophisticated plumbing.
10. Since the spectrometer sample lift is difficult to use with the tubing attached to the NMR tube, rather tie a string to the NMR tube just under the screw-cap neck for lowering the tube into the spectrometer. Adjust the sample spinner.
11. It is a good precaution to add one or two safety suction tubes, one on top of the septum, a possible second one in (or on top of) the sample spinner for protecting the probehead in case of a leakage.
12. Insert the NMR tube in the magnet.

3.2 Spectrometer Setup: Lock, Shim

If the NMR magnet drift is not too important, the NMR spectra can be acquired without field lock (sweep off) (*see* **Note 7**). Otherwise, depending on the sensitivity of the lock channel, 1–2 % of the medium water has to be replaced by D_2O which might slightly modify kinetics in metabolism and which renders mass spectrometric isotopomer analysis of the biomass after the in vivo NMR experiment very difficult. An alternative is to use a closed capillary filled with D_2O inserted into the sample. However the resulting loss of sample volume is an inconvenient for narrow NMR tubes.

1. Tune and match the probehead for ^{1}H and for ^{31}P or ^{13}C (see below).
2. Shim the magnet while continuously observing the ^{1}H spectrum. The water peak is most useful for coarse shimming and improving the lineshape. However its natural linewidth is quite large, hardly below 10 Hz for in vivo NMR. Therefore for fine

shimming, it is useful to select another intense well resolved peak that can easily be observed in one scan like an anomeric sugar peak or a methyl for example of alanine. If automatic gradient shimming on the ^{1}H signal is available this can be useful to accelerate and improve the coarse shimming, the fine shimming however should be verified manually. Note that the ^{2}H signal is unsuitable for shimming especially if it stems from a capillary.

3.3 ^{31}P Spectrum

If the probehead can be tuned to ^{31}P (otherwise skip this step), it is a good idea to record a 15–30 min ^{31}P spectrum to ascertain that the oxygenation of the cells/tissue is satisfactory as reflected in their energetic state. During this time, the sample also continues to acclimate to the physiological conditions in the NMR tube. Choose rapid pulsing conditions (e.g., a 45° pulse angle, 0.5 s repetition time and power-gated ^{1}H decoupling) to maximize the signal from the fast relaxing NTP and NDP nucleotides. In well oxygenated cells/tissues, there should be visible three clear NTP signals: γ-NTP at −5.5 ppm, α-NTP at −10.6 ppm, and β-NTP at −19.3 ppm, whereas NDP (β-NDP at −6.3 ppm) should not be detectable (or be about ten times smaller than NTP). Furthermore, the position of cytoplasmic P_i is indicative of cytoplasmic pH: ideally it should be at 2.3 ppm (i.e., at -2.0 ± 0.1 ppm from Glc6P at 4.3 ppm). Under hypoxic conditions, the cytoplasm acidifies and the P_i peak shifts to the right ("highfield").

3.4 ^{1}H and ^{13}C Reference Spectra

Before adding the ^{13}C label it is important to record reference spectra at natural abundance. This timeslot gives also the opportunity to make the necessary spectrometer calibrations without time pressure.

1. Tune and match the probehead for ^{13}C and ^{1}H if ^{31}P spectra were recorded.
2. Carefully calibrate the 90° pulses for both nuclei, especially if you plan to perform 2D NMR (*see* **Note 8**).
3. Record a ^{1}H spectrum and note the absolute frequency of the water peak. Use this frequency as irradiation frequency for ^{1}H spectra with presaturation (*see* **Note 9**).
4. Optimize the presaturation power for presaturation sequence(s) and adjust the receiver gain so as to leave room for doubling of the residual water peak which can happen due to changes in magnetic homogeneity or magnetic field drift.
5. Record reference spectra of the type chosen for following the label kinetics.

3.5 Preparation of NMR Files for Kinetics

Prepare in advance the NMR files for the kinetics experiment. Set up a unit cycle containing a short 1D ^{1}H spectrum for spectrometer and sample control, a 1D ^{1}H water presaturation spectrum and/or a ^{1}H-TOCSY spectrum for recording changes in pool sizes and one or several spectra sensitive to ^{13}C label kinetics (1D ^{13}C and/or 1D ^{13}C INEPT or DEPT and/or 2D ^{13}C-^{1}H HSQC or HMQC). Spectra should have the shortest possible duration, just enough to have the sensitivity to detect natural abundance peaks of the more abundant metabolites as this determines the time resolution of the kinetics. For less abundant metabolites, several spectra can be summed during data processing (*see* **Note 10** for an example). Spectral parameters should be optimized for maximizing the signal to noise ratio. Copy the unit cycle as many times as the projected total experimental time (*see* **Notes 11** and **12**).

3.6 Label Introduction and Kinetics Data Acquisition

1. Fill a second reservoir with at least two volumes of medium without the carbon source, and place the aeration or oxygenation system in it.
2. Position the entrance tube in the second reservoir thus progressively replacing the unlabeled medium with the medium without the carbon source.
3. When the volume in the new reservoir has been lowered to the level of the old one, place also the return tube in the new reservoir thus reestablishing the circulation system.
4. Now add the labeled carbon source and start the NMR kinetics experiment after the lapse of time necessary for the substrate to reach the sample in the NMR tube (*see* **Note 13**).
5. During the course of the experiment, check regularly that the perfusion system works well (no leak, i.e., constant liquid height in the reservoir; constant perfusion rate), that there is no degradation of magnetic homogeneity due for instance to an air bubble. Also regularly inspect the NMR spectra to make sure that the carbon supply is sufficient and the medium pH is fairly stable (a pH change can be seen by shifted signals especially of the pH buffer compound).

3.7 End of NMR Experiment

In vivo NMR experiments for metabolic flux analysis typically last for between 12 and 48 h when isotopic steady state is reached for most compounds. Indeed for such durations, plant tissues can be maintained rather simply in a suitable metabolic steady state in the NMR tube, whereas longer experiments need more precautions in terms of sterility and medium homeostasis.

1. At the end of the experiment, steady state of energy metabolism can be controlled with a rapid ^{31}P NMR spectrum that can be compared with the initial one. Afterwards it might be advisable to perform a last unit cycle of the kinetic experiment.
2. Directly after the last NMR recording, take the sample out of the tube.

3. Rinse it rapidly with water (*see* **Note 14**).
4. Freeze it in liquid nitrogen for subsequent extraction and analysis (as for isotopic steady state MFA).

3.8 Data Analysis

1. Separate the NMR files according to spectrum type.
2. Fourier-transform each series with the same parameters.
3. Correct the baseline automatically or, if necessary, manually.
4. Use a prominent and well resolved peak for chemical shift calibration.
5. Identify peaks from the literature (especially [8]) or by comparison with extracts.
6. Integrate all peaks and export results in tabular form for further analysis (*see* **Note 12**). Note with each spectrum the exact time point as available from the status parameters. Relative timecourses can now be charted and analyzed. An example is shown in Fig. 2.
7. The last time point of each curve for different peaks can be calibrated by quantifying the corresponding peaks in the extract thus giving a scale to the whole curve.
8. The determination of initial slopes from ^{13}C NMR (left slope in Fig. 2) together with knowledge of the precursor pool enrichment allows to determine unidirectional synthesis rates [2]. Steady state slopes from ^{1}H or ^{13}C NMR (right slope in Fig. 2) give flux information for accumulating metabolites [2].

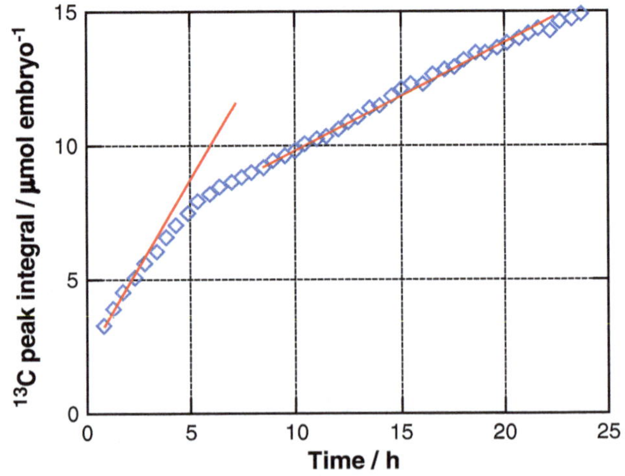

Fig. 2 In vivo timecourse of the ^{13}C signal from C1 of the fructosyl moiety of sucrose in developing rapeseed (*Brassica napus*) embryos perfused with [1-^{13}C] glucose. The vertical scale was established from the corresponding tissue extract. The continuos lines suggest the initial and the steady state slopes

4 Notes

1. For cell or callus cultures, Mn should already be omitted from the growth medium at least for the last subculture before NMR analysis which is usually no problem for heterotrophic cultures [5].
2. For developing oilseed embryos (rapeseed, linseed) we use sorbitol on physiological reasons. However one of the six sorbitol ^{13}C signals coincides with the signal from the carbon 6 of the fructosyl moiety of sucrose rendering this peak only quantifiable after subtraction of the sorbitol peak thus generating a substantial uncertainty.
3. Depending on the physiological parameters associated with the experiment, the reservoir can be extended to become a small bioreactor with for example pH control, pO_2 control, measurement of O_2 consumption rate, temperature control, and reinforced sterility [9]. However this goes with an increase of the minimum volume of the circulation system.
4. If tissues float they probably contain air spaces and should be vacuum infiltrated before in vivo NMR to improve magnetic homogeneity [5]. For vacuum infiltration, immerse the tissue in medium in a Büchner flask, prevent it from floating (e.g., with a meshed wire plate) and submit it to reduced pressure with a vacuum pump or simply a water aspirator. The tissue will visibly degas. A degassing run should not exceed 2 or 3 min in order to limit hypoxic injury of the tissue; if necessary rather repeat the procedure after a recovery phase.
5. Pull on the central capillary and push with a rod. A bicycle spoke does this quite nicely.
6. For measuring the circulation rate, direct the return flow during 30 or 60 s into a small graduated cylinder, note the volume and pour the medium back into the reservoir.
7. When using pulsed field gradients (e.g., for HSQC or HMQC) in unlocked mode, be careful to use the appropriate gradient pulse commands without lockhold in the pulse program.
8. Pulse calibration is best performed with the 360° method. First record a spectrum with an approximate 90° pulse (actually any pulse length <180 is fine) and phase it carefully. Then record a spectrum with your estimate of a 360° pulse using the phase values previously determined. If you need several scans (^{13}C) reduce the interpulse delay to virtually zero. If the spectrum is positive, reduce your estimated 360° pulse and restart; if the spectrum is negative increase your estimation. The null passage is rather sharp and well defined and gives the 360° pulse, and divided by four the 90° pulse.

9. For ^{1}H spectra, use a spectral width of at least 20 ppm which allows for a more robust phase correction.
10. A typical 1D cycle might contain a very short 1D ^{1}H spectrum (four scans), a short 1D ^{1}H presaturation spectrum (2 min) and four 1D ^{13}C spectra of 14.5 min each giving a total cycle length of 60 min.
11. Rather prepare more than less unit cycle copies since it is easier to delete empty files in the end than having to restart an acquisition.
12. Shell scripts for copying and separating kinetics files as well as AU-programs for data acquisition and processing on Bruker spectrometers are available on request.
13. Be careful to take into account a possible delay between the operators' start command and the actual start of data acquisition (e.g., the standard Bruker multizg utility will first calculate the total experimental time which can easily take a minute).
14. Keep the rinsing water and add it to the medium when concentrating the medium for further analysis.

References

1. Ratcliffe RG, Shachar-Hill Y (2001) Probing plant metabolism with NMR. Annu Rev Plant Physiol Plant Mol Biol 52:499–526
2. Troufflard S, Roscher A, Thomasset B et al (2007) *In vivo* ^{13}C-NMR determines metabolic fluxes and steady state in linseed embryos. Phytochem 68:2341–2350
3. Noack S, Nöh K, Moch M et al (2011) Stationary versus non-stationary ^{13}C-MFA: a comparison using a consistent dataset. J Biotechnol 154:179–190
4. Fan TW-M, Higashi RM, Lane AN (1986) Monitoring of hypoxic metabolism in superfused plant tissues by *in vivo* ^{1}H NMR. Arch Biochem Biophys 251:674–687
5. Ratcliffe RG, Roscher A, Shachar-Hill Y (2001) Plant NMR spectroscopy. Prog Nucl Magn Reson Spectrosc 39:267–300
6. Massou S, Nicolas C, Letisse F et al (2007) Application of 2D-TOCSY NMR to the measurement of specific ^{13}C-enrichments in complex mixtures of ^{13}C-labeled metabolites. Metab Eng 9:252–257
7. Shachar-Hill Y, Pfeffer PE, Germann MW (1996) Following plant metabolism *in vivo* and in extracts with heteronuclear two-dimensional nuclear magnetic resonance spectroscopy. Anal Biochem 243:110–118
8. Fan TW-M (1996) Recent advances in profiling plant metabolites by multi-nuclear and multi-dimensional NMR. In: Shachar-Hill Y, Pfeffer PE (eds) Nuclear magnetic resonance in plant biology. American Society of Plant Physiologists, Rockville, MD, USA, pp 181–254
9. Roscher A, Emsley L, Raymond P et al (1998) Unidirectional steady-state rates of central metabolism enzymes measured simultaneously in a living tissue. J Biol Chem 273: 25053–25061

Part II

From Labeling Data to Flux Maps (from the Labeling Data to a Metabolic Map)

Chapter 11

Steady-State and Instationary Modeling of Proteinogenic and Free Amino Acid Isotopomers for Flux Quantification

Yuting Zheng and Ganesh Sriram

Abstract

Metabolic flux analysis (MFA) is a powerful tool for exploring and quantifying carbon traffic in metabolic networks. Accurate flux quantification requires (1) high-quality isotopomer measurements, usually of biomass components including proteinogenic/free amino acids or central carbon metabolites, and (2) a mathematical model that relates the unknown fluxes to the measured isotopomers. Modeling requires a thorough knowledge of the structure of the underlying metabolic network, often available from many databases, as well as the ability to make reasonable assumptions that will enable simplification of the model. Here we describe a general methodology underlying computer-aided mathematical modeling of a flux–isotopomer relationship and some of the accompanying data-processing steps. One of two modeling strategies will need to be employed, depending on the type of isotope labeling experiment performed. These strategies—steady-state modeling and instationary modeling—have different experimental and computational demands. We discuss the concepts underlying these two types of modeling and demonstrate steady-state modeling in a step-by-step manner. Our methodology should be applicable to most isotope-assisted MFA applications and should serve as a general framework applicable to many realistic metabolic networks with little modification.

Key words Mathematical modeling, Isotope labeling experiment, Amino acids, Mass spectrometry, Steady-state model, Instationary model

1 Introduction

Metabolic flux analysis (MFA) is a methodology for quantifying carbon traffic through chemical reactions that interconvert metabolites in a cell or a biological system [1]. Being a powerful tool to assess the metabolic impacts of genetic and environmental perturbations, MFA provides valuable guidance toward metabolic engineering. MFA involves performing mass balances on

Yuting Zheng and Ganesh Sriram conceived the chapter. Yuting Zheng wrote the first draft and prepared data displays. Ganesh Sriram critically edited the draft and prepared a final version. Both authors approved the final version of the chapter.

Martine Dieuaide-Noubhani and Ana Paula Alonso (eds.), *Plant Metabolic Flux Analysis: Methods and Protocols*, Methods in Molecular Biology, vol. 1090, DOI 10.1007/978-1-62703-688-7_11, © Springer Science+Business Media New York 2014

metabolites, the solving of which by linear algebraic methods ultimately results in a metabolic flux map. For implementation on simple metabolic networks, this procedure only requires growth rates, measurements of extracellular fluxes, and fluxes toward biomass as experimental inputs.

For complex biological systems such as plants, this approach may by itself not be sufficient to unambiguously determine the values of all or most fluxes. In such cases, tracing isotope labels through the metabolic network provides critical information toward completing the flux map. MFA involving tracing of isotope labels is called isotope-assisted MFA (isotope MFA). This approach is especially beneficial for parallel pathways with common, intracellular start and end points as well as for metabolic cycles [2]. The central idea of this approach is that different pathways feature different carbon atom rearrangements, so that any given pathway will likely process a mixture of carbon isotopes (e.g., ^{13}C and ^{12}C) uniquely. Thus, when an appropriately designed mixture of carbon isotopes is supplied to a set of pathways, the resulting isotopic mixture will reflect the metabolic "signatures" of the pathways that contributed to it. Figure 1 illustrates a simple scenario. Here, a mixture of labeled and naturally abundant (NA, 98.8 % unlabeled) version of metabolite A is fed to a network that converts metabolite A to metabolite D through two available pathways. In Fig. 1,

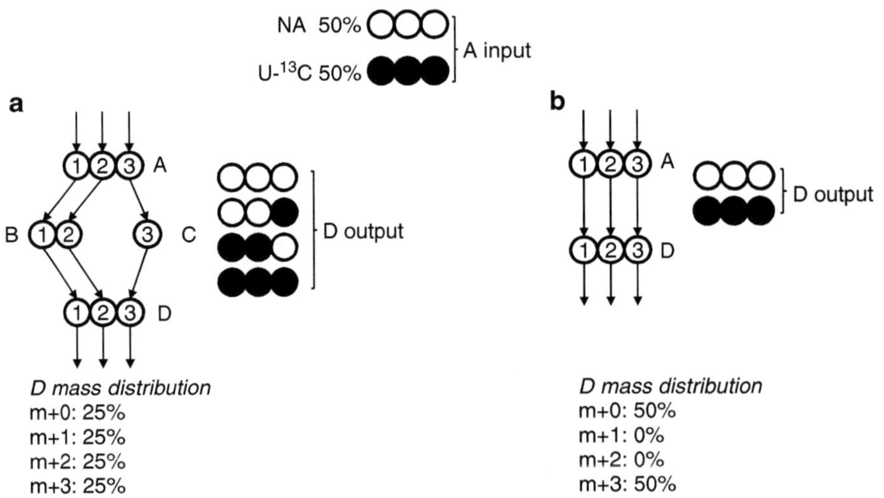

Fig. 1 A simple network with two parallel pathways differing in carbon atom rearrangement. Carbon isotopes and isotopomers are notated using *empty circles* for ^{12}C and *filled circles* for ^{13}C. The network has two available pathways between metabolites A and D. Feeding a mixture of 50 % natural abundant (NA, nearly unlabeled) and 50 % uniformly ^{13}C-labeled (U-^{13}C) A to this network will result in different isotopomers of D, whose ^{13}C–^{12}C patterns will depend on the pathway that contributed to their synthesis. These isotopomers will generally differ in mass. Thus, each of the two pathways in this example has a signature mass distribution for D. (**a**) shows the theoretical isotopomer combination of D when the intermediate products C and D are formed from A en route to D, and (**b**) shows the other case involving no carbon atom rearrangements

pathway (b) is "straightforward" and does not involve carbon atom rearrangements, whereas pathway (a) effects a carbon atom rearrangement by splitting metabolite A into metabolites B and C that condense to form D. Because the two pathways differ in their carbon atom rearrangements and because enzymes exhibit no macroscopic preference toward ^{12}C and ^{13}C (a very reasonable assumption in MFA), each pathway synthesizes different ^{12}C–^{13}C patterns or isotopomers of D. For instance, pathway (a) in Fig. 1 uniquely synthesizes a ^{12}C–^{12}C–^{13}C and a ^{13}C–^{13}C–^{12}C isotopomer. The isotopomers synthesized by different pathways generally differ in their masses or magnetic properties. In Fig. 1, pathway (a) alone produces isotopomers of mass $m + 1$ and $m + 2$ (in addition to isotopomers of mass $m + 0$ and $m + 3$), where m is the molecular weight of metabolite D. Contrastingly, pathway (b) only produces isotopomers of mass $m + 0$ and $m + 3$. The mass spectra of metabolites such as D, or of biomass components derived from them, can be measured via gas chromatography-mass spectrometry (GC-MS) to not only identify which metabolic pathways are active but also evaluate the relative fluxes through each pathway.

Several experimental steps are necessary to obtain isotopomer abundance measurements from a cell or a tissue after it reaches metabolic steady state. These include timely cell or tissue harvesting, extraction and processing of biomass components (proteinogenic and free amino acids, fatty acids, sugars, and soluble metabolites), GC-MS or nuclear magnetic resonance (NMR) sample preparation and analysis, and post-analysis processing of data. These procedures are described in the PART I in this volume. This chapter principally discusses modeling of isotopomers of biomass components that are measureable by GC-MS. Readers should note that NMR is an alternative method to measure isotopomer abundances and provides information complementary to MS. The use of NMR in isotope MFA is demonstrated in, e.g., refs. 3, 4 and reviewed in, e.g., refs. 5, 6.

Mathematical modeling of an isotope labeling experiment can involve significant complexity as determined by the structural sophistication of the underlying metabolic network, the amount of isotopomer abundance data available from labeling experiments, and whether isotope MFA is to be performed at isotopic steady state or unsteady state. The first paragraph of Subheading 3 discusses the situations under which the steady-state or instationary methodologies are appropriate.

Several methods broadly based on mass balancing enable isotope label-free, stoichiometric, or kinetic analysis of metabolic networks. These include flux balance analysis (FBA) [7–9], genome-scale metabolic modeling [10–12], and ensemble modeling [13]. Genome-scale modeling involves the construction of an organism-specific stoichiometric model inferred from the organism's annotated genome or from a metabolic pathway database that is based on the genome annotation. FBA models a metabolic network as a system of linear, stoichiometric equations constrained by extracellular flux measurements and thermodynamic irreversibilities of reactions. Optimization of different objective functions (e.g., maximization of biomass production) via this model results in metabolic flux distributions or maps.

Akin to the aforementioned methods, models for isotope labeling experiments involve mass balances of metabolites and isotopomers. However, these balances can be bilinear [14], making solution strategies iterative and complex. Researchers have developed several model frameworks and methods to reduce this complexity. This includes the use of compartmental matrix techniques such as cumulative isotopomers (cumomers) [15], bond isomers (bondomers) [16, 17], or elementary metabolite units (EMUs) [18]. All these methods involve condensing a nonlinear set of isotopomer balance equations to a cascade of linear equations that can be solved by straightforward matrix computations [15]. Although these techniques were developed for analysis at isotopic steady state, they have now been successfully extended to isotopically instationary situations [19, 20].

The mathematical methods listed above have progressively improved the efficiency and speed of isotopomer data analysis while not sacrificing the information contained in isotope labeling datasets. Despite their availability, the workload of data processing and flux evaluation can be tremendous. Currently, many computer-aided flux evaluation tools are available that significantly shorten the time between data acquisition and evaluation of fluxes and confidence intervals. These tools are especially useful in instationary flux analysis. Programming platforms used by these tools are wide ranging, with C/C++ and MATLAB being favored.

This chapter describes general principles that should be followed to construct both steady-state and instationary models, as well as the experimental and computational differences between them of which practitioners of flux analysis need to be aware.

2 Materials

2.1 Metabolic Network Databases

Detailed information on metabolic pathways is now available through several metabolic pathway databases including KEGG [21] and MetaCyc [22]. KEGG and MetaCyc have online interfaces (www.genome.jp/kegg/kegg2.html and www.metacyc.org, respectively); additionally, many of their functionalities are available off-line. These databases provide detailed information on metabolites, reactions, pathways, and genomic information (e.g., gene loci) across various species, often quite comprehensively (*see* **Note 1**). Additionally, because most information in these databases is curated from previous research, they host significantly more information on model species such as Arabidopsis. Information on relatively unexplored species is best obtained by searching the literature, especially via Google Scholar or PubMed with appropriate keywords. An alternative method is to begin with canonical pathways or pathways known to exist in related species and to iteratively improve the model on the basis of discrepancies observed while fitting the model to isotopomer measurements.

2.2 Collecting Information from Isotope Labeling Experiments

Isotope labeling experiments also provide information toward constructing a metabolic network. Particularly, an initial metabolic network model can be iteratively refined on the basis of isotope labeling patterns that are not satisfactorily explained by it [4, 5, 23]. Additionally, labeling data can help narrow down feasible pathways when (re)constructing a network. Appropriately designing labels of the carbon source is crucial for this to be effective [24–26].

2.3 Computational Platforms

Isotope labeling experiments may be modeled on several computational platforms, depending on the level of sophistication of the underlying metabolic network and the number and types of labeling measurements available. For simple networks such as those in Fig. 1, it is easy to manually obtain algebraic equations relating isotopomers and fluxes. Somewhat more complicated networks can be solved by using the Solver function on Microsoft Excel. More complex networks will need to be modeled on MATLAB, C, C++, or other platforms by writing ad hoc programs or by using programs specifically suited for modeling isotope labeling experiments, such as 13CFlux2 [27], OpenFLUX [28], Metran [18], NMR2Flux+ [3], or FiatFlux [29]. Although we have historically used NMR2Flux+ [3, 30] and continue to do for routine isotope MFA [26], we use MATLAB as a demonstrative tool in the classroom or for illustration. MATLAB is an excellent platform to initially execute the modeling examples presented in this chapter, principally because of its user-friendly interface and flexibility. Furthermore, with its versatile optimization and parallel computing toolboxes, MATLAB is adequate for steady-state flux analyses of many realistic metabolic networks. However, development of more robust tools may require programming languages such as C or C++.

3 Methods

After collecting all the necessary data from either metabolic network databases or previous iterations of isotope labeling experiments, different modeling strategies may be employed depending on the nature of the labeling experiment and the supplied carbon source. For many cases, especially if (1) multiple carbon sources or a carbon source containing covalently bonded carbon atoms are (is) supplied and (2) labeling data is collected at a single time point after isotopic steady state is attained, then steady-state modeling should be employed. In other cases, if (1) a single carbon source containing exactly one carbon atom or multiple, non-covalently bonded carbon atoms is supplied or (2) the labeling data is collected transiently at multiple time points, then an instationary model is indispensable. The type of model employed also determines which metabolites need to be sampled and analyzed to measure isotope labeling. The typical length of time over which steady-state experiments are conducted enables the accumulation of substantial amounts of label in biomass components including proteinogenic amino acids, fatty acids, sugars, and nucleic acids. Therefore, it is not surprising that steady-state experiments involve quantification of isotopomer abundances in these compounds (e.g., refs. 3, 31) rather than in intracellular metabolites, which are typically present in low amounts in the biomass. However, instationary experiments are conducted over time frames so short that sufficient label does not accumulate in biomass components. Therefore, it is advisable to isolate and measure labeling in intracellular metabolites such as those present in glycolysis, the tricarboxylic acid (TCA) cycle, and other central carbon metabolic pathways. Metabolic steady state, i.e., the constancy of the fluxes to be measured, is necessary for both types of experiments and can be partially verified by examining if extracellular flux measurements are constant during the labeling experiment.

3.1 Assumptions Underlying Isotopomer Network Models

The following assumptions are usually implicit in the construction of isotopomer network models. *See* **Note 2** for tips on simplifying steady-state models.

1. Carbon participates only in the reactions included in the model. For this reason, it is essential that the model include all reactions known to account for more than ~1 % of the total carbon flux.
2. The system (cell or tissue) being modeled is at metabolic steady state; therefore, (1) the metabolic flux values, directions of net fluxes, and reaction reversibilities are constant and (2) the pool sizes (concentrations) of the metabolites are constant.
3. Enzymes do not distinguish between ^{12}C and ^{13}C. Although in reality enzymes do show a slight preference for one isotope over another (e.g., ribulose-1,5-bisphosphate carboxylase oxygenase, RuBisCO [32]), this isotopic discrimination is almost negligible in isotope MFA wherein substantial amounts of label are fed.
4. In steady-state models, the system has reached isotopic steady state and the isotope labeling patterns of metabolites do not vary with time. This assumption does not apply to instationary models. Since the objective of isotope MFA is to measure (constant) flux values, the kinetics of the underlying enzymatic reactions or the regulatory effects governing them are

irrelevant to the analysis. Typically, the overall flux of a metabolic reaction varies with the activity of the catalyzing enzyme and the concentrations of the participating metabolites. However, because these quantities are constant at metabolic steady state, the overall flux is constant and its dependence on enzyme activities or metabolite concentrations can be ignored. This argument enables the decoupling of kinetics from flux analysis, so the only unknown parameters in the flux model are the unknown fluxes and (in case of instationary models) metabolite concentrations. Furthermore, comparative isotope MFA of two environmental conditions, chemical treatments, or genetic variants can be used as a tool to probe into the effects of conditions or genetic interventions on kinetics or regulatory mechanisms.

3.2 Mass Fragmentography and Interpretation of Mass Spectral Data

Although Chapters 7–9 in this volume cover the use of mass spectrometry for quantifying isotopomer abundances, we provide a brief overview as background for the following sections on modeling. MS can detect both entire molecules and more often fragments of molecules and distinguish isotopomers of these molecules or fragments by mass. The fragmentography of a particular analyte depends on its chemical structure, whether it is derivatized prior to analysis (often the case if GC-MS is used), and the particular derivatization chemistry used. For instance, MS detects two distinct metabolic fragments of the *t*-butyldimethylsilyl (tBDMS) derivative of the three-carbon amino acid alanine (Ala). One fragment, corresponding to the ions with $m/z = 158$ or 232, contains carbon atoms C-2 and C-3 of Ala. We therefore notate this fragment as "Ala[23]". Another fragment, corresponding to an ion with $m/z = 260$, contains all three carbon atoms of Ala and we therefore notate it as "Ala[123]". MS, however, detects three distinct metabolic fragments of another three-carbon amino acid, serine (Ser). Using the notation above, these fragments are Ser[12] ($m/z = 302$), Ser[23] ($m/z = 288$ or 362), and Ser [123] ($m/z = 390$).

A close examination of these fragments will reveal a mass isotopomer distribution (MID). For example, the Ala[123] fragment could exhibit several isotopomers of masses m, $m + 1$, $m + 2$, etc., where m is the molecular weight of the fragment. The abundances of these mass isotopomers depend on (1) the metabolically generated labeling pattern of the three carbon atoms constituting the Ala[123] fragment (e.g., a ^{12}C–^{12}C–^{12}C fragment will have a mass 3 amu lesser than a ^{13}C–^{13}C–^{13}C fragment) and (2) natural isotopic distributions of atoms other than metabolic carbon in the fragment. The effect in (2) can easily be decoupled from the effect of interest in (1) by using algorithms described elsewhere (e.g., refs. 30, 33). While processing MID data of biomass components such as amino acids, two important problems need to be solved: (1) relating amino acid MIDs to the isotopomer distributions of their central carbon metabolic precursors and (2) mathematically relating isotopomer distributions of central carbon metabolic precursors to pathway fluxes. The first problem is relevant in steady-state experiments, where amino acid MIDs are usually measured as surrogates for isotopomer distributions of central carbon metabolic precursors. The second problem is relevant in both steady-state and instationary labeling experiments. The following sections discuss methods to solve both these problems.

3.3 Retrobiosynthetic Deduction of Isotope Labeling of Central Carbon Metabolites from Labeling Measurements of Biomass Components

To keep flux–isotopomer models reasonably small in size, it is convenient to restrict them to central carbon metabolism. Such models will, therefore, not simulate isotopomers of biomass components (e.g., amino acids) because the pathways leading from central carbon metabolism to biomass synthesis are not directly incorporated into the model. However, especially in steady-state isotope MFA, labeling patterns of biomass components are the ones that are best measured. Therefore, it is necessary to relate the labeling patterns of biomass components to the isotopomers of central carbon metabolites. This requires information on the biosynthesis of proteinogenic amino acids from central carbon metabolic precursors, which is available from the literature [34] as well as metabolic pathway databases such as MetaCyc. Amino acids are synthesized from one or more metabolic precursors through almost universally valid pathways. For example, Ala is synthesized from pyruvate (Pyr) with a one-to-one correspondence between its three carbon atoms and those of Pyr, whereas isoleucine (Ile) is synthesized from one molecule each of Pyr and oxaloacetate (OAA) (Fig. 2). Some central carbon metabolites are precursors for the

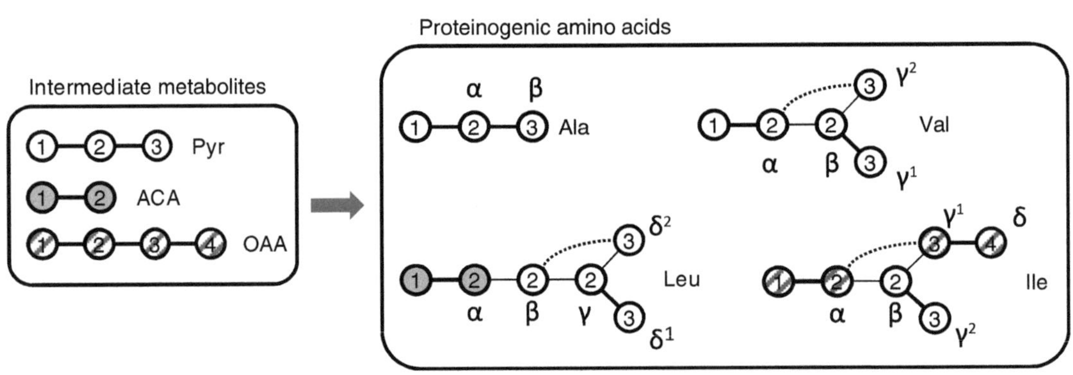

Fig. 2 Relationships between precursor metabolites and their derived biomass components, such as proteinogenic amino acids. *Circles* filled with different patterns denote carbon atoms of different precursors from central carbon metabolism. *Bold solid lines* in an amino acid molecule denote intact carbon–carbon bonds in its precursor molecule(s) that were not broken during the synthesis of the amino acid from the precursor(s). *Thin straight lines* denote biosynthetic bonds that were formed by reassembling carbon atoms from two different precursor molecules. *Dashed lines* connect carbon atoms of amino acid molecules that originated from the same precursor molecule and were covalently bonded in the precursor molecule but are not covalently bonded in the amino acid molecule. Abbreviations: ACA, acetyl-coA; Ala, alanine; Ile: isoleucine; Leu: leucine; OAA, oxaloacetate; Pyr: pyruvate; Val: valine

Table 1
Retrobiosynthetic calculation of the isotopomer distribution of an intracellular metabolite (Pyr) from the MIDs of biomass components (amino acids Ala and Val) known to be synthesized from it

Fragment	Mass ($m + n$)	MID	SD	Predicted formula	Predicted value	χ^2	Pyr isotopomer	Optimized abundance
Ala[23]	0	0.69	0.04	=[000] + [100]	0.69	0.00	000	0.63
	1	0.30	0.04	=[010] + [110] + [001] + [101]	0.30	0.00	100	0.06
	2	0.01	0.00	=[011] + [111]	0.01	0.43	010	0.09
Ala[123]	0	0.63	0.03	=[000]	0.63	0.01	110	0.02
	1	0.34	0.03	=[100] + [010] + [001]	0.34	0.00	001	0.19
	2	0.03	0.00	=[110] + [101] + [011]	0.03	0.39	101	0.00
	3	0.00	0.00	=[111]	0.00	0.05	011	0.01
Val[12]	0	0.81	0.01	=[000] + [001]	0.82	0.00	111	0.00
	1	0.16	0.01	=[100] + [010] + [101] + [011]	0.16	0.00	$\chi^2 = \frac{(\text{Prediction} - \text{MID})^2}{\text{SD}^2}$	
	2	0.03	0.00	=[110] + [111]	0.02	1.05	Total $\chi^2 = 1.93$	

The amino acids Ala and Val are known to be synthesized from Pyr. We employ an optimization algorithm to convert the MIDs of the fragments Ala [23], Ala [123], and Val [12] to the eight isotopomers of Pyr. Toward this we minimize a statistical χ^2 criterion, whose expression is shown in this table. Abundances of Pyr isotopomers are simulated and optimized as shown in the last two columns. Isotopomers are represented using the notation 1 = ^{13}C, 0 = ^{12}C. *SD*, measurement standard deviation

synthesis of multiple amino acids. For instance Ala, valine (Val), as well as parts of leucine (Leu) and Ile originate from Pyr (Fig. 2). Pyr, being a three-carbon compound, has eight (= 2^3) different isotopomers as shown in Table 1 using the notation 0 = ^{12}C, 1 = ^{13}C. Because Ala, Val, Leu, and Ile are entirely or partially synthesized from Pyr, their mass isotopomer distributions should reflect the abundances of the eight Pyr isotopomers. However, the Pyr isotopomer abundances cannot be directly back-calculated from the MID of any single amino acid derived from it. For example, the two available mass fragments of Ala, Ala[23] and Ala[123], do not distinguish between labeling at C-2 and C-3 of Ala and therefore do not provide complete information on mass isotopomers of Pyr. A mass spectrum of tBDMS-derivatized Val contains the fragments Val[COOH · C^{α}] (or Val[12], which corresponds to Pyr[12]) and Val[C^{α}·C^{β}·$C^{\gamma 1}$·$C^{\gamma 2}$] (or Val[2344], which corresponds to Pyr[2233]). As with Ala, the two Val fragments do not, by themselves, provide information to back-calculate the abundances of the eight Pyr isotopomers. However, the MIDs of the two Ala fragments Ala [23] and Ala[123] and one Val fragment Val [COOH · C^{α}] (or Val[12]) can be combined to provide sufficient information to infer the abundances of the eight Pyr isotopomers. Table 1 summarizes our method for accomplishing this. Given the experimentally observed MIDs of three fragments Ala[23], Ala [123], and Val[23], the abundances of any seven Pyr isotopomers are collectively guessed (the abundance of the eighth isotopomer is automatically determined as the eight isotopomer abundances of Pyr should sum up to unity), and the three MIDs are predicted on the basis of the guess. Then, the guesses for the Pyr isotopomer abundances are successively refined until a set of abundances is obtained that best matches the experimental MIDs. We use the χ^2 criterion as a metric for the discrepancy between experimental and predicted MIDs:

$$\chi^2 = \sum_j \left(\frac{I_j - I_j^x}{\sigma_j} \right)^2 \tag{1}$$

where I_j^x and I_j are, respectively, the experimentally measured and predicted abundances of the *j*th mass isotopomer and σ_j is the experimental standard deviation of this isotopomer. Thus, this problem can be cast as an optimization problem with the objective of minimizing χ^2. The optimization can be readily solved using the Solver utility in Microsoft Excel or by writing a small MATLAB script. Ultimately, a robust method of determining the isotopomer distribution of a central carbon metabolite such as Pyr is to estimate it statistically by combining information from the MIDs of all amino acids of which Pyr is a precursor. Furthermore, because many central carbon metabolites are precursors of multiple amino acids and many amino acids have multiple precursors, we usually solve a comprehensive version of the problem illustrated in Table 1, wherein we simultaneously solve for the isotopomer abundances of all amino acid precursors by using all available MIDs of amino acid fragments. Amino acids contain redundant metabolic information, and the above method enables leveraging this redundancy toward estimating statistics (means and standard deviations) for the isotopomer distributions of the central carbon metabolite precursors. Readers should note that compartmentalization of amino acid biosynthesis in plants should be considered while performing the analyses described above. *See* **Note 3** regarding compartmentalization of amino acid synthesis in plants.

3.4 Quantification of Flux Ratios at Metabolic Branchpoints via Steady-State Isotope MFA

3.4.1 Relating Pathway Fluxes to Isotopomers of Central Carbon Metabolites

In this subsection, we illustrate modeling of flux–isotopomer relationships for a simplified, unicompartmental version of glycolysis and the pentose phosphate pathway (PPP) (Fig. 3). An important metabolic branchpoint in this network is at glucose-6-phosphate (G6P), where carbon traffic splits between glycolysis and the PPP. The flux partitioning at this branchpoint is difficult to measure on the basis of extracellular measurements alone, making isotope labeling indispensable for accurate evaluation of fluxes through this network.

For the purpose of demonstration, we assume that all reactions in this network are irreversible in the directions indicated in Fig. 3

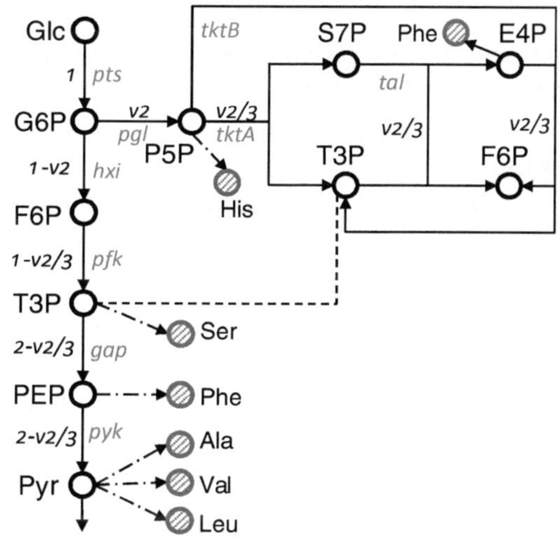

Fig. 3 A simplified metabolic network including glycolysis and the pentose phosphate pathway (PPP). *Open circles* denote carbon atoms of metabolic precursors and *gray hashed circles* denote carbon atoms of biomass components (in this case, proteinogenic amino acids) derived from the precursor(s). *Arrows* denote net fluxes, with *solid lines* representing pathway fluxes *and dashed-dotted lines* representing fluxes toward biomass synthesis. *Dashed lines* connect multiple occurrences of the same metabolite. Reaction names are marked in *gray italic letters. Numbers* alongside pathway flux arrows denote expressions for the corresponding fluxes in terms of the *pgl* flux (v_2), derived by performing mass balances on the metabolites. Abbreviations: Ala, alanine; E4P, erythrose-4-phosphate; F6P, fructose-6-phosphate; G6P, glucose-6-phosphate; Glc, glucose; His, histidine; Leu, leucine; PEP, phosphoenolpyruvate; Phe, phenylalanine; Pyr, pyruvate; S7P, sedoheptulose-7-phosphate; Ser, serine; T3P, consolidated triose-3-phosphates; Val, valine

and that natural abundance of ^{13}C is negligible. However, the realistic version of this problem that relaxes these assumptions and uses appropriately compartmentalized pathways can also be solved by extending the concepts explained here to a computational algorithm. The general methodology for this is explained in Subheadings 3.5 and 3.6, and a full treatment is available in another publication from our research group [26]. In Fig. 3, the fluxes in the network are expressed in terms of two variables: the glucose uptake flux (arbitrarily set to 1 unit because isotope MFA can only evaluate relative and not absolute fluxes [35]) and the flux of *pgl* reaction, which we choose as a free flux and designate as v_2. By performing mass balances on the intracellular metabolites in the network, all other fluxes in the network can be expressed in terms of v_2 as shown in Fig. 3. Below, we discuss how the isotopomers of metabolites in this network, specifically the consolidated triose-3-phosphates (T3P), can be expressed as functions of v_2. This enables (1) estimation of v_2 given the measurements of these isotopomers

Table 2
Isotopomer abundances in the glycolysis–PPP network of Fig. 3 resulting from feeding 1-^{13}C or 6-^{13}C glucose, expressed in terms of the flux v_2

Fed 1-^{13}C glucose			Fed 6-^{13}C glucose		
Metabolite	**Isotopomer**	**Abundance**	**Metabolite**	**Isotopomer**	**Abundance**
G6P	$\underline{1}23456$	100 %	G6P	$12345\underline{6}$	100 %
P5P	12345	100 %	P5P	$1234\underline{5}$	100 %
S7P	1234567	100 %	S7P	$123456\underline{7}$	100 %
E4P	1234	100 %	E4P	$123\underline{4}$	100 %
F6P	$\underline{1}23456$	$\frac{3-3v_2}{3-v_2}$	F6P	$12345\underline{6}$	$\frac{6-2v_2}{6-v_2}$
	$12345\underline{6}$	$\frac{3v_2(1-v_2)}{(3-v_2)(6-v_2)}$		123456	$\frac{v_2}{6-v_2}$
	123456	$\frac{v_2(9+v_2)}{(3-v_2)(6-v_2)}$			
T3P/PEP/Pyr	$12\underline{3}$	$\frac{3-3v_2}{6-v_2}$	T3P/PEP/Pyr	$12\underline{3}$	$\frac{3}{6-v_2}$
	123	$\frac{2v_2+3}{6-v_2}$		123	$\frac{3-v_2}{6-v_2}$

Labeled carbon atoms are *underlined*. Carbon atom numbering follows IUPAC rules

and (2) assessment of the relative merit of one labeled carbon source over another on the basis of sensitivities of the isotopomer expressions to changes in v_2.

Feeding certain labeled varieties of glucose to the network in Fig. 3 will result in isotopomer distributions that depend on the flux split ratio at the G6P branchpoint. For example, manual isotopomer delineation shows that feeding either 1-^{13}C or 6-^{13}C glucose will result in the isotopomers listed in Table 2. To derive expressions for the isotopomers in terms of v_2, we proceed as follows:

1. Comprehensively delineate the isotopomers. This will generate the isotopomer list shown in Table 2.
2. Perform mass balances on the individual isotopomers by accounting for all generation and consumption fluxes. For example, when 6-^{13}C glucose is fed, the balance equation for isotopomer T3P $12\underline{3}$ is as follows (underline denotes ^{13}C, no underline denotes ^{12}C):

$$\left(1-\frac{v_2}{3}\right)[F6P\ 12345\underline{6}]+\frac{2v_2}{3}[P5P\ 1234\underline{5}] - \left(2-\frac{v_2}{3}\right)[T3P\ 12\underline{3}]-\frac{v_2}{3}[T3P\ 12\underline{3}]=0 \tag{2}$$

where square brackets represent abundance. In Eq. (2), the first two terms are generation fluxes for T3P $12\underline{3}$: this isotopomer is generated from F6P 12345$\underline{6}$ through the *pfk* reaction whose flux is $(1 - v_2/3)$ as well as from P5P 1234$\underline{5}$ through the *tktA* and *tktB* reactions whose combined flux is $2v_2/3$. The next (last) two terms in Eq. (2) are consumption fluxes for T3P 12$\underline{3}$, which is consumed by the *gap* and *tal* reactions, whose fluxes are $(2 - v_2/3)$ and $v_2/3$, respectively. Along these lines, isotopomer balance equations can be written for all the other isotopomers in Table 2. Solving these balances simultaneously will result in the final expressions for the isotopomers in terms of v_2, as displayed in Table 2.

3.4.2 Designing Labels: Choice of Labeled Carbon Source(s)

Accurate identification of fluxes often requires the use of designed isotope labels. In this important step in modeling, researchers determine the sensitivities of isotopomer abundances to changes in flux values and use this information in deciding if one labeled version of a carbon source provides more flux information than another labeled variety [15, 26, 36, 37]. As a simple example, we can use the isotopomer expressions derived above to analyze whether an experiment using 1-^{13}C glucose or one using 6-^{13}C glucose provides superior information on the flux partitioning at the G6P branchpoint. Assuming that we can only measure the isotopomers of T3P (either directly or via amino acids or biomass components synthesized from T3P), we examine its isotopomer abundances for extreme values of the free flux v_2. For instance, when $v_2 = 0$ (all flux through glycolysis), the isotopomer distribution of T3P is {123: 0.5, 12$\underline{3}$: 0.5} irrespective of whether 1-^{13}C glucose or 6-^{13}C glucose is fed. When $v_2 = 1$ (all flux through the PPP), feeding 1-^{13}C glucose will result in a distribution of {123: 1.0, 12$\underline{3}$: 0.0}, whereas feeding 6-^{13}C glucose will result in a distribution of {123: 0.40, 12$\underline{3}$: 0.60}. Thus, if T3P is the only metabolite whose labeling state can be measured, 1-^{13}C glucose is superior to 6-^{13}C glucose because it results in isotopomer distributions that are more sensitive to the flux partitioning between glycolysis and the PPP. However, this situation may change significantly if the isotopomers of both T3P and F6P are measurable, or if the isotopomers of T3P, F6P, and P5P are simultaneously measurable. We have recently explored this problem in detail in the context of compartmentalized glycolysis and PPP in plants [26]. In this context, it is important to note that biomass components derived from T3P (certain amino acids and glycerol) are different types of macromolecules compared to those derived from F6P (certain carbohydrates) or P5P (nucleotide sugars). Therefore, procedures for isolating all of these biomass components can be time consuming. The incremental benefit of isolating and measuring isotopomers of additional types of biomass components can be solved by the experimental design procedure described here. Also *see* **Note 4**.

Table 3
Correspondence between the carbon atoms of the amino acids shown in Fig. 3 and the carbon atoms of their central carbon metabolic precursors

Amino acid	Amino acid fragments	Metabolite source
His	His 12345	P5P 54321
Phe	Phe 123 Phe 45 Phe 5667	PEP 123 PEP 23 E4P 1234
Ser	Ser 123	T3P 123
Ala	Ala 123	Pyr 123
Val	Val 124 Val 34	Pyr 123 Pyr 23
Leu	Leu 12 Leu 35 Leu 45	Pyr 23 Pyr 23 Pyr 23

Repeated numbers in amino acid fragments indicate chemically identical carbons

3.4.3 Relating Pathway Fluxes to Amino Acid Mass Isotopomers

In the previous discussion we related a flux split ratio at a branchpoint in a simple metabolic network to the isotopomer distribution of the central carbon metabolite T3P. However, as measuring labeling in amino acids and other biomass components is more practical for steady-state isotope MFA, we need to establish relationships between pathway fluxes and amino acid MIDs. For this, we employ the methodology described in Subheading 3.3 and obtain information on the correspondence between the carbon atoms of their central carbon metabolic precursors and the carbon atoms of the proteinogenic amino acids of Fig. 3. This relationship is nearly universal and can be obtained either from the literature [34] or, in case of rare organisms with unusual metabolic pathways, by performing targeted labeling experiments. The relationships pertaining to the network of Fig. 3 are illustrated in Table 3. Using this information, we worked out relationships between the flux variable v_2 of Fig. 2 and amino acid MIDs. Figure 4 depicts, for an experiment in which 1-^{13}C glucose is fed, the MIDs of the amino acid fragments Ala[23] and Leu[23455] for $v_2 = 0$ (only glycolysis), $v_2 = 1$ (only PPP), and an intermediate value of $v_2 = 0.6$. Because all isotopomer abundances and MIDs are dependent on one flux (v_2) in this example, measuring the relevant amino acid MIDs will enable direct determination of v_2, the carbon partitioning at the G6P branchpoint. However, a realistic network will contain several flux variables akin to v_2 of the network just discussed, and multiple MIDs will be dependent on these flux variables. To model such a circumstance, the concepts used in this simple model need to be generalized and automated, as explained below.

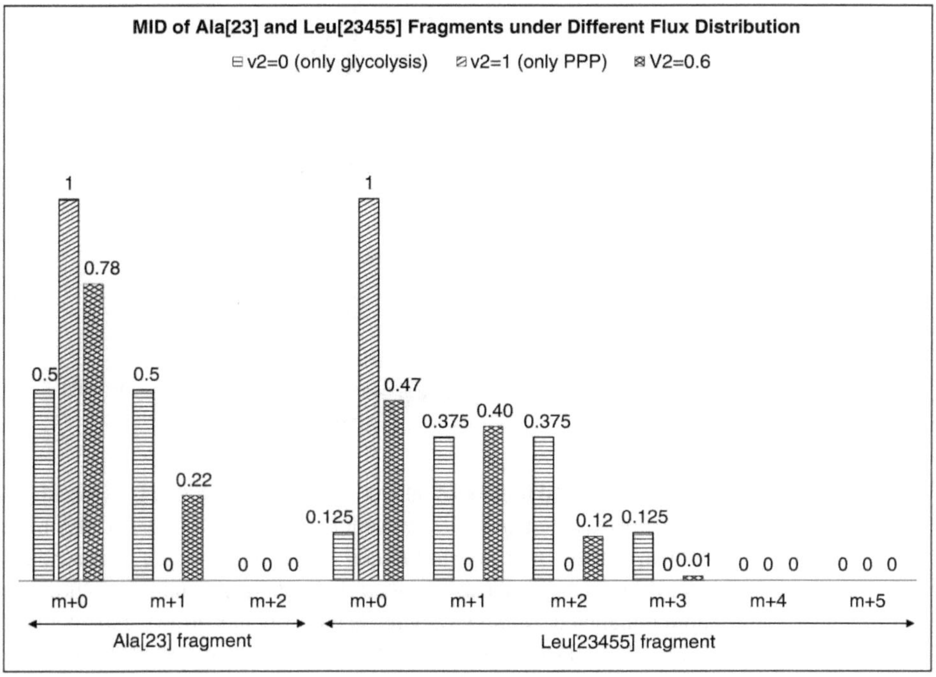

Fig. 4 The predicted MIDs of Ala[23] and Leu[23455] fragments under different flux scenarios. MIDs are calculated on the basis of information provided in Table 2 and Table 3

3.5 General Procedures for Construction of Steady-State Models

This section generalizes two modeling steps used in the previous glycolysis–PPP example: (1) determining how many "free flux" variables (e.g., v_2 in the previous example) are necessary, (2) expressing fluxes in the network in terms of measured fluxes and free flux variables, and (3) writing and solving balances for the isotopomers. To illustrate these generalizations, we use a new network that contains the TCA cycle and the glyoxylate shunt (Fig. 5).

3.5.1 Determining the Number of Free Flux Variables and Using Mass Balances to Derive Expressions for Fluxes

The assumption of metabolic steady state implies that no metabolite accumulates or depletes in the system (cell or tissue) under consideration. Therefore, for each metabolite, we have the mass balance:

$$\frac{\mathrm{d}[M]}{\mathrm{d}t} = 0 = \left\{\Sigma v_{\mathrm{in}} + \Sigma v_{\mathrm{generation}}\right\} - \left\{\Sigma v_{\mathrm{out}} + \Sigma v_{\mathrm{consumption}}\right\} \quad (3)$$

where, at metabolic steady state, $[M]$ is the time-invariant concentration of the metabolite being balanced, $\sum v_{\mathrm{in}}$ is the sum of all extracellular fluxes that transport M into the system across its boundary (e.g., the flux v_1 in the network of Fig. 5 transports acetyl-CoA into the system), and $\sum v_{\mathrm{generation}}$ is the sum of all fluxes that generate M through chemical reactions. Likewise, $\sum v_{\mathrm{out}}$ is the sum of all extracellular fluxes that transport M out of the system across its boundary (e.g., the flux v_{10} in the network of Fig. 5 transports malate [Mal] out of the system) and $\sum v_{\mathrm{consumption}}$ is

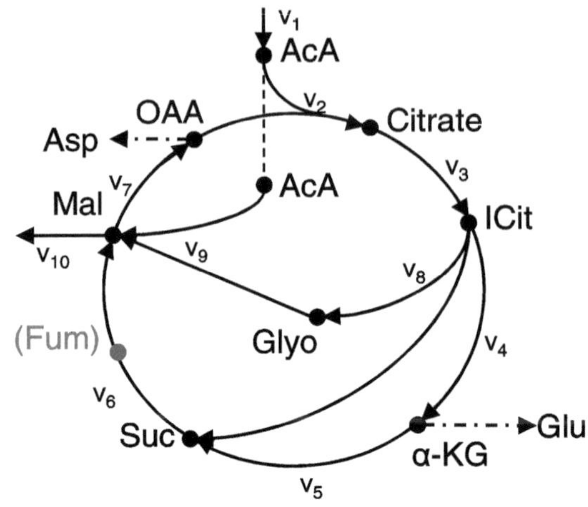

Fig. 5 A simplified metabolic network including the tricarboxylic acid (TCA) cycle and the glyoxylate shunt. *Arrows* denote net fluxes, with solid lines representing pathway fluxes and *dashed-dotted lines* representing (negligible) effluxes toward biomass synthesis. *Dashed lines* connect multiple occurrences of the same metabolite. *Numbers* alongside pathway flux arrows denote flux variables. Metabolites shown in *gray* are present in the real metabolic network but are eliminated in the steady-state model for simplifying it without loss of information. Abbreviations: AcA, acetyl-coA; Asp, aspartic acid; Fum, fumarate; Glu, glutamic acid; Glyo, glyoxylate; ICit, isocitrate; Mal, malate; OAA, oxaloacetate; Suc, succinate; α-KG, α-ketoglutarate

the sum of all fluxes that consume M through chemical reactions. Biomass synthesis fluxes, if significant, are conveniently modeled as extracellular fluxes. In this simple network, the two biomass synthesis fluxes (dashed-dotted lines in Fig. 5) are assumed to be negligible and are therefore not included in the model.

Mass balances for all metabolites in a network can be collected into a matrix equation:

$$\mathbf{G} \cdot \mathbf{v} = \mathbf{0} \tag{4}$$

where **G** is called the stoichiometric matrix for the network. Table 4 depicts **G** for the network of Fig. 5.

The matrix **G** represents the mathematical interdependency of the fluxes in the network. Algebraic processing of **G** enables the expression of a large number of dependent fluxes in the network in terms of a few independent or free fluxes whose values are known, measured, or assumed. For example, in the glycolysis–PPP network discussed in Fig. 3 and Subheading 3.4.1, we expressed all fluxes in terms of two independent fluxes: the glucose uptake flux (a measurement; 1 unit) and the flux variable v_2 (a free flux). In isotope MFA, this expression permits a reduction of the problem, so the identification of only a few independent fluxes on the basis of the

Table 4
Stoichiometric matrix obtained from mass balance equations

	v_1	v_2	v_3	v_4	v_5	v_6	v_7	v_8	v_9	v_{10}
AcA	1	-1	0	0	0	0	0	0	-1	0
Cit	0	1	-1	0	0	0	0	0	0	0
ICit	0	0	1	-1	0	0	0	-1	0	0
α-KG	0	0	0	1	-1	0	0	0	0	0
Suc	0	0	0	0	1	-1	0	1	0	0
Mal	0	0	0	0	0	1	-1	0	1	-1
OAA	0	-1	0	0	0	0	1	0	0	0
Glyo	0	0	0	0	0	0	0	1	-1	0

The group of numbers not shaded in gray represents the stoichiometric matrix G of the simple metabolic network shown in Fig. 5

labeling measurements can be followed by the algebraic steps explained below to obtain values and statistics for all fluxes.

To express dependent fluxes in terms of independent fluxes, it is first necessary to determine a quantity called the degrees of freedom (DOF) of the network. This quantity is defined as

$$\begin{aligned}\text{DOF} = {} & \text{Total number of fluxes in the network} \\ & - \text{Number of intracellular metabolites} \\ & - \text{Number of independent extracellular} \\ & \quad \text{and biomass synthesis fluxes in the network}\end{aligned} \tag{5}$$

This expression is valid only if there are no linearly dependent rows (metabolite balances) in **G**. If there are linearly dependent metabolite balances, only an independent set of balances should be counted in Eq. (5).

A zero DOF indicates that all unknown fluxes in the network can be calculated from the measured extracellular or biomass synthesis fluxes via Eq. (4). A positive DOF indicates that there are certain free fluxes whose values cannot be unequivocally determined from the measured fluxes. Isotope MFA guesses the values of all free fluxes that best account for a given set of isotope labeling measurements (e.g., amino acid MIDs).

Once DOF has thus been determined for a network, the flux vector $\mathbf{v}$ can be split into two vectors $\mathbf{v_m}$ and $\mathbf{v_x}$ such that $\mathbf{v_m}$ contains the independent fluxes, i.e., the (measured) extracellular and biomass synthesis fluxes as well as the free fluxes, and $\mathbf{v_x}$ contains the other, dependent fluxes. The matrix **G** can be correspondingly split into matrices $\mathbf{G_m}$ and $\mathbf{G_x}$. Applying this partitioning, Eq. (4) can be rewritten as

$$\mathbf{G}\cdot\mathbf{v} = \mathbf{G_m}\cdot\mathbf{v_m} + \mathbf{G_x}\cdot\mathbf{v_x} = \mathbf{0} \tag{6}$$

or

$$\mathbf{v_x} = (\mathbf{G_x})^{-1}(\mathbf{G_m}\cdot\mathbf{v_m}) \tag{7}$$

which is an expression for the dependent fluxes in terms of the measured and free fluxes (*see* **Note 2**).

For the network of Fig. 5, the stoichiometric matrix **G** (Table 4) has eight rows (corresponding to metabolites) and ten columns (corresponding to fluxes). The rank of matrix (= 8) equals the number of rows, indicating that none of the metabolite balances are linearly dependent on each other. The DOF for this network, calculated using Eq. (5), equals 2, indicating that the network has two independent fluxes. A typical scenario will involve one extracellular measurement (e.g., the flux of acetyl-CoA uptake, v_1) and one free flux. We chose v_8, the flux into the glyoxylate shunt, as the free flux. Certain other fluxes could be valid choices for the free flux; however, not every flux in the network may be a valid choice. A valid choice of free fluxes will render $\mathbf{G_x}$ non-singular, assuming that **G** did not have linearly dependent rows (metabolite balances).

We now split the flux vector **v** into $\mathbf{v_m}$ (containing only the independent fluxes v_1 and v_8) and $\mathbf{v_x}$ (containing the other fluxes), and correspondingly split the stoichiometric matrix **G** into matrices $\mathbf{G_m}$ and $\mathbf{G_x}$. This partitioning is shown in Table 5.

To obtain the unknown $\mathbf{v_x}$, we need to make use of isotopomer abundance measurements or MIDs, as explained in the following sections.

3.5.2 Simulating Isotopomer Abundances from Guessed or Given Flux Values

Isotopomer balance equations can often be bilinear, which require iterative methods for direct solution. However, the bilinearity can be circumvented by expressing the labeling state in terms of transformed variables such as cumomers [15], bondomers [16, 17] (when applicable), or EMUs [18]. In this chapter, we focus on cumomers. Whereas an isotopomer representation expresses labeling states of individual atoms as labeled ("1") or unlabeled ("0"), a cumomer representation expresses these labeling states as labeled ("1") and undetermined or irrelevant ("x"). For example, a three-carbon metabolite has eight isotopomers (000, 100, 010, 110, 001, 101, 011, 111) and eight cumomers (xxx, 1xx, x1x, 11x, xx1, 1x1, x11, 111). Each cumomer is the sum of a certain number of isotopomers, and the cumomer and isotopomer abundance distributions of a metabolite are easily interconvertible [15].

Table 5
Splitting of the stoichiometric matrix G on the basis of known fluxes (v_m) and unknown fluxes (v_x); the corresponding matrices are G_m and G_x

	v_2	v_3	v_4	v_5	v_6	v_7	v_9	v_{10}	v_1	v_8
	G_x								G_m	
AcA	-1	0	0	0	0	0	-1	0	1	0
Cit	1	-1	0	0	0	0	0	0	0	0
ICit	0	1	-1	0	0	0	0	0	0	-1
α-KG	0	0	1	-1	0	0	0	0	0	0
Suc	0	0	0	1	-1	0	0	0	0	1
Mal	0	0	0	0	1	-1	1	-1	0	0
OAA	-1	0	0	0	0	1	0	0	0	0
Glyo	0	0	0	0	0	0	-1	0	0	1

v_x	v_2
	v_3
	v_4
	v_5
	v_6
	v_7
	v_9
	v_{10}
v_m	v_1
	v_8

A corresponding split of the flux vector v is shown on the right

The order of a cumomer is the number of atoms in it that are definitively labeled. For example, the cumomer xx1 is order 1, 1x1 is order 2, and 111 is order 3.

The advantage of the cumomer representation is that bilinear isotopomer balance equations are transformed into cascades of linear equations. For example, the balances for the first-order cumomers in a network can be written as [15]

$$\mathbf{diag}[X]\cdot\frac{d\,{}^1\mathbf{x}}{dt} = \mathbf{0} = {}^1\mathbf{A}(\mathbf{v})\cdot{}^1\mathbf{x} + {}^1\mathbf{b}\left(\mathbf{v}, {}^1\mathbf{x}_{\text{inp}}\right) \quad (8)$$

where $\mathbf{diag}[X]$ is a diagonal matrix containing metabolite concentrations, ${}^1\mathbf{x}$ is a vector containing all first-order cumomers in the network, ${}^1\mathbf{A}(\mathbf{v})$ is a constant matrix whose entries depend on the flux vector $\mathbf{v}$, and ${}^1\mathbf{b}(\mathbf{v}, {}^1\mathbf{x}_{\text{inp}})$ is a constant vector whose entries depend on the first-order cumomers of the carbon source fed to the network (${}^1\mathbf{x}_{\text{inp}}$). At isotopic steady state, Eq. (8) is linear with ${}^1\mathbf{x}$ as the only unknown, so it can be solved by matrix inversion or decomposition of ${}^1\mathbf{A}(\mathbf{v})$:

$$\begin{aligned} \mathbf{0} &= {}^1\mathbf{A}(\mathbf{v})\cdot{}^1\mathbf{x} + {}^1\mathbf{b}\left(\mathbf{v}, {}^1\mathbf{x}_{\text{inp}}\right) \\ {}^1\mathbf{x} &= -\left\{{}^1\mathbf{A}(\mathbf{v})\right\}^{-1}\cdot{}^1\mathbf{b}\left(\mathbf{v}, {}^1\mathbf{x}_{\text{inp}}\right) \end{aligned} \quad (9)$$

Upon solution of the first-order cumomer balances, the second-order cumomer balances become linear at isotopic steady state:

$$\mathbf{0} = {}^{2}\mathbf{A}(\mathbf{v})\cdot{}^{2}\mathbf{x} + {}^{2}\mathbf{b}\left(\mathbf{v}, {}^{2}\mathbf{x}_{\mathrm{inp}}, {}^{1}\mathbf{x}\right) \tag{10}$$

where ${}^{2}\mathbf{x}$ is a vector containing all second-order cumomers, ${}^{2}\mathbf{A}(\mathbf{v})$ is a constant matrix whose entries depend on the flux vector **v**, and ${}^{2}\mathbf{b}(\mathbf{v}, {}^{2}\mathbf{x}_{\mathrm{inp}}, {}^{1}\mathbf{x})$ is a constant vector whose entries depend on the known second-order cumomers of the carbon source fed to the network (${}^{2}\mathbf{x}_{\mathrm{inp}}$) and the already solved first-order cumomer abundances ${}^{1}\mathbf{x}$. Again, Eq. (10) is linear with ${}^{2}\mathbf{x}$ as the only unknown and can be solved by matrix inversion of ${}^{2}\mathbf{A}(\mathbf{v})$.

This balance can be generalized to *n*th-order cumomers:

$$\mathbf{0} = {}^{n}\mathbf{A}(\mathbf{v})\cdot{}^{n}\mathbf{x} + {}^{n}\mathbf{b}\left(\mathbf{v}, {}^{n}\mathbf{x}_{inp}, {}^{1}\mathbf{x}, {}^{2}\mathbf{x}, \ldots, {}^{n-1}\mathbf{x}\right) \tag{11}$$

where the only unknown is ${}^{n}\mathbf{x}$. As before, this balance can be solved by inverting ${}^{n}\mathbf{A}(\mathbf{v})$ as above [15].

3.5.3 An Overall Modeling Methodology

The following overall modeling methodology summarizes the steps described above:

1. Assemble the stoichiometric matrix **G** for the network. Determine DOF for the network and choose a valid set of free fluxes accordingly. On the basis of this choice, split the flux vector **v** into $\mathbf{v_m}$ and $\mathbf{v_x}$ and correspondingly split **G** into $\mathbf{G_m}$ and $\mathbf{G_X}$.
2. Write cumomer balances from order 1 to the highest order that needs to be simulated in the network.
3. Assume values for the free fluxes.
4. Using Eq. (7) and the current guesses for free fluxes, calculate the values of all fluxes in the network.
5. Solve for the cumomer abundances corresponding to the current flux values calculated in **step 4**, and employing the methodology described in Subheading 3.5.2. *See*, e.g., ref. 15.
6. Convert the cumomer abundances first to isotopomer abundances and then to MIDs of biomass components or directly to MIDs of biomass components.
7. Compare the predicted MIDs of **step 6** with the experimentally obtained MIDs and compute a χ^2 metric:

$$\chi^2 = \sum_j \left(\frac{I_j - I_j^x}{\sigma_j}\right)^2 \tag{12}$$

where I_j^x and I_j are, respectively, the experimentally measured and predicted abundances of the *j*th mass isotopomer and σ_j is the experimental standard deviation of this isotopomer.

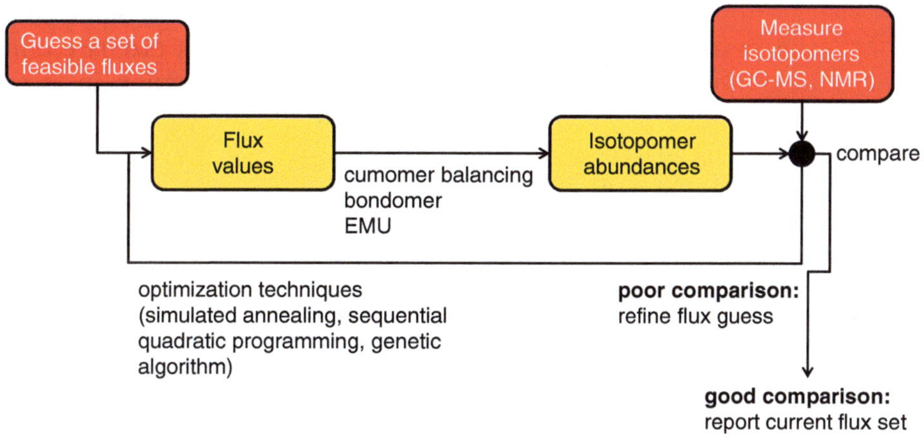

Fig. 6 Iterative methodology for evaluating fluxes from isotopomer data. Adapted from [2]

8. If the value of the χ^2 metric is acceptable, report the current flux values. Else, use an optimization algorithm to minimize the χ^2 metric, thereby refining the previous free flux guesses. Then return to **step 4** and iterate. In our experience, optimization algorithms require ~10,000 or more iterations to converge.

   The iterative steps of this methodology are depicted in Fig. 6.

   Steady-state isotope MFA is essentially a parameter estimation problem wherein the unknown (free) fluxes are parameters that are estimated from isotope labeling data. Therefore, global optimization algorithms are enormously beneficial in refining previous free flux guesses to efficiently arrive at a flux guess that reasonably account for the labeling data. Examples of previously used optimization algorithms include simulated annealing [38], which we use in our software NMR2Flux+ [16]; sequential quadratic programming [2]; and genetic algorithms. Often, it is helpful to use local optimization algorithms to fine-tune the results of a global search. Programs such as 13CFLUX2 [27] offer several optimization libraries that users can choose from.

3.6 Construction of Instationary Models

As explained at the beginning of Subheading 3, instationary models differ from steady-state ones in experimental and computational aspects. Experimentally, it is essential to obtain transient measurements of labels and, if feasible, metabolite concentrations. Computationally, it is necessary to solve the isotopomer balance equations as functions of time. Thus, the steady-state n-other cumomer balance now becomes an instationary balance:

$$\mathbf{diag}[X]\cdot\frac{\mathrm{d}\,^{n}\mathbf{x}}{\mathrm{d}t} = {}^{n}\mathbf{A}(\mathbf{v})\cdot{}^{n}\mathbf{x} + {}^{n}\mathbf{b}\left(\mathbf{v},{}^{n}\mathbf{x}_{inp},{}^{1}\mathbf{x},{}^{2}\mathbf{x},\ldots,{}^{n-1}\mathbf{x}\right) \tag{13}$$

The cascades of algebraic equations encountered in the steady-state model now become cascades of ordinary differential equations. The extensive nature of computation in instationary models makes flux evaluation a lengthy process. Despite this, flux evaluation for relatively small networks can be performed by using homegrown MATLAB or C code (*see* **Note 5**). While methods for detailed flux evaluation have been presented in detail [19, 20, 39], we list a few points from our experience:

1. For instationary experiments involving a single carbon source containing exactly one carbon atom (e.g., $^{13}CO_2$), much of the labeling information is often available in the first- and second-order cumomers. Therefore, it is unadvisable to expend computational resources to simulate cumomer abundances beyond the second order.
2. It is important that the matrices $^{n}\mathbf{A}(\mathbf{v})$ are well conditioned, with condition numbers between 1 and 100. This prevents complex eigenvalues from occurring during matrix decomposition.

4 Notes

1. Metabolic databases provide detailed information on metabolites, reactions, and pathways often so comprehensively that for the purpose of steady-state isotope MFA, it may be advisable to reduce the size of the model. Also, *see* **Note 2**.
2. While building steady-state models, it is helpful to consolidate metabolites to reduce the size of the model. Two or more metabolites may be consolidated into a single "lumped" metabolite as long as (1) the influxes and generation fluxes for all the metabolites lead into the lumped metabolite, (2) the effluxes and consumption fluxes for all the metabolites lead out of the lumped metabolite, and (3) any intracellular flux that only connects the metabolites that were lumped is eliminated from the network. For instance, in a steady-state model of glycolysis and the PPP, a number of three-carbon metabolites (glyceraldehyde-3-phosphate, dihydroxyacetone phosphate, 1,3-bisphosphoglycerate, 3-phosphoglycerate, and 2-phosphoglycerate) can be lumped into a single metabolite named T3P (Fig. 3). Likewise, in our model of the TCA cycle and glyoxylate shunt (Fig. 5), fumarate (Fum; grayed out in Fig. 5) can be trivially eliminated from a steady-state model without loss of information because there are no branch fluxes leading into or out of it. Furthermore, the sequence of reactions oxaloacetate + acetyl-CoA → citrate → aconitate → isocitrate → α-ketoglutarate + CO_2 in the TCA cycle involves the branched metabolites citrate, aconitate, and isocitrate. In most circumstances, this sequence of reactions can

be written bypassing the branched compounds as oxaloacetate + acetyl CoA → → α-ketoglutarate, with a consolidated carbon atom rearrangement and without loss of information. However, because instationary models involve metabolite pool sizes, it may not be wise to consolidate metabolites in them.

3. Amino acid biosynthesis in plants is compartmentalized, with aromatic and branched-chain amino acids (e.g., Phe, Tyr, Trp, Val, Leu, Ile) synthesized almost exclusively in the plastid and some other amino acids (e.g., Ala, Asp) synthesized in the cytosol, plastid, and mitochondrion. Information on compartmentation is available from the literature [40] and to an extent from metabolic databases. This information must be carefully considered while deriving isotopomer abundances of central carbon metabolites from the MIDs of amino acids. See [3] and [41] for illustrations.
4. Designing labels for isotope labeling experiments may identify certain powerful labels that may be too expensive to feed in pure form. In such cases, it may be desirable to reduce cost by diluting such labels with naturally abundant versions of the same carbon source. However, mixing different isotopically labeled versions of the same carbon source (e.g., 1-^{13}C glucose and 6-^{13}C glucose) in the same experiment is not recommended, since labels can often interfere with each other. If two labels are found to provide equally powerful, orthogonal information in evaluating fluxes, it may be advisable to perform parallel labeling experiments with these labels. *See*, e.g., ref. 26.
5. In an instationary model containing cumomers no higher than second order, solving for ^{13}C enrichments (first-order cumomers) is desirable instead of solving cascades of ordinary differential equations for different orders of cumomers. Often, this will save significant simulation time with minimal loss of information.

Acknowledgments

This work was funded by the National Science Foundation (award numbers CBET 1134115 and IOS 0922650).

References

1. Stephanopoulos G (1999) Metabolic fluxes and metabolic engineering. Metab Eng 1:1–11
2. Wiechert W (2001) ^{13}C metabolic flux analysis. Metab Eng 3:195–206
3. Sriram G, Fulton DB, Iyer VV et al (2004) Quantification of compartmented metabolic fluxes in developing soybean embryos by employing biosynthetically directed fractional ^{13}C labeling, two-dimensional [^{13}C, ^{1}H] nuclear magnetic resonance, and comprehensive isotopomer balancing. Plant Physiol 136:3043–3057
4. Sriram G, Fulton DB, Shanks JV (2007) Flux quantification in central carbon metabolism of

Catharanthus roseus hairy roots by ^{13}C labeling and comprehensive bondomer balancing. Phytochemistry 68:2243–2257

5. Iyer V, Sriram G, Shanks JV (2007) Metabolic flux maps of central carbon metabolism in plant systems. In: Wurtele ES, Nikolau BJ (eds) Concepts in plant metabolomics. Springer, Dordrecht, The Netherlands, pp 125–144
6. Nargund S, Joffe ME, Tran D, Tugarinov V, Sriram G (2013) Nuclear magnetic resonance methods for metabolic fluxomics. In: Alper HS (ed) Systems metabolic engineering. Humana, New York, NY, pp 335–351
7. Edwards JS, Palsson BO (2000) Metabolic flux balance analysis and the in silico analysis of *Escherichia coli* K-12 gene deletions. BMC Bioinformatics 1:1
8. Lee JM, Gianchandani EP, Papin JA (2006) Flux balance analysis in the era of metabolomics. Brief Bioinform 7:140–150
9. Kauffman KJ, Prakash P, Edwards JS (2003) Advances in flux balance analysis. Curr Opin Biotechnol 14:491–496
10. Resendis-Antonio O, Reed JL, Encarnación S, Collado-Vides J, Palsson BØ (2007) Metabolic reconstruction and modeling of nitrogen fixation in Rhizobium etli. PLoS Comput Biol 3: e192
11. AbuOun M, Suthers PF, Jones GI et al (2009) Genome scale reconstruction of a Salmonella metabolic model: comparison of similarity and differences with a commensal *Escherichia coli* strain. J Biolog Chem 284:29480–29488
12. Feist AM, Henry CS, Reed JL et al (2007) A genome-scale metabolic reconstruction for *Escherichia coli* K-12 MG1655 that accounts for 1260 ORFs and thermodynamic information. Mol Syst Biol 3:121
13. Tran LM, Rizk ML, Liao JC (2008) Ensemble modeling of metabolic networks. Biophys J 95:5606–5617
14. Wiechert W, Siefke C, de Graaf AA, Marx A (1997) Bidirectional reaction steps in metabolic networks: II. Flux estimation and statistical analysis. Biotechnol Bioeng 55:118–135
15. Wiechert W, Möllney M, Isermann N, Wurzel M, De Graaf AA (1999) Bidirectional reaction steps in metabolic networks: III. Explicit solution and analysis of isotopomer labeling systems. Biotechnol Bioeng 66:69–85
16. Sriram G, Shanks JV (2004) Improvements in metabolic flux analysis using carbon bond labeling experiments: bondomer balancing and Boolean function mapping. Metab Eng 6:116–132
17. van Winden WA, Heijnen JJ, Verheijen PJT (2002) Cumulative bondomers: a new concept in flux analysis from 2D [^{13}C, ^{1}H] COSY NMR data. Biotechnol Bioeng 80:731–745
18. Antoniewicz MR, Kelleher JK, Stephanopoulos G (2007) Elementary metabolite units (EMU): a novel framework for modeling isotopic distributions. Metab Eng 9:68–86
19. Nöh K, Wiechert W (2006) Experimental design principles for isotopically instationary ^{13}C labeling experiments. Biotechnol Bioeng 94:234–251
20. Young JD, Walther JL, Antoniewicz MR, Yoo H, Stephanopoulos G (2008) An elementary metabolite unit (EMU) based method of isotopically nonstationary flux analysis. Biotechnol Bioeng 99:686–699
21. Masoudi-Nejad A, Goto S, Endo TR, Kanehisa M (2008) KEGG bioinformatics resource for plant genomics research [Internet]. In: Edwards D (ed) Plant bioinformatics. Humana, New York, NY, pp 437–458
22. Zhang P, Foerster H, Tissier CP et al (2005) MetaCyc and AraCyc. Metabolic pathway databases for plant research. Plant Physiol 138:27–37
23. Schwender J, Goffman F, Ohlrogge JB, Shachar-Hill Y (2004) Rubisco without the Calvin cycle improves the carbon efficiency of developing green seeds. Nature 432:779–782
24. Yang TH, Heinzle E, Wittmann C (2005) Theoretical aspects of ^{13}C metabolic flux analysis with sole quantification of carbon dioxide labeling. Comput Biol Chem 29:121–133
25. Wittmann C, Heinzle E (2001) Modeling and experimental design for metabolic flux analysis of lysine-producing Corynebacteria by mass spectrometry. Metab Eng 3:173–191
26. Nargund S, Sriram G (2013) Designer labels for plant metabolism: statistical design of isotope labeling experiments for improved quantification of flux in complex plant metabolic networks. Mol Biosyst 9:99–112
27. Weitzel M, Nöh K, Dalman T et al (2012) 13CFLUX2—high-performance software suite for ^{13}C-metabolic flux analysis. Bioinformatics 29:143–145
28. Quek L-E, Wittmann C, Nielsen LK, Krömer JO (2009) OpenFLUX: efficient modelling software for ^{13}C-based metabolic flux analysis. Microb Cell Fact 8:25
29. Zamboni N, Fischer E, Sauer U (2005) FiatFlux - a software for metabolic flux analysis from ^{13}C-glucose experiments. BMC Bioinformatics 6:209
30. Sriram G, Rahib L, He J-S et al (2008) Global metabolic effects of glycerol kinase overexpression in rat hepatoma cells. Mol Genet Metab 93:145–159

31. Masakapalli SK, Lay PL, Huddleston JE et al (2010) Subcellular flux analysis of central metabolism in a heterotrophic *Arabidopsis thaliana* cell suspension using steady-state stable isotope labeling. Plant Physiol 152: 602–609
32. O'Leary MH (1988) Carbon isotopes in photosynthesis. BioScience 38:328–336
33. Winden WAV, Wittmann C, Heinzle E, Heijnen JJ (2002) Correcting mass isotopomer distributions for naturally occurring isotopes. Biotechnol Bioeng 80:477–479
34. Szyperski T (1998) ^{13}C-NMR, MS and metabolic flux balancing in biotechnology research. Q Rev Biophys 31:41–106
35. Isermann N, Wiechert W (2003) Metabolic isotopomer labeling systems. Part II: structural flux identifiability analysis. Math Biosci 183:175–214
36. Libourel IGL, Gehan JP, Shachar-Hill Y (2007) Design of substrate label for steady state flux measurements in plant systems using the metabolic network of *Brassica napus* embryos. Phytochemistry 68:2211–2221
37. Crown SB, Antoniewicz MR (2012) Selection of tracers for ^{13}C-metabolic flux analysis using elementary metabolite units (EMU) basis vector methodology. Metab Eng 14:150–161
38. Onbaşoğlu E, Özdamar L (2001) Parallel simulated annealing algorithms in global optimization. J Global Optim 19:27–50
39. Young JD, Shastri AA, Stephanopoulos G, Morgan JA (2011) Mapping photoautotrophic metabolism with isotopically nonstationary ^{13}C flux analysis [Internet]. Metab Eng 13:656–665
40. Singh BK (1998) Plant amino acids (Books in soils, plants, & the environment). CRC, USA
41. Allen DK, Laclair RW, Ohlrogge JB, Shachar-Hill Y (2012) Isotope labelling of Rubisco subunits provides in vivo information on subcellular biosynthesis and exchange of amino acids between compartments. Plant Cell Environ 35:232–1244

Chapter 12

Isotopically Nonstationary MFA (INST-MFA) of Autotrophic Metabolism

Lara J. Jazmin, John P. O'Grady, Fangfang Ma, Doug K. Allen, John A. Morgan, and Jamey D. Young

Abstract

Metabolic flux analysis (MFA) is a powerful approach for quantifying plant central carbon metabolism based upon a combination of extracellular flux measurements and intracellular isotope labeling measurements. In this chapter, we present the method of isotopically nonstationary ^{13}C MFA (INST-MFA), which is applicable to autotrophic systems that are at metabolic steady state but are sampled during the transient period prior to achieving isotopic steady state following the introduction of $^{13}CO_2$. We describe protocols for performing the necessary isotope labeling experiments, sample collection and quenching, nonaqueous fractionation and extraction of intracellular metabolites, and mass spectrometry (MS) analysis of metabolite labeling. We also outline the steps required to perform computational flux estimation using INST-MFA. By combining several recently developed experimental and computational techniques, INST-MFA provides an important new platform for mapping carbon fluxes that is especially applicable to autotrophic organisms, which are not amenable to steady-state ^{13}C MFA experiments.

Key words Metabolic flux analysis, Isotopomer analysis, Nonaqueous fractionation, Mass spectrometry, Elementary metabolite unit, Isotopomer modeling, Autotrophic metabolism

1 Introduction

Metabolic engineering of plants has great potential for the production of novel compounds and chemical feedstocks at low costs. However, in order to take full advantage of this potential, a better understanding of plant central carbon metabolism is needed. The ability to quantitatively map intracellular carbon fluxes using isotope tracers and metabolic flux analysis (MFA) is critical for identifying pathway bottlenecks and elucidating network regulation in plants, especially transgenic varieties that have been engineered to alter their native metabolic capacities [1, 2]. Typically, MFA relies on the assumption of both metabolic and isotopic steady state. Achieving this situation experimentally involves (1) equilibrating the system in a stable metabolic state, (2) introducing an

Martine Dieuaide-Noubhani and Ana Paula Alonso (eds.), *Plant Metabolic Flux Analysis: Methods and Protocols*, Methods in Molecular Biology, vol. 1090, DOI 10.1007/978-1-62703-688-7_12, © Springer Science+Business Media New York 2014

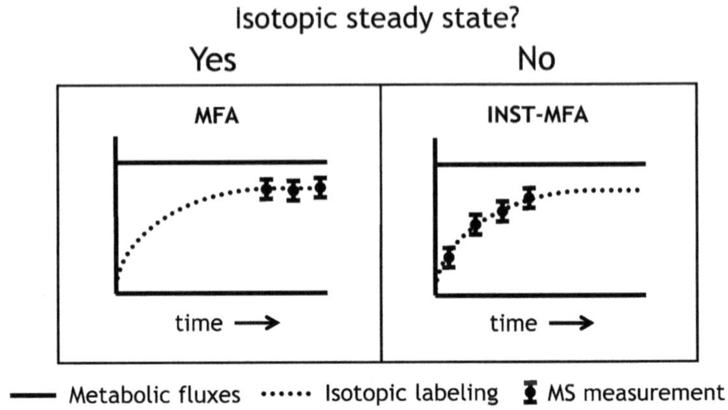

Fig. 1 Comparison between steady-state and nonstationary MFA methodologies. The relative speed of metabolic and isotopic dynamics will influence the type of MFA study performed. The *left panel* shows the conventional MFA approach under both metabolic and isotopic steady state. The *right panel* shows INST-MFA at metabolic steady state but not isotopic steady state

isotopically labeled substrate without perturbing the metabolic steady state, (3) allowing the system to establish a new isotopic steady state that is dictated by the underlying metabolic fluxes, and (4) measuring isotopic labeling in the fully equilibrated system as shown in Fig. 1. Depending on the relative speed of metabolic versus isotopic dynamics, however, other experimental scenarios can be envisioned. If labeling occurs slowly but metabolism is maintained in a fixed state, isotopically nonstationary MFA (INST-MFA; Fig. 1) can be applied to determine fluxes from transient isotope labeling measurements [3]. This requires isotopomer measurements to be collected at multiple time points during the pre-steady-state labeling period followed by least-squares fitting of dynamic isotopomer balance equations to identify best-fit estimates of flux parameters.

Although ^{13}C is the preferred isotope tracer for quantifying central carbon metabolism in heterotrophic systems, autotrophic organisms assimilate carbon solely from CO_2 and therefore produce a uniform steady-state ^{13}C-labeling pattern that is insensitive to fluxes (Fig. 2). This makes conventional steady-state ^{13}C-MFA ineffective for quantifying autotrophic metabolism [4]. However, transient measurements of isotope incorporation following a step change from natural CO_2 to $^{13}CO_2$ can be used to determine fluxes by application of INST-MFA. Furthermore, INST-MFA has the ability to quantify metabolite pool sizes based solely on their labeling dynamics [5, 6], thus providing a potential framework for integrating metabolomic datasets with MFA. Despite its advantages, however, the increased complexity of INST-MFA introduces

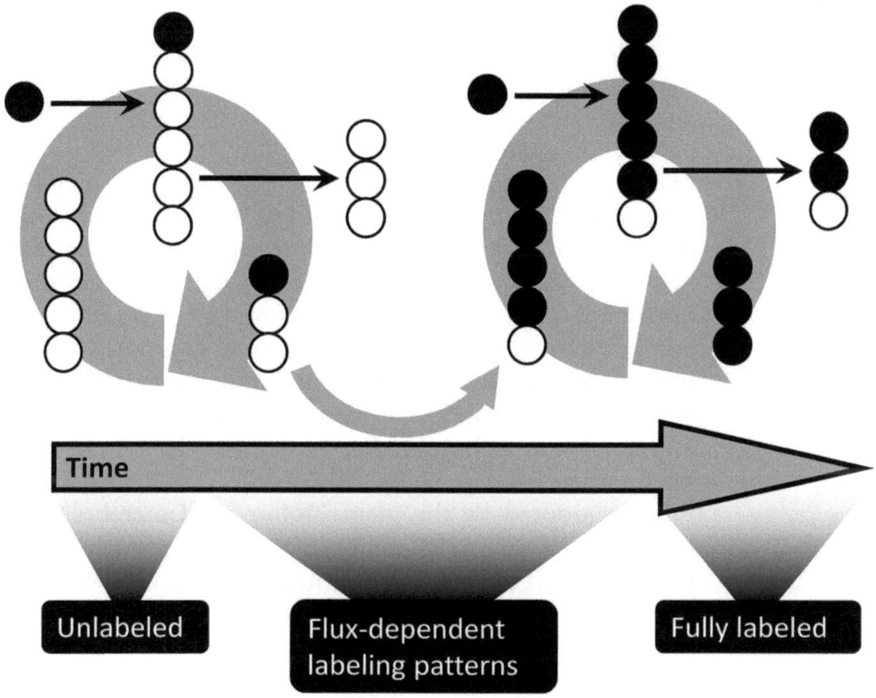

Fig. 2 Example of carbon labeling in an autotrophic system. Following the introduction of $^{13}CO_2$ to the Calvin cycle, intracellular metabolites become gradually labeled over time. Once steady-state labeling is achieved, all metabolites are uniformly ^{13}C enriched irrespective of fluxes and intracellular pool sizes. Labeling patterns observed during the isotopically transient period, however, can be computationally analyzed to determine fluxes

additional difficulties at both the computational and experimental levels that must be addressed. First, the solution of large-scale ordinary differential equation (ODE) models poses a substantial challenge to efficiently simulate transient isotope labeling experiments. The application of EMU decomposition to INST-MFA has greatly reduced this computational burden and has enabled determination of fluxes and accurate confidence intervals in biologically relevant networks [7, 8]. Second, the requirement for isotopically nonstationary measurements adds further complexity to experimental design, including selection of sampling time points and metabolite concentration measurements. Finally, rapid sampling and metabolite quenching must be applied in order to obtain meaningful isotopomer data from rapidly labeled intracellular metabolites. The field of metabolomics has witnessed considerable progress in this area, and some of these measurement techniques have already been successfully adapted for isotopomer studies in plants [9–12].

In this contribution, we present up-to-date protocols for performing INST-MFA on photosynthetic tissues under conditions of metabolic steady state, which is enabled by recent advancements in both experimental techniques for obtaining transient isotopomer measurements and computational tools for nonstationary flux analysis. Specifically, we describe experimental protocols for performing transient isotope labeling experiments, sample collection and quenching, extraction of intracellular metabolites, and mass spectrometry (MS) analysis of metabolite labeling, We also describe the important technique of nonaqueous fractionation (NAF), which can be used to obtain compartment-specific information on isotopic enrichment [13]. We then discuss the computational steps required to simultaneously integrate mass isotopomer distribution (MID) measurements from multiple pathways and compartments to reconstruct comprehensive flux maps of autotrophic metabolism. We anticipate that these methods will become part of a growing MFA toolbox, which has already been rapidly adopted by the plant biology and metabolic engineering communities due to its ability to quantify network-wide metabolic phenotypes in seeds and other non-photosynthetic tissues.

2 Materials

2.1 ^{13}C Labeling Experiment and Quenching

1. Plants at metabolic steady state (*see* **Note 1**).
2. Glove bag (Glas-Col, 17″ × 17″ × 11″) or other labeling chamber.
3. ^{13}C labeled CO_2 gas; $^{13}CO_2/N_2/O_2$ gas mix, ratio 0.033/78/21.967 (Sigma).
4. Liquid nitrogen.

2.2 CO_2 Consumption and Production Measurement

1. Li-COR 6400XT portable photosynthesis system (Lincoln, NE, USA).

2.3 Starch Production Measurement

1. 75 % (v/v) methanol.
2. α-amylase.
3. Amyloglucosidase.
4. Megazyme Kit (K-TSTA).
5. UV–Vis spectrophotometer.

2.4 Nonaqueous Fractionation

1. Tetrachloroethylene.
2. Heptane.
3. Centrifuge (capable of at least 13,000 × *g*).
4. Inert centrifuge tubes, preferably transparent.

2.5 Enzymatic Assays

2.5.1 Phosphoenolpyruvate Carboxylase Assay Buffer

1. 90 μL of 110 mM Tris sulfate adjusted to pH 8.5 at 25 °C with NaOH (all volumes are per sample).
2. 5 μL of 300 mM $MgSO_4{\cdot}7H_2O$.
3. 15 μL of 100 mM $NaHCO_3$.
4. 5 μL of 6 mM β-NADH.
5. 15 μL of dioxane.
6. 5 μL of 300 mM dithioerythritol (DTE).
7. 0.5 μL of 600 units/mL malic dehydrogenase (MDH).

2.5.2 Prephenate Aminotransferase Storage Buffer

1. 10 wt% glycerol.
2. 5 wt% of 1 M sodium phosphate, adjusted to pH 8.0 at 25 °C with NaOH.
3. 1 wt% of 10 mM pyridoxal 5-phosphate (PLP).

2.5.3 Prephenate Aminotransferase Assay Solution

1. 1.25 μL of 1 M sodium phosphate, adjusted to pH 8.0 at 25 °C with NaOH (all volumes are per sample).
2. 0.5 μL of 10 mM PLP.
3. 5 μL of 100 mM aspartic acid.
4. 2.5 μL of 10 mM prephenic acid.

2.5.4 Prephenate Aminotransferase Assay Derivatizing Solution

1. 5 mg Phthaldialdehyde (all volumes are per sample).
2. 900 μL of 0.5 M boric acid, adjusted to pH 10.2 at 20 °C with NaOH.
3. 125 μL Methanol.
4. 5 μL β-mercaptoethanol.

2.5.5 α-Mannosidase Assay

1. 50 mM sodium citrate buffer adjusted to a pH 4.5 at 25 °C.
2. 5 mM *p*-nitrophenol-α-D-mannoside.
3. 0.2 M boric buffer adjusted to pH 9.8 at 25 °C.
4. PD-10 protein desalting columns.
5. Microplate reader.
6. HPLC system equipped with a UV–Vis detector.

2.6 Metabolite Extraction

1. Cellulose acetate spin-X columns.
2. Methanol.
3. Chloroform.
4. DI water.
5. Bench-top centrifuge (capable of at least 8,000 × *g*).

2.7 Mass Spectrometry Analysis and Data Processing

1. GC-MS and/or LC-MS/MS.
2. Derivatization agents, vials, heating blocks, and nitrogen evaporator for GC-MS sample preparation (*see* **Note 2**).
3. Vials, columns, gases, buffers, solvents, and other consumables for GC-MS or LC-MS/MS.
4. Computer equipped with either (1) freeware MS analysis software (*see* **Note 3**) or (2) commercial software for searching and integrating mass spectra.
5. Mass spectral library for compound verification, such as the NIST/EPA/NIH Mass Spectral Database [14], Golm Metabolome Database [15], FiehnLib [16], Metlin [17], HMDB [18], or MassBank [19].

2.8 Isotopically Nonstationary Metabolic Flux Analysis

1. Computer equipped with research code or publically available software capable of performing INST-MFA, such as Isotopomer Network Compartmental Analysis (INCA; http://mfa.vueinnovations.com/), which runs through the computing environment of MATLAB.

3 Methods

3.1 ^{13}C Labeling Experiment and Quenching

The ^{13}C labeling experiment should be initialized once the plants are producing maximum biomass without production of reproductive organs (i.e., flowers and seeds) and approaching a metabolic (pseudo) steady state (*see* **Note 1**).

1. To maintain steady state metabolism, phototrophs should be isotopically labeled in the chamber near midday or 3–4 h after exposure to light.
2. Remove samples from the chamber at multiple time points (~5–15) prior to reaching isotopic steady state and quench immediately.
3. Initiate quenching by immediately pouring liquid nitrogen onto plant tissue in a way that minimizes shading.
4. Grind using a mortar and pestle.

3.2 CO_2 Consumption and Production Measurement

1. Measure leaf CO_2 assimilation in an incubator via a Li-COR 6400XT portable photosynthesis system and 2×3 cm standard clear top or red/blue light source leaf chamber with the following settings:
   (a) Photon flux density (PFD): 200 μmol/(m^2 s).
   (b) Air temperature: 24 ± 1 °C.
   (c) Reference CO_2 concentration: ambient level.

3.3 Starch Production Measurement

1. Collect approximately 100 mg of fresh tissue.
2. Add tissue to 1.4 mL of 75 % (v/v) methanol at 70 °C for 30 min.
3. Add 750 μL chloroform (−20C) and 1500 μL DI water (4C) and vortex
4. Spin down the sample to fraction the nonpolar metabolites into the chloroform.

   Remove the methanol/water phase for further analysis of water-soluble metabolites.
5. Re-extract the remaining residue with 1.5 mL 80 % (v/v) ethanol at 80 °C three times to remove all the remaining water-soluble metabolites.
6. Dry down the remaining residue.
7. Digest the pellet with α-amylase and amyloglucosidase to hydrolyze starch into maltodextrins and then into glucose.
8. Determine the starch concentration spectrophotometrically as glucose equivalents using the Megazyme Kit.

3.4 Nonaqueous Fractionation

NAF is an optional, yet invaluable, technique for separating metabolites from different subcellular compartments in an ordered manner so that they can be characterized and their relative quantities at each locus can be identified. This procedure, first outlined for plants by Gerhardt and Heldt [20], begins with the rapid freezing and lyophilizing of the tissue. This arrests metabolism, and the removal of water prevents polar metabolites from being able to diffuse—instead they remain with the membrane of the compartments in which they were localized while all enzymes are retained compartmentally but inactive. At this point the cellular material can be stratified by compartment using a nonaqueous density gradient. This separation is quantified by extracting the enzymes of each compartment and analyzing their relative activity in each fraction of the density gradient. Using information on how the compartments have been separated, coupled with a thorough analysis of the metabolite profile of each fraction of the density gradient, the metabolic profiles of each compartment can be deconvoluted and the resulting data employed for further analysis (*see* **Note 5**).

1. Grind the leaf tissue. It is crucial that frequency and time be optimized to get small particles.
2. Filter the leaf powder by size before suspension in the loading fraction of the gradient.
3. Prepare the steps in the gradient by mixing together various ratios of tetrachloroethylene and heptanes. The NAF gradient is prepared to achieve the maximum separation of biomass (*see* **Note 6**).

4. Carefully layer the steps in the gradient on top of each other in the centrifuge tube, allowing only minimal mixing of the layers. A loading layer and density well are prepared for the top and bottom layers, respectively.
5. Place the loading layer with the sample suspended in it on top of the gradient, once it is created.
6. Centrifuge the tube for 2 h at 13,000 × *g* to ensure full separation of the cellular fragments throughout the gradient.
7. Remove the tube after centrifugation and draw out the sample in fractions. The sample is drawn from the bottom of the tube using a syringe of glass or other nonreactive material. Fraction volumes are selected to help normalize the amount of biomass in each fraction. Generally, six or seven fractions are drawn.
8. While still suspended, further separate the fractions in order to accommodate the number of assays that will be performed (i.e., if three compartmental markers will be analyzed and each requires suspension of cellular material in a separate buffer, then the fraction will be separated into four parts—three for the different assays and one for metabolite analysis).
9. Dry down the fractions under a stream of N_2 to remove the nonaqueous solvent before subsequent analysis.

3.5 Enzymatic Assays

Enzymatic assays are used to determine the distribution of each subcellular compartment throughout the fractions. Because the cellular fragments frequently contain parts of multiple compartments, this information is necessary to deconvolute the labeling patterns observed in the metabolite profiles of the fractions. The enzymes assayed are selected because they are easily and reliably quantifiable and they are known to be present in only one compartment. The assays outlined here are for the determination of cytosolic (phospho*enol*pyruvate carboxylase), plastidial (prephenate aminotransferase), and vacuolar (α-mannosidase) distribution.

Mitochondrial activity has been determined by others using the method of Bergmeyer [21], although, due to the large standard deviation of the assay, the compartment has not been unambiguously delineated for any plant species [22]. Activity measurements for each protein assay are normalized by their protein content. Protein content is measured using a Pierce BCA protein assay kit from Thermo Fisher Scientific (Rockford, IL).

3.5.1 Phosphoenolpyruvate Carboxylase Assay

1. Resuspend samples in 0.5 mL of Tris sulfate buffer and allow to incubate for 30–60 min at 4 °C to effectively extract the protein.
2. Spin down the samples at 17,000 × *g* for 30 s.
3. Aliquot 10 μL of supernatant from each sample into the wells of a 96-well plate.

4. Following the procedure prescribed by Sigma-Aldrich [23], add to the extract phospho*enol*pyruvate carboxylase (PEPc) assay buffer, as described in Subheading 2.5.
5. Allow the wells to equilibrate at 25 °C for 5 min.
6. Add 5 μL of 30 mM PEP solution to each well, except for the last one, which will serve as a blank. Mix thoroughly for 1 min.
7. Measure the OD_{340} of each well periodically over the next 10 min to determine relative enzyme activity.

3.5.2 Prephenate Aminotransferase Assay

A modified form of the method presented by Maeda et al. [24] is used to measure prephenate aminotransferase (PAT) activity.

1. Resuspend the samples in 0.4 mL of PAT storage buffer. Allow the samples to incubate for 30–60 min at 4 °C to effectively extract the protein.
2. Spin down at 18,000 × *g* for 30 s.
3. Load the supernatant onto PD-10 columns, which have been prewashed with the PAT storage buffer solution.
4. Elute fractions from the PD-10 columns 0.25 mL at a time.
5. Measure the OD_{280} of each PD-10 column fraction to determine the relative protein content. The set of subfractions with the highest absorbance will be used in the PAT assay.
6. Perform the assay by adding PAT assay buffer to 15 μL of the desalted enzyme solution.
7. Incubate the samples at 37 °C for 30–60 min.
8. Quench the samples with 50 μL of methanol and store at −20 °C overnight to precipitate the protein.
9. Centrifuge the samples at 18,000 × *g* for 5 min to remove all precipitated protein. The supernatant is analyzed by high-performance liquid chromatography (HPLC).
10. Derivatize the samples immediately prior to injection using a PAT assay derivatizing solution.
11. 10 μL samples are derivatized with 10 μL of derivatizing agent in the injection needle for 2.5 min prior to sample separation. This derivatizing solution should be changed every 12 h.
12. Run sample on an Agilent Eclipse XDB-C8 column (4.6 × 150 mm) maintained at 35 °C. The mobile phase is a ratio of 0.1 % ammonium acetate:methanol that varies according to Table 1 with a constant flow rate of 0.4 mL/min. Derivative concentrations were measured using a UV/Vis detector observing absorbance at 336 nm with a reference of 390 nm. Figure 3 shows an example of the chromatogram.

3.5.3 α-Mannosidase Assay

1. Extract protein samples in 50 mM sodium citrate buffer.
2. Allow samples to incubate for 30–60 min at 4 °C to effectively extract the protein.

Table 1
HPLC mobile-phase gradient for PAT assay

Time (min)	% MeOH
0	15
15	65
25	65
27	15
32	15

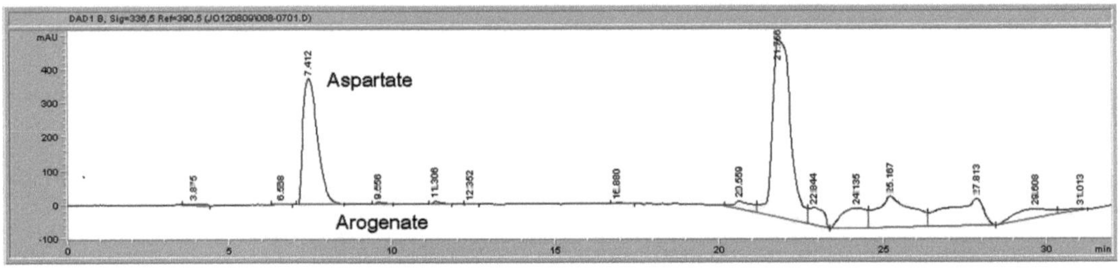

Fig. 3 Sample HPLC chromatogram of the PAT assay using phthaldialdehyde as a derivatizing agent

3. Following the method of Li and Primate [25], add the extracted enzyme to 5 mM *p*-nitrophenol-α-D-mannoside solution at 25 °C in a 10:1 ratio.
4. Terminate the reaction after 10 min with a volume of 0.2 M borate buffer at pH 9.8 equal to twice the reaction mixture.
5. Measure the OD_{400} of each sample to determine enzyme activity.

3.6 Metabolite Extraction

1. Weigh 200 mg tissue powder (kept in liquid N_2) to a plastic 2 mL tube.
2. Add 3 mL 7:3 methanol–chloroform (−20 °C); after vigorous shaking, the mixture is incubated at −20 °C for 2 h with occasional shaking.
3. Water-soluble metabolites are extracted from the chloroform phase by adding 2.4 mL H_2O; after vigorous shaking, centrifuge for 5 min at 420 × *g*.
4. Remove the upper, aqueous methanol–H_2O phase to a glass reaction tube and repeat **step 3**; mix the second upper phase with the first one; the final volume of the extract should be about 6.5 mL.
5. Dry the extract with N_2 at room temperature.

6. Redissolve the dry extract with 200 μL H_2O; remove the viscous, high-molecular-mass components from the samples by applying extracts to 0.45 μm cellulose acetate centrifuge tube filters.

3.7 Mass Spectrometry Analysis

Divide the extracted sample from each fraction into three different vials. One vial will be used for LC-MS/MS analysis and the remaining two vials will be dried down the rest of the way for amino acid and organic acid analyses using GC-MS.

3.7.1 GC-MS Analysis of Amino Acids

GC-MS analysis of amino acids is most readily achieved following conversion to *tert*-butyldimethylsilyl (TBDMS) derivatives [26]. TBDMS derivatives are approximately 1,000 times more stable than trimethylsilyl (TMS) derivatives and produce characteristic $[M\text{-}57]^+$, $[M\text{-}85]^+$, and $[M\text{-}157]^+$ fragment ions that facilitate identification. Ion masses and relative retention times are detailed by Wittman et al. [27].

1. Dissolve dried sample in 100 μL anhydrous pyridine in a fume hood.
2. Add 70 μL MTBSTFA + 1 % TBDMCS.
3. Sonicate for 5 min.
4. Incubate for 2 h at 45 °C on dry block heater.
5. Transfer supernatant to injection vial.
6. Run sample on single quadrupole GC-MS with the following settings:
   (a) Injection volume: 1 μL.
   (b) Split injection mode, 10:1 ratio (*see* **Note 7**).
   (c) Column: DB-5MS (or similar), 30 m × 0.25 mm ID, 0.25 μm film.
   (d) Column flow: He at 1 mL/min (*see* **Note 8**).
   (e) Inlet temp: 270 °C.
   (f) Interface temp: 300 °C.
   (g) Temp profile: 135 °C for 3 min, ramp at 3 °C/min to 280 °C, hold at 280 °C for 2 min (53.3-min run time).
   (h) MS settings: Electron impact (EI) ionization, full-scan detection (100–500 *m/z*), 5-min solvent delay.

3.7.2 GC-MS Analysis of Organic Acids

Organic acids can be converted to their TBDMS derivatives and analyzed in much the same way as amino acids. However, it is recommended to first convert all ketone and aldehyde functional groups to their associated methyloxime derivatives. This prevents keto-enol tautomerization, which would otherwise result in multiple TBDMS derivatives. Ion masses are detailed by Roessner et al. [28].

1. Dissolve dried sample in 100 μL pyridine with 20 mg/mL methoxyamine hydrochloride MOX reagent in a fume hood.
2. Sonicate for 5 min.
3. Incubate for 90 min at 30 °C on dry block heater.
4. Add 100 μL MTBSTFA + 1 % TBDMCS.
5. Incubate for 60 min at 37 °C on dry block heater.
6. Remove from heating block and incubate overnight at room temperature (*see* **Note 9**).
7. Centrifuge for 5 min at 18,000 × *g* to remove solid debris.
8. Transfer liquid to injection vial.
9. Run sample on single quadrupole GC-MS with the following settings:
   (a) Injection volume: 1 μL.
   (b) Purged splitless mode, set to activate at 1 min (*see* **Note 7**).
   (c) Column: DB-5MS (or similar), 30 m × 0.25 mm ID, 0.25 μm film.
   (d) Column flow: He at 1 mL/min (*see* **Note 8**).
   (e) Inlet temp: 270 °C.
   (f) Interface temp: 300 °C.
   (g) Temp profile: 70 °C for 3 min, ramp at 4 °C/min to 280 °C, hold at 280 °C for 2 min (75-min run time).
   (h) MS settings: Scan mode (100–550 *m/z*), 5-min solvent delay.

3.7.3 LC-MS/MS Analysis of Sugar Phosphates

Although analysis of labeling in macromolecule components such as protein, starch, and lipids has dominated the MFA literature due to their high abundance in tissues, macromolecules label too slowly for INST-MFA and recent advances in MS technology have facilitated direct analysis of free intracellular metabolites. This approach enables quantification of MIDs in sugar phosphate intermediates that participate directly in glycolysis and Calvin cycle reactions, providing comprehensive and dynamic information on flux through these important pathways. However, for intermediates found in multiple subcellular compartments (e.g., glucose-6-phosphate in both the cytosol and chloroplast), this will only provide the *average* labeling among the various intracellular pools, unless NAF is applied to generate enriched subfractions first.

The vast majority of sugar phosphate analysis has been performed using LC-MS/MS to avoid thermal degradation of these nonvolatile analytes [29]. The LC-MS/MS conditions provided are based on the method developed by Luo et al. [30], which has been subsequently modified and applied to analyze labeling in cyanobacteria extracts by Shastri et al. [31].

Table 2
LC-MS/MS ion transitions and method parameters

Analyte	Retention time (min)	MS1	MS2	DP	EP	CE	CXP	CUR	IS	TEM	GS1	GS2	CAD
Glycolate	6.2	75	43	−52	−10	−15	−7	40	−4,500	550	85	90	High
G6P/F6P	12.4/13.3	259	97	−50	−2	−23	−10	40	−4,500	550	85	90	High
GAP/DHAP	12.2/16.9	169	97	−50	−3	−12	−8.5						
S7P	12.9	289	97	−26	−11	−14	−4.5						
R5P/X5P/RU5P	12.7/14.2/14.5	229	97	−79	−6	−17	−15						
PYR	15.7	87	43	−33	−10	−13	−7						
SUC	22.8	117	73	−50	−3	−15	−2	30	−4,500	550	90	90	9
MAL	24.4	133	73	−50	−9	−23	−11						
αKG	25.5	145	101	−43	−5	−11	−9.6						
FUM	26.4	115	71	−37	−4	−12	−11						
3PGA	26.5	185	79	−42	−5	−46	−11						
FBP	27	339	97	−66	−8	−26	−11						
CIT/ICT	27.4	191	75	−60	−10	−25	−10						
RUBP	27.6	309	97	−68	−6	−28	−12						
PEP	27.8	167	79	−55	−5	−21	−7						

Labeling of selected metabolites measured by monitoring transitions from selected precursor ions (MS1) to daughter ions (MS2) in multiple reaction monitoring (MRM) mode. Compound specific MS parameters are abbreviated as follows: declustering potential (DP), entrance potential (EP), collision energy (CE), collision exit potential (CXP), curtain gas (CUR), ion spray velocity (IS), temperature (TEM), ion source gas 1 (GS1), ion source gas 2 (GS2), and collision (CAD). List of analyte abbreviations can be found in Subheading 4 (*see* **Note 15**)

1. Transfer sample to LC injection vial.
2. Run sample on linear ion-trap triple quadrupole LC-MS/MS with the following settings:
   (a) Injection volume: 10 μL.
   (b) Column: Phenomenex Synergi Hydro-RP (or similar), 150 mm × 2.1 mm ID, 4 μm particle size.
   (c) Column flow: 0.3 mL/min.
   (d) Column temp: 25 °C.
   (e) Eluent A: Solution of 10 mM tributylamine + 15 mM acetic acid (*see* **Note 10**).
   (f) Eluent B: HPLC-grade methanol.
   (g) Gradient profile: 0 % B (0 min), 8 % B (8 min), 22 % B (18 min), 40 % B (28 min), 60 % B (32 min), 90 % B (34 min), 90 % B (37 min), 0 % B (39 min), 0 % B (49 min).
   (h) MS settings: Negative-mode electrospray ionization (ESI), MRM mode (*see* Table 2 and **Note 11**).

3.8 MS Data Processing

Analysis of MS data requires (1) identification of chromatographic peaks and fragment ions associated with target analytes of interest, (2) integration of ion chromatograms over time to quantify relative abundance of specific isotope peaks, and (3) assessment of measurement standard errors.

1. Identify the chromatographic peaks associated with the analytes of interest based on both the retention time (RT) and the MS fingerprint of the peak.
2. Identify ions to be used for mass isotopomer analysis and determine their molecular composition. The best GC-EI-MS fragment ions are highly abundant ions with masses greater than 150 Da, since these are less likely to be contaminated by interfering fragment ions of similar mass.
3. Integrate the mass isotopomer peaks using either custom or commercial software.
4. Correct MIDs for natural isotope abundance (optional). The method of Fernandez et al. [32] can be applied to perform the correction.
5. Calculate the mean and standard error of MIDs for each metabolite at each time point.

3.9 Isotopically Nonstationary Metabolic Flux Analysis (INST-MFA)

A flow chart of a typical INST-MFA process is shown in Fig. 4. INST-MFA is concerned with solving an "inverse problem" where fluxes and pool sizes are estimated from measured labeling patterns and extracellular rates through the means of an iterative least-squares fitting procedure. At each iteration, a "forward problem" is solved where an isotopomer model is used to simulate labeling measurements for a given metabolic network and a given set of parameter estimates. The discrepancy between the simulated and measured labeling patterns is then assessed, and the parameter estimates are updated to achieve an improving fit. Once convergence to the best-fit solution is obtained, the procedure terminates and the optimal flux and pool size estimates are returned.

1. *Build an isotopomer model for INST-MFA*. In order to perform INST-MFA, it is necessary to reconstruct a metabolic network from biochemical literature and the annotated genome of the organism of interest. This network must prescribe both (1) the stoichiometry of all enzymatic reactions under consideration and (2) atom transitions for each reaction (*see* **Note 12**). Reactions must also be classified as either reversible or irreversible.
2. *Construct the stoichiometric matrix S*. Fluxes are required to satisfy the stoichiometric constraint

$$S \cdot v = 0 \tag{1}$$

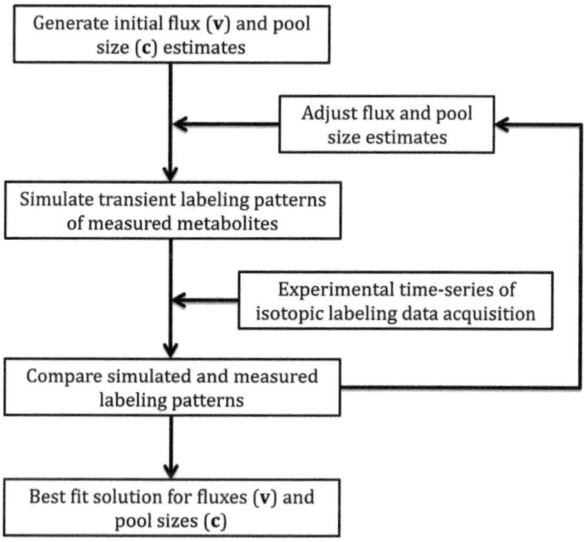

Fig. 4 Flow chart showing the overall schematic of ^{13}C INST-MFA. Following the labeling experiment and MS analysis of the measured metabolites, computational analysis of the dynamic changes in isotope labeling patterns can be used to estimate metabolic pathway fluxes and pool sizes. This involves solving an inverse problem whereby the vectors of flux (**v**) and pool size (**c**) parameters are iteratively adjusted until the mismatch between simulated and experimentally measured data sets is minimized

where S is the stoichiometric matrix and v is the flux vector. In Eq. 1, the stoichiometric matrix S is a $k \times j$ matrix, with k metabolites and j fluxes. Reversible reactions should be modeled as separate forward and backward fluxes so that all fluxes are nonnegative. Appendix shows a simple network model example, which illustrates the process of setting up the stoichiometric matrix as well as subsequent steps discussed below.

3. *Identify the free fluxes of the network and the null space matrix.* From a computational standpoint, it is more convenient to work with "free" fluxes rather than the "true" network fluxes [33, 34]. Free fluxes can be obtained from the general solution to Eq. 1:

$$v = N \cdot u \tag{2}$$

where N is the null space matrix of S and u is the vector of free fluxes. There are many methods to calculate a valid null space matrix, and generally there is not a unique null space matrix for any given stoichiometric matrix [35]. The size of the null space matrix and the number of independent flux variables are determined by the rank of the stoichiometric matrix. With $r = \text{rank}(S) \leq k$, the null space matrix is a $j \times j - r$ matrix and the number of free fluxes is $j - r$.

4. *Identify the minimal set of EMUs required to simulate the MIDs of the measured metabolites.* In INST-MFA, the isotopomer balances are described by a system of ordinary differential equations, which is significantly more difficult to solve than the algebraic systems that describe steady-state labeling. Due to this additional difficulty, algorithms for solving the forward problem of INST-MFA need to be carefully designed so that computational expense does not become prohibitive. The most efficient approach involves first decomposing the isotopomer network into elementary metabolite units (EMUs) [7, 8]. By only solving for the isotopomer distributions of EMUs that contribute to the available measurements, this approach minimizes the number of ODEs that need to be integrated and thereby enables the forward problem to be solved thousands of times faster than previous methods. This, in turn, increases the efficiency of solving the inverse problem of INST-MFA because each iteration of the parameter estimation procedure can be completed in minimal time.

   An EMU is defined as a distinct subset of a metabolite's atoms and can exist in a variety of mass states depending on its isotopic composition. In its lowest mass state, an EMU is referred to as M0, while an EMU that contains one additional atomic mass unit (e.g., as a result of a ^{13}C atom in place of ^{12}C atom) is referred to as M1, with higher mass states described accordingly. An MID is a vector that contains the fractional abundance of each mass state of an EMU. To solve the forward problem of simulating metabolite labeling in INST-MFA, the isotopomer network is first systematically searched to enumerate all EMUs that contribute to measurable MS fragment ions [8]. The main advantage of the EMU decomposition is that metabolites are never broken into smaller pieces than is strictly required to describe the labeling state of the selected metabolites (*see* Appendix).

5. *Set up EMU balances and simulate the labeling distribution.* The EMU reactions identified from network decomposition form the new basis for generating system equations. In INST-MFA, these EMUs are grouped into mutually dependent blocks using a Dulmage–Mendelsohn decomposition [36, 37] (*see* **Note 13**). Therefore, by definition, all EMUs within a particular block have the same number of atoms and must be solved simultaneously and not sequentially. The decoupled blocks can be arranged into a cascaded system of ODEs with the following form:

$$\mathbf{C}_n \cdot \frac{\mathrm{d}\mathbf{X}_n}{\mathrm{d}t} = \mathbf{A}_n \cdot \mathbf{X}_n + \mathbf{B}_n \cdot \mathbf{Y}_n \tag{3}$$

   Level n of the cascade represents the network of EMUs within the nth block. The rows of the state matrix $\mathbf{X}_n$ correspond to MIDs of EMUs within the nth block. The input matrix $\mathbf{Y}_n$ is analogous but with rows that are MIDs of EMUs that are previously calculated inputs to the *n*th block (or MIDs of source EMUs that are unbalanced). The concentration matrix $\mathbf{C}_n$ is a diagonal matrix whose elements are pool sizes corresponding to EMUs represented in $\mathbf{X}_n$. The system matrices $\mathbf{A}_n$ and $\mathbf{B}_n$ describe the network as follows:

$$\mathbf{A}_n(i,j) = \begin{cases} -\text{sum of fluxes consuming } i\text{th EMU in } \mathbf{X}_n\ i=j \\ \text{flux to } i\text{th EMU in } \mathbf{X}_\text{n} \text{ from } j\text{th EMU in } \mathbf{X}_n\ i \neq j \end{cases} \quad (4)$$

$$\mathbf{B}_n(i,j) = \{\text{flux to } i\text{th EMU in } \mathbf{X}_n \text{ from } j\text{th EMU in } \mathbf{Y}_n \quad (5)$$

The system matrices $\mathbf{A}_n$ and $\mathbf{B}_n$ can be evaluated directly once the "true" flux vector (v) has been determined from the free fluxes (u) and the null space matrix (N) (*see* Appendix).

6. *Optimize parameters.* Fluxes and pool sizes are estimated by minimizing the difference between measured and simulated data according to the following equation [5, 8]:

$$\begin{aligned} \min_{u,c}\phi &= [m(u,c,t) - \widehat{m}(t)]^T \cdot \textstyle\sum_{\mathbf{m}}^{-1} \cdot [m(u,c,t) - \widehat{m}(t)] \\ &\text{s.t. } N \cdot u \geq 0, c \geq 0 \end{aligned} \quad (6)$$

   where ϕ is the objective function to be minimized, u is a vector of free fluxes, c is a vector of metabolite concentrations, t is time, $m(u,c,t)$ is a vector of simulated measurements, $\widehat{m}(t)$ is a vector of observed measurements, $\sum_{\mathbf{m}}$ is the measurement covariance matrix, and N is the null space of the stoichiometric matrix. A reduced gradient method can be implemented to handle the linear constraints of this problem within a Levenberg–Marquardt nonlinear least-squares solver [38, 39]. Alternatively, gradient-free optimization approaches have been applied by Noh et al. [5]. **Steps** 7 and **8** do not need to be solved when using a gradient-free method.

7. *Differentiate the isotopomer balances with respect to each free flux to generate the sensitivity equations. Solve the sensitivity equations to calculate the first derivatives of the measured MIDs with respect to the free parameters.* Estimation of both the unknown fluxes and pool sizes using INST-MFA is accomplished by finding a best-fit solution to the inverse problem. Efficient solution of this problem typically relies on optimization algorithms that choose the search direction based on the gradient of the least-squares objective function (*see* Eq. 6) with respect to all adjustable parameters. The most accurate and least expensive way to obtain the required gradient information is to integrate a system of sensitivity equations whose solution describes how the calculated MIDs vary in response to changes in the model parameters. Implicit differentiation of Eq. 3 yields the following sensitivity equation:

$$\frac{\mathrm{d}}{\mathrm{d}t}\frac{\partial \mathbf{X}_n}{\partial p} = \mathbf{C}_n^{-1} \cdot \mathbf{A}_n \cdot \frac{\partial \mathbf{X}_n}{\partial p} + \frac{\partial(\mathbf{C}_n^{-1} \cdot \mathbf{A}_n)}{\partial p} \cdot \mathbf{X}_n + \mathbf{C}_n^{-1} \cdot \mathbf{B}_n \cdot \frac{\partial \mathbf{Y}_n}{\partial p} + \frac{\partial(\mathbf{C}_n^{-1} \cdot \mathbf{B}_n)}{\partial p} \cdot \mathbf{Y}_n \quad (7)$$

where p is the vector of adjustable flux and pool size parameters. This system of equations can be solved in tandem with those of Eq. 3, and the time-dependent sensitivities can be used to evaluate the objective function gradient during each iteration of the INST-MFA inverse problem.

8. *Calculate the Hessian and gradient of the objective function.* Differentiation of the objective function with respect to the flux and pool size parameters will yield equations for the Hessian matrix **H** and the gradient vector **J**:

$$\mathbf{J} = \left(\frac{\partial m}{\partial p}\right)^{\mathrm{T}} \cdot \sum\nolimits_m^{-1} \cdot (m - \widehat{m}) \quad (8)$$

$$\mathbf{H} = \left(\frac{\partial m}{\partial p}\right)^{\mathrm{T}} \cdot \sum\nolimits_m^{-1} \cdot \left(\frac{\partial m}{\partial p}\right) \quad (9)$$

where $(\partial m/\partial p)$ is the matrix of sensitivities of the simulated measurements with respect to the parameters. Both the Hessian and gradient will be evaluated at each iteration if a gradient-based optimization algorithm is employed.

9. *Evaluate the goodness of fit.* Flux and pool size estimation is initiated with random values for all fluxes and pool sizes. The estimation algorithm will continue until no further improvements are made to the sum of squared residuals (SSRES) of the objective function. However, the fact that INST-MFA yields a set of fluxes that minimize the difference between the observed and simulated measurements does not mean that the flux model is adequate. Testing the goodness of fit will determine whether the optimal solution is statistically acceptable based on the minimized SSRES. At convergence, the minimized variance-weighted SSRES is a stochastic variable drawn from a chi-square distribution with n–p degrees of freedom (DOF), where n is the number of independent measurements and p is the number of estimated parameters. The SSRES that is calculated should therefore be in the interval $\left[\chi^2_{\alpha/2}, \chi^2_{1-\alpha/2}\right]$, where α is a chosen threshold value corresponding to the desired confidence level (e.g., 0.05 for 95 % confidence or 0.01 for 99 % confidence). The model fit is accepted when the SSRES falls within the limits of the expected chi-square range [40]. Additionally, the distribution of residuals should be assessed for normality. The standard deviation-weighted residuals should be normally distributed with a mean of zero and standard deviation of one.

One approach that can be used to evaluate the hypothesis that the residuals are normally distributed is the Lilliefors test [41]. Various plots can also be constructed to assess normality of the residuals.

10. *Calculate the 95 % confidence intervals using either continuation methods or Monte Carlo analysis.* Once an optimal solution has been obtained, nonlinear confidence intervals on the fitted parameters should be computed using robust, global methods instead of relying solely upon local standard errors. The local standard errors can be easily obtained from the parameter covariance matrix at the optimal solution; however, they do not accurately reflect changing sensitivities at points removed from the optimal solution. Parameter continuation can be performed to calculate accurate upper and lower bounds on the 95 % confidence interval for each flux or pool size parameter [40]. This determines the sensitivity of the minimized SSE to varying a single parameter away from its optimal value while allowing the remaining parameters to adjust in order to minimize $\Delta\phi$. Large confidence intervals indicate that the flux cannot be estimated precisely. On the other hand, small confidence intervals indicate that the flux is well determined. Monte Carlo simulation can also be used to calculate the 95 % confidence intervals. This method is typically more expensive than the parameter continuation approach but is expected to yield similar results.
11. *Report the flux values and flux uncertainties.* Once an acceptable fit to the experimental measurements has been achieved and confidence intervals have been computed for all parameters, the results are best summarized visually in the form of a flux map. Figure 5 shows an example plant flux map under photoautotrophic growth conditions using INST-MFA, involving multiple subcellular compartments. Several software tools have been recently developed, which aid in the construction of these maps (*see* **Note 14**).

4 Notes

1. For labeling experiments involving *Arabidopsis thaliana*, it is suggested that 4-week-old plants be used. This is because 4-week-old plants at metabolic steady state provide the maximum biomass, RuBisCO content, and chlorophyll content without the production of reproductive organs (i.e., flowers and seeds). Thus, the 4-week-old plants are active and vegetative. Additionally, metabolic steady state can be determined by measuring the sucrose and starch production of leaves; if the production of sucrose and starch remains consistent, then metabolic steady state can be established.

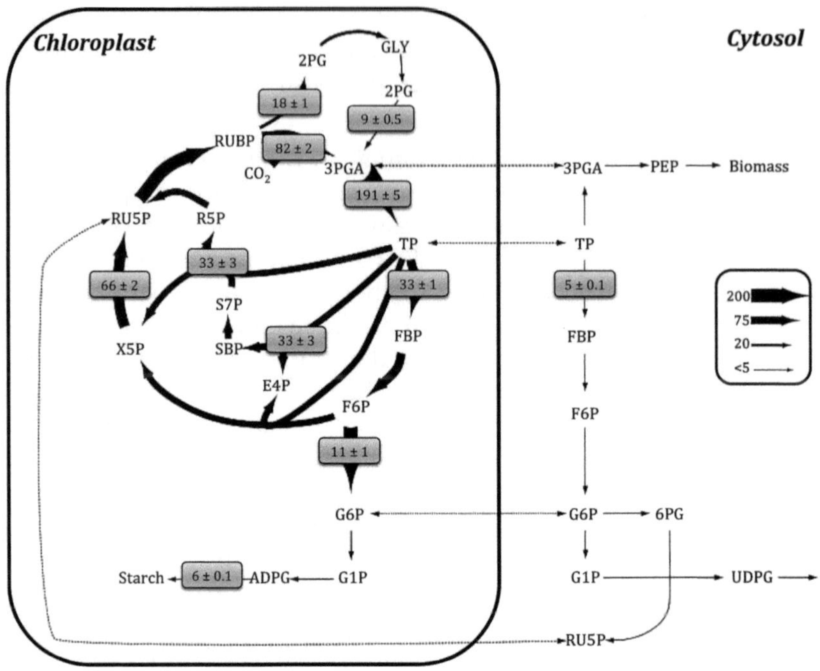

Fig. 5 Example flux map for a plant INST-MFA study under photoautotrophic growth conditions. This flux map shows a hypothetical flux map associated with a plant INST-MFA study involving multiple subcellular compartments. Values are represented as $M \pm$ SE, where M is the median of the flux confidence interval and SE is the estimated standard error of M. *Arrow* thickness is scaled proportional to net flux. *Dotted arrows* indicate transport reactions between compartments

2. Derivatization agents such as methoxyamine (MOX), trimethylsiloxane (TMS), or *tert*-butyl dimethylsiloxane (TBDMS) are typical for GC-MS analysis. The MOX reaction protects ketone and aldehyde functional groups and thereby prevents the formation of multiple TMS or TBDMS derivatives. This step is unnecessary if no ketone or aldehyde functional groups are present in the analytes of interest. TMS and TBDMS derivatives produce several characteristic fragment ions that facilitate identification [42]. Huege et al. [10] provide a list of some metabolites derivatized with TMS used for isotopomer analysis via GC-MS analysis. Ahn and Antoniewicz [43] provide a similar list for TBDMS-derivatized metabolites for GC-MS analysis.

3. Available GC-MS freeware include AMDIS (http://chemdata.nist.gov/mass-spc/amdis/) and Wsearch32 (http://www.wsearch.com.au/wsearch32/wsearch32.htm). Two popular freeware programs for LC-MS/MS data analysis are MZmine and XCMS, the latter of which runs in the R statistical programming environment. Both programs require the user to convert raw data files into a nonproprietary format such as mzXML, NetCDF, or mzData. Conversion to mzXML format

can be accomplished using one of several instrument-specific software tools developed and maintained by the Seattle Proteome Center (http://tools.proteomecenter.org/software.php).

4. Samples should be collected more frequently near the beginning of the tracer experiment, as the isotopic labeling will be changing most rapidly during this initial time period. For example, Wiechert et al. [5] have recommended using an approach where the length of each time interval increases exponentially (e.g., 1, 2, 4, 8, 16) following an initial period where uniformly spaced samples are collected at the maximum rate.

5. The results from the fractional labeling measurements are not immediately employable in their acquired form but must be deconvoluted to determine the actual MIDs for the metabolite pools in each compartment. To do this we first describe the contribution of a metabolite to each fraction. This is a function of the metabolite pool size and the presence of the compartment in the fraction, as described by Eq. 10:

$$F_{i,j} = A_i \cdot E_{i,j} \tag{10}$$

where $F_{i,j}$ is the contribution of metabolite A of compartment i to fraction j, A_i is the pool size of the metabolite in compartment i, and $E_{i,j}$ is the relative compartmental activity of compartment i in fraction j. The fractional contribution can then be normalized such that

$$B_{i,j} = \frac{A_i \cdot E_{i,j}}{\sum_i A_i \cdot E_{i,j}} \tag{11}$$

Using this relationship we can write calculated fractional MIDs as a function of $B_{i,j}$ and the actual compartmental MIDs:

$$\mathrm{MID}_{\mathrm{calc},j} = \sum_i B_{i,j} \cdot \mathrm{MID}_i \tag{12}$$

This problem can then be solved for all MID_i and A_i simultaneously using a nonlinear least-squares fitting to minimize the difference between the calculated fractional MIDs for all fractions and those observed experimentally, and confidence intervals can be obtained accordingly.

6. For *A. thaliana*, the NAF gradient range to achieve maximum separation of biomass is 1.43–1.62 g/cm^3 [22]. For soybeans, this range has been found to be 1.28–1.59 g/cm^3 [44].

7. The split ratio (in split mode) or purge activation time (in splitless mode) can be varied to achieve total ion counts in the desired range. Split mode is appropriate for concentrated samples while splitless mode is most appropriate for dilute samples. Typical split ratios vary from 1:5 to 1:100, with 1:10 representing a good initial value. Typical purge times range from 0.5 to 2 min, with 1 min representing a good initial value. Note that GC inlet liners are specific for either split or splitless operation and are not

interchangeable. Also note that when using splitless mode, the initial column temperature should be near or below the boiling point of the solvent in which the sample is dissolved [42].

8. The optimal linear velocity is in the range 20–40 cm/s for helium. Agilent's FlowCalc tool or the instrument control software can help determine what the linear velocity will be for a particular combination of flow rate, column diameter, temperature, and pressure.
9. This step may be necessary for dilute samples in order to ensure complete conversion. However, it may be skipped for concentrated samples.
10. In order to dissolve tributylamine completely in water, the tributylamine and acetic acid are first mixed together in a dry flask before the requisite amount of ultrapure water is added. The solution is then filtered through a 0.45 μm membrane prior to use. The final pH should be 4.5–5.
11. To increase sensitivity, the method should be divided into multiple time segments with different MRM transitions scanned in each interval. Dwell time of each transition should be optimized such that the total cycle time in each time segment does not exceed 2 s. This will provide at least 10–15 scans of each chromatographic peak as it elutes from the column.
12. Networks used for MFA typically include glycolysis, pentose phosphate pathway, amino acid metabolism, TCA cycle, and various amphibolic pathways that interact with the TCA cycle. This backbone of central metabolic pathways may be further augmented by additional reactions of interest to a particular investigation. Some helpful online public libraries include Kyoto Encyclopedia of Genes and Genomes (KEGG; http://www.genome.jp/kegg/), BioCyc (http://biocyc.org/), metaTIGER (http://www.bioinformatics.leeds.ac.uk/metatiger/), ENZYME (http://enzyme.expasy.org/), and BRENDA (http://www.brenda-enzymes.info/).
13. Blocks are defined by sets of EMUs whose MIDs are mutually dependent within the context of the EMU reaction network. The EMUs are arranged into blocks where the EMU reaction network is regarded as a directed graph, where the nodes represent EMUs and edges represent EMU reactions. An $N \times N$ adjacency matrix is constructed for the directed graph, where N is the total number of EMUs. A nonzero entry $a(i, j)$ of the adjacency matrix indicates the dependence of the ith EMU's MID on the jth EMU's MID. A Dulmage–Mendelsohn decomposition is performed on the adjacency matrix, returning an upper block triangular matrix from which the diagonal blocks are extracted. Blocks can be arranged

so that each is a self-contained subproblem that depends on the outputs of previously solved blocks, creating a cascaded system.

14. Several tools have been recently developed for flux visualization in the context of metabolic networks, such as FluxMap [45], FluxViz [46], fa-BINA [47], Omix [48], BioCyc Omics Viewer [49], Reactome Skypainter [50], Pathway Projector [51], MetaFluxNet [52], and OptFlux [53].

15. Abbreviations: *3PGA*, 3-phosphoglycerate; *αKG*, alpha-ketoglutarate; *CIT*, citrate; *DHAP*, dihydroxyacetone phosphate; *F6P*, fructose-6-phosphate; *FBP*, fructose 1,6-bisphosphate; *FUM*, fumarate; *G6P*, glucose 6-phosphate; *GAP*, glyceraldehyde 3-phosphate; *ICT*, isocitrate; *MAL*, malate; *PEP*, phosphoenol-pyruvate; *PYR*, pyruvate; *R5P*, ribose 5-phosphate; *RU5P*, ribulose 5-phosphate; *RUBP*, ribulose 1,5-bisphosphate; *S7P*, sedoheptulose 7-phosphate; *SUC*, succinate; *X5P*, xylulose 5-phosphate.

Acknowledgements

This work was supported by NSF EF-1105249 and DOE DE-SC0008118. LJJ was supported by DOE SCGF DE-AC05-06OR23100 and GAANN P200A090323.

Appendix: Simple Network Example for INST-MFA Calculations

A simple metabolic network appears in Fig. 6 as an example of how to construct the stoichiometric matrix *S*, identify the set of EMUs required to simulate MIDs of measured metabolites, and set up

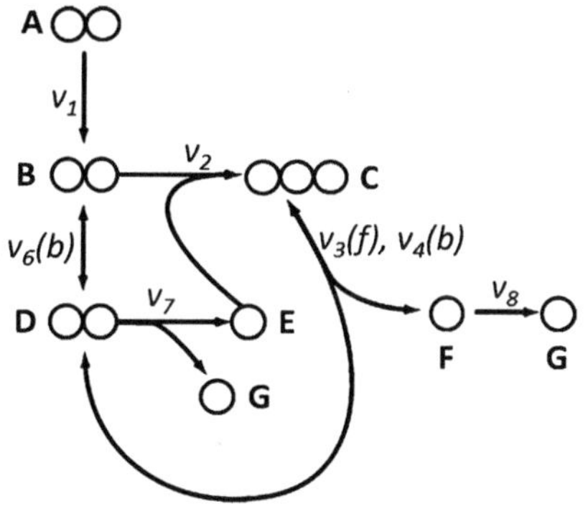

Fig. 6 Simple metabolic network used to illustrate the decomposition into EMUs. Atom transitions for the reactions in this model are given in Table 3. The network fluxes are assumed to be constant since the system is at metabolic steady state. Extracellular metabolite A is assumed to be at a fixed state of isotopic labeling to which intracellular metabolites B, C, D, E, F, and G adapt over time

Table 3
Stoichiometry and atom transitions for the reactions in the example metabolic network

Reaction number	Reaction stoichiometry	Atom transitions
1	A → B	ab → ab
2	B + E → C	ab + c → abc
3 and 4	C ↔ D + F	abc ↔ cb + a
5 and 6	B ↔ D	ab ↔ ab
7	D → E + G	ab → b + a
8	F → G	a → a

dynamic isotopomer balances on these EMUs. Table 3 delineates the atom transitions for the network. In this network example, metabolite A is the sole substrate and metabolite G is the only final product. The intermediary metabolites B, C, D, E, and F are assumed to be at metabolic steady state but isotopically nonstationary.

The stoichiometric matrix S is shown below, which has $k = 5$ intermediary metabolites and $j = 8$ fluxes, resulting in a 5×8 matrix:

$$S = \begin{bmatrix} 1 & -1 & 0 & 0 & -1 & 1 & 0 & 0 \\ 0 & 1 & -1 & 1 & 0 & 0 & 0 & 0 \\ 0 & 0 & 1 & -1 & 1 & -1 & -1 & 0 \\ 0 & -1 & 0 & 0 & 0 & 0 & 1 & 0 \\ 0 & 0 & 1 & -1 & 0 & 0 & 0 & -1 \end{bmatrix}$$

Therefore, $S{\cdot}v = 0$ is expressed in vector form as

$$\begin{bmatrix} v_1 - v_2 - v_5 + v_6 \\ v_2 - v_3 + v_4 \\ v_3 - v_4 + v_5 - v_6 - v_7 \\ -v_2 + v_7 \\ v_3 - v_4 - v_8 \end{bmatrix} = 0$$

A systematic method of EMU network decoupling in which metabolite units are grouped into mutually dependent blocks is described through this simple network example. For this example, we will set up the simplest possible model to simulate the MID of metabolite C, i.e., EMU C_{123}. First, we need to identify all the possible EMUs that contribute to the formation of C_{123}—in this reaction model, C_{123} is formed from the condensation of B_{12} + E_1 and D_{12} + F_1 in reactions 2 and 4, respectively. This is recorded and the process is then repeated for all new EMUs, starting with the largest EMU in size; in this case, all EMUs of size 3 have already been identified. Next, the process is repeated to determine all the

Table 4
Complete list of EMU reactions generated for metabolite C

Reaction number	EMU reaction	EMU reaction size balance	Block
2	$B_{12} + E_1 \rightarrow C_{123}$	$2 + 1 = 3$	6
4	$D_{12} + F_1 \rightarrow C_{123}$	$2 + 1 = 3$	6
2	$B_2 + E_1 \rightarrow C_{23}$	$1 + 1 = 2$	5
4	$D_{12} \rightarrow C_{23}$	$2 = 2$	5
3	$C_{23} \rightarrow D_{12}$	$2 = 2$	5
5	$B_{12} \rightarrow D_{12}$	$2 = 2$	5
6	$D_{12} \rightarrow B_{12}$	$2 = 2$	5
1	$A_{12} \rightarrow B_{12}$	$2 = 2$	5
2	$B_1 \rightarrow C_1$	$1 = 1$	4
4	$F_1 \rightarrow C_1$	$1 = 1$	4
3	$C_1 \rightarrow F_1$	$1 = 1$	4
2	$E_1 \rightarrow C_3$	$1 = 1$	3
4	$D_1 \rightarrow C_3$	$1 = 1$	3
3	$C_3 \rightarrow D_1$	$1 = 1$	3
5	$B_1 \rightarrow D_1$	$1 = 1$	3
6	$D_1 \rightarrow B_1$	$1 = 1$	3
1	$A_1 \rightarrow B_1$	$1 = 1$	3
7	$D_2 \rightarrow E_1$	$1 = 1$	2
2	$B_2 \rightarrow C_2$	$1 = 1$	1
4	$D_2 \rightarrow C_2$	$1 = 1$	1
1	$A_2 \rightarrow B_2$	$1 = 1$	1
6	$D_2 \rightarrow B_2$	$1 = 1$	1
5	$B_2 \rightarrow D_2$	$1 = 1$	1
3	$C_2 \rightarrow D_2$	$1 = 1$	1

Subscripts denote atoms that are part of their respective EMUs. The EMU reactions are also divided into their respective blocks after Dulmage–Mendelsohn decomposition has been applied

EMUs of size 2 that were previously identified, starting with D_{12}. D_{12} is formed from two different reactions—from B_{12} in reaction 5 and from C_{23} in reaction 3. Following this, we determine which reactions form C_{23}; C_{23} is formed from D_{12} in reaction 4 and $B_2 + E_1$ in reaction 2. Finally, we need to determine which reactions form B_{12}. B_{12} is formed from A_{12} and D_{12} in reactions 1 and 6, respectively. A_{12} is a network substrate and is not produced by any other reactions and D_{12} has already been considered in the previous step. Therefore, all EMU reactions of size 2 have been identified. The process is repeated once again for EMUs of size 1, until all the EMUs have been traced back to network substrates or previously identified EMUs. Table 4 shows the complete EMU decomposition of this system, which involves 24 EMU reactions connecting 16 EMUs.

After EMU decomposition, the reaction network can be further decoupled into blocks, which group together minimal sets of mutually dependent metabolite units that must be solved simultaneously. Figure 7 shows the EMU network decomposition for the simple network example after block decoupling. The blocks are arranged so that each one is a self-contained subproblem, which will depend on the outputs of the previously solved blocks. Therefore, EMUs in block 1 should first be solved, then block 2, etc.

The EMU reactions obtained from network decomposition and block decoupling form the new basis for generating system equations. The decoupled blocks can be arranged into a cascaded system of ODEs with the following form, as described in Subheading 3.9:

$$\mathbf{C}_n \cdot \frac{\mathrm{d}\mathbf{X}_n}{\mathrm{d}t} = \mathbf{A}_n \cdot \mathbf{X}_n + \mathbf{B}_n \cdot \mathbf{Y}_n$$

The concentration matrix $\mathbf{C}_n$ is a diagonal matrix whose elements are pool sizes corresponding to EMUs represented in $\mathbf{X}_n$. $\mathbf{X}_n$ comprises row vectors that represent the MIDs of each EMU and $\mathrm{d}\mathbf{X}_n/\mathrm{d}t$ is the time derivative of $\mathbf{X}_n$. Analogously, the input matrix $\mathbf{Y}_n$ also comprises row vectors that represent MIDs of EMUs that have been previously calculated. The system matrices $\mathbf{A}_n$ and $\mathbf{B}_n$ come from calculating the "true" flux vectors (v) based on the chosen free fluxes (u) and null space matrix (N). Furthermore, in the decoupled blocks, the full MID of products formed from condensation reactions can be obtained from the convolution (or Cauchy product, denoted by "×") of MIDs of preceding EMUs. In the case of C_{123}, these MIDs are B_{12} and E_1 or D_{12} and F_1, i.e., $C_{123} = B_{12} \times E_1$ or $C_{123} = D_{12} \times F_1$. The following equations represent the system of ODEs for the simple network example:

$$\begin{bmatrix} C_C & 0 & 0 \\ 0 & C_B & 0 \\ 0 & 0 & C_D \end{bmatrix} \begin{bmatrix} \frac{\mathrm{d}C_2}{\mathrm{d}t} \\ \frac{\mathrm{d}B_2}{\mathrm{d}t} \\ \frac{\mathrm{d}D_2}{\mathrm{d}t} \end{bmatrix} = \begin{bmatrix} -v_2 - v_4 & v_2 & v_4 \\ 0 & -v_1 - v_6 & v_6 \\ v_3 & v_5 & -v_3 - v_5 \end{bmatrix} \begin{bmatrix} C_2 \\ B_2 \\ D_2 \end{bmatrix} + \begin{bmatrix} 0 \\ v_1 \\ 0 \end{bmatrix} [A_2]$$

$$[C_E]\left[\frac{\mathrm{d}E_1}{\mathrm{d}t}\right] = [-v_7][E_1] - [v_7][D_2]$$

$$\begin{bmatrix} C_C & 0 & 0 \\ 0 & C_D & 0 \\ 0 & 0 & C_B \end{bmatrix} \begin{bmatrix} \frac{\mathrm{d}C_3}{\mathrm{d}t} \\ \frac{\mathrm{d}D_1}{\mathrm{d}t} \\ \frac{\mathrm{d}B_1}{\mathrm{d}t} \end{bmatrix} = \begin{bmatrix} v_2 + v_4 & -v_4 & 0 \\ -v_3 & v_3 + v_5 & -v_5 \\ 0 & -v_6 & v_1 + v_6 \end{bmatrix} \begin{bmatrix} C_3 \\ D_1 \\ B_1 \end{bmatrix} + \begin{bmatrix} v_2 & 0 \\ 0 & 0 \\ 0 & v_1 \end{bmatrix} \begin{bmatrix} E_1 \\ A_1 \end{bmatrix}$$

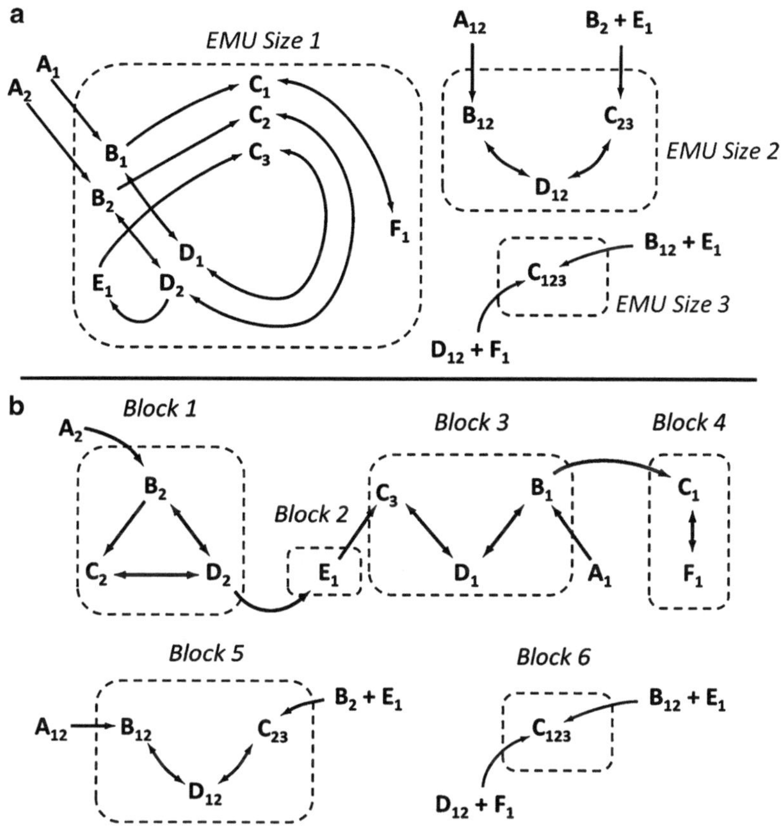

Fig. 7 (**a**) EMU network decomposition for simple example network (Fig. 6) generated to simulate the labeling of metabolite C. The EMU network was decoupled based on EMU size and network connectivity. (**b**) EMU network decomposition for the same network using block decoupling. *Subscripts* refer to the atoms that are contained within the EMU

$$\begin{bmatrix} C_C & 0 \\ 0 & C_F \end{bmatrix} \begin{bmatrix} \frac{dC_1}{dt} \\ \frac{dF_1}{dt} \end{bmatrix} = \begin{bmatrix} -v_2 - v_4 & v_4 \\ v_3 & -v_3 \end{bmatrix} \begin{bmatrix} C_1 \\ F_1 \end{bmatrix} + \begin{bmatrix} v_2 \\ 0 \end{bmatrix} [B_1]$$

$$\begin{bmatrix} C_C & 0 & 0 \\ 0 & C_D & 0 \\ 0 & 0 & C_B \end{bmatrix} \begin{bmatrix} \frac{dC_{23}}{dt} \\ \frac{dD_{12}}{dt} \\ \frac{dB_{12}}{dt} \end{bmatrix} = \begin{bmatrix} -v_2 - v_4 & v_4 & 0 \\ v_3 & -v_3 - v_5 & v_5 \\ 0 & v_6 & -v_1 - v_6 \end{bmatrix} \begin{bmatrix} C_{23} \\ D_{12} \\ B_{12} \end{bmatrix} + \begin{bmatrix} v_2 & 0 \\ 0 & 0 \\ 0 & v_1 \end{bmatrix} \begin{bmatrix} B_2 \times E_1 \\ A_{12} \end{bmatrix}$$

$$[C_C]\left[\frac{dC_{123}}{dt}\right] = [-v_2 - v_4][C_{123}] + [v_2 \quad v_4]\begin{bmatrix} B_{12} \times E_1 \\ D_{12} \times F_1 \end{bmatrix}$$

Solving this system of ODEs will simulate the EMU labeling trajectories needed to calculate the time-dependent MID of metabolite C. The flux and pool size parameters can then be adjusted iteratively using an optimization search algorithm to converge on parameter values that minimize the lack of fit with experimental mass isotopomer data.

References

1. Wiechert W (2001) 13C Metabolic flux analysis. Metab Eng 3:195–206
2. Sauer U (2006) Metabolic networks in motion: 13C-based flux analysis. Mol Syst Biol 2:62
3. Nöh K, Grönke K, Luo B et al (2007) Metabolic flux analysis at ultra short time scale: isotopically non-stationary 13C labeling experiments. J Biotechnol 129:249–267
4. Shastri AA, Morgan JA (2007) A transient isotopic labeling methodology for 13C metabolic flux analysis of photoautotrophic microorganisms. Phytochemistry 68:2302–2312
5. Nöh K, Wahl A, Wiechert W (2006) Computational tools for isotopically instationary 13C labeling experiments under metabolic steady state conditions. Metab Eng 8:554–577
6. Wiechert W, Nöh K (2005) From stationary to instationary metabolic flux analysis. Adv Biochem Eng Biotechnol 92:145–172
7. Young JD, Walther JL, Antoniewicz MR et al (2008) An elementary metabolite unit (EMU) based method of isotopically nonstationary flux analysis. Biotechnology 99:686–699
8. Antoniewicz MR, Kelleher JK, Stephanopoulos G (2007) Elementray metabolite units (EMU): a novel framework for modeling isotopic distributions. Metab Eng 9:68–86
9. Arrivault S, Guenther M, Ivakov A et al (2009) Use of reverse-phase liquid chromatography, linked to tandem mass spectrometry, to profile the Calvin cycle and other metabolic intermediates in Arabidopsis rosettes at different carbon dioxide concentrations. Plant J 59: 826–839
10. Huege J, Sulpice R, Gibon Y et al (2007) GC-EI-TOF-MS analysis of in vivo carbon-partitioning into soluble metabolite pools of higher plants by monitoring isotope dilution after 13CO2 labelling. Phytochemistry 68:2258–2272
11. Römisch-Margl W, Schramek N, Radykewicz T et al (2007) 13CO2 as a universal metabolic tracer in isotopologue perturbation experiments. Phytochemistry 68:2273–2289
12. Sekiyama Y, Kikuchi J (2007) Towards dynamic metabolic network measurements by multi-dimensional NMR-based fluxomics. Phytochemistry 68:2320–2329
13. Geigenberger P, Tiessen A, Meurer J (2011) Use of non-aqueous fractionation and metabolomics to study chloroplast function in Arabidopsis. Methods Mol Biol 775:135–160
14. Ausloos P, Clifton CL, Lias SG et al (1999) The critical evaluation of a comprehensive mass spectral library. J Am Soc Mass Spectrom 10:287–299
15. Kopka J, Schauer N, Krueger S et al (2005) GMD@CSB.DB: the golm metabolome database. Bioinformatics 21:1635–1638
16. Kind T, Wohlgemuth G, Lee DY et al (2009) FiehnLib: mass spectral and retention index libraries for metabolomics based on quadrupole and time-of-flight gas chromatography/mass spectrometry. Anal Chem 81: 10038–10048
17. Smith CA, O'Maille G, Want EJ et al (2005) METLIN: a metabolite mass spectral database. Ther Drug Monit 27:747–751
18. Wishart DS, Tzur D, Knox C et al (2007) HMDB: the human metabolome database. Nucleic Acids Res 35:D521–D526
19. Horai H, Arita M, Kanaya S et al (2010) MassBank: a public repository for sharing mass spectral data for life sciences. J Mass Spectrom 45:703–714
20. Gerhardt R, Heldt HW (1984) Measurement of subcellular metabolite levels in leaves by fractionation of freeze-stopped material in nonaqueous media. Plant Physiol 14:542–547

21. Bergmeyer H (1983) Methods of enzymatic analysis. Verlag Chemie, Deerfield Beach, FL
22. Krueger S, Giavalisco P, Krall L et al (2011) A topological map of the compartmentalized Arabidopsis thaliana leaf metabolome. PLoS One 6:e17806
23. Sigma-Aldrich Enzymatic assay of phospho (enol)pyruvate carboxylase. Presented at the (1994).
24. Maeda H, Yoo H (2011) Prephenate aminotransferase directs plant phenylalanine biosynthesis via arogenate. Nat Chem Biol 7:19–21
25. Li Y-T, Primate DR (1967) Studies on the glycosidases in jack bean meal. J Biol Chem 242:5474–5480
26. Antoniewicz MR, Kelleher JK, Stephanopoulos G (2007) Accurate assessment of amino acid mass isotopomer distributions for metabolic flux analysis. Anal Chem 79:7554–7559
27. Wittmann C, Hans M, Heinzle E (2002) In vivo analysis of intracellular amino acid labelings by GC/MS. Anal Biochem 307:379–382
28. Roessner U, Wagner C, Kopka J et al (2000) Technical advance: simultaneous analysis of metabolites in potato tuber by gas chromatography-mass spectrometry. Plant J 23:131–142
29. Yang L, Kasumov T, Yu L et al (2006) Metabolomic assays of the concentration and mass isotopomer distribution of gluconeogenic and citric acid cycle intermediates. Metabolomics 2:85–94
30. Luo B, Groenke K, Takors R et al (2007) Simultaneous determination of multiple intracellular metabolites in glycolysis, pentose phosphate pathway and tricarboxylic acid cycle by liquid chromatography-mass spectrometry. J Chromatogr A 1147:153–164
31. Young JD, Shastri AA, Stephanopoulos G et al (2011) Mapping photoautotrophic metabolism with isotopically nonstationary (13)C flux analysis. Metab Eng 13:656–665
32. Fernandez CA, Des Rosiers C, Previs SF et al (1996) Correction of 13C mass isotopomer distributions for natural stable isotope abundance. J Mass Spectrom 31:255–262
33. Schmidt K, Carlsen M, Nielsen J et al (1997) Modeling isotopomer distributions in biochemical networks using isotopomer mapping matrices. Biotechnol Bioeng 55:831–840
34. Wiechert W, Siefke C, de Graaf AA et al (1997) Bidirectional reaction steps in metabolic networks: II. Flux estimation and statistical analysis. Biotechnol Bioeng 55:118–135
35. Foster LV (1986) Rank and null space calculations using matrix decomposition without column interchanges. Linear Algebra Appl 74:47–71
36. Dulmage AL, Mendelsohn NS (1958) Coverings of bipartite graphs. Can J Math 10: 517–534
37. Pothen A, Fan CJ (1990) Computing the block triangular form of a sparse matrix. ACM Trans Math Softw 16:303–324
38. Gill PE, Murray W, Wright MH (1981) Practical optimization. Academic, London
39. Madsen K, Nielsen HB, Tigleff O (2004) Methods for non-linear least squares problems. http://www2.imm.dtu.dk/pubdb/views/publication_details.php?id=3215
40. Antoniewicz MR, Kelleher JK, Stephanopoulos G (2006) Determination of confidence intervals of metabolic fluxes estimated from stable isotope measurements. Metab Eng 8:324–337
41. Conover WJ (1999) Practical nonparameteric statistics. Wiley, New York
42. Kitson FG, Larsen BS, McEwen CN (1996) Gas chromatography and mass spectrometry: a practical guide. Academic, San Diego
43. Ahn WS, Antoniewicz MR (2011) Metabolic flux analysis of CHO cells at growth and non-growth phases using isotopic tracers and mass spectrometry. Metab Eng 13:598–609
44. Benkeblia N, Shinano T, Osaki M (2007) Metabolite profiling and assessment of metabolome compartmentation of soybean leaves using non-aqueous fractionation and GC-MS analysis. Metabolomics 3:297–305
45. Rohn H, Hartmann A, Junker A et al (2012) FluxMap: a VANTED Add-on for the visual exploration of flux distributions in biological networks. BMC Syst Biol 6:33
46. König M, Holzhütter H-G (2010) Fluxviz—cytoscape plug-in for visualization of flux distributions in networks. Genome informatics. International Conference on Genome Informatics. 24:96–103
47. Hoppe A, Hoffmann S, Gerasch A et al (2011) FASIMU: flexible software for flux-balance computation series in large metabolic networks. BMC Bioinformatics 12:28
48. Droste P, Miebach S, Niedenführ S et al (2011) Visualizing multi-omics data in metabolic networks with the software Omix: a case study. Biosystems 105:154–161
49. Paley SM, Karp PD (2006) The pathway tools cellular overview diagram and Omics Viewer. Nucleic Acids Res 34:3771–3778
50. Matthews L, Gopinath G, Gillespie M et al (2009) Reactome knowledgebase of human biological pathways and processes. Nucleic Acids Res 37:D619–D622

51. Kono N, Arakawa K, Ogawa R et al (2009) Pathway projector: web-based zoomable pathway browser using KEGG atlas and Google Maps API. PLoS One 4:e7710
52. Lee SY, Lee D-Y, Hong SH et al (2003) MetaFluxNet, a program package for metabolic pathway construction and analysis, and its use in large-scale metabolic flux analysis of Escherichia coli. Genome informatics. International Conference on Genome Informatics. 14:23–33
53. Rocha I, Maia P, Evangelista P et al (2010) OptFlux: an open-source software platform for in silico metabolic engineering. BMC Syst Biol 4:45

Chapter 13

Simulating Labeling to Estimate Kinetic Parameters for Flux Control Analysis

Amy Marshall-Colon, Neelanjan Sengupta, David Rhodes, and John A. Morgan

Abstract

An important aspect of kinetic modeling is the ability to provide predictive information on network control and dynamic responses to genetic or environmental perturbations based on innate enzyme kinetics. In a top-down approach to model assembly, unknown kinetic parameters are calculated using experimental data such as metabolite pool concentrations and transient labeling patterns after supply of an isotopically labeled substrate. These kinetic parameters can then be used to calculate flux control coefficients for every reaction in a network, which aids in the identification of enzymatic reactions that exert the most control over the network as a whole. This chapter describes a modeling approach to estimate kinetic parameters which are then used to perform metabolic control analysis. An example is provided for the benzenoid network of *Petunia hybrida*; however, the methodologies can be applied to any small segment of metabolism.

Key words Kinetic modeling, Metabolic control analysis, Metabolic flux analysis, Isotopic labeling, Computer simulation, Network decomposition

1 Introduction

The increasing need for systems biology in agricultural sciences has encouraged the use of multiple "omics" methodologies to construct a comprehensive model of whole organism responses to perturbations, in order to avoid off target effects of genetic engineering. These unforeseen effects can potentially be avoided by mathematical modeling of biological systems, which can help to predict in vivo changes in transcript, protein and metabolite levels in response to network modification. Due to the nonstationary functioning of a plant cell the most appropriate modeling strategy is kinetic modeling [1–3], which requires extensive knowledge of the kinetic expressions of the constituent pathway enzymes [2].

Three major utilities of kinetic models were recently described in an excellent review by [1]. Kinetic modeling makes it possible to (1) perform control analysis of plant metabolic systems, (2) identify

Martine Dieuaide-Noubhani and Ana Paula Alonso (eds.), *Plant Metabolic Flux Analysis: Methods and Protocols*, Methods in Molecular Biology, vol. 1090, DOI 10.1007/978-1-62703-688-7_13,

targets for biotechnological engineering, and (3) determine the regulatory importance of different enzymes. The majority of kinetic models are constructed with previously determined enzymatic parameters that can be found in databases such as BRENDA [4]. However, these parameters are often based on in vitro experiments that are frequently not representative of cellular dynamics [2]. Additionally, many networks, especially in secondary metabolism, are poorly characterized with a large number of unknown steps. To address this, one approach is to collapse multiple biochemical steps into a single enzymatic reaction, which can be appropriate for simple linear pathways but frequently leads to incomplete or non-physiological models in complex networks that are highly branched or under multiple levels of control. An alternative approach is to solve for the unknown kinetic parameters, where they are fitted through iterative adjustment and optimization to experimental data from dynamic isotopic labeling experiments.

In this chapter we describe a method to determine unknown kinetic parameters by fitting empirical kinetic equations to dynamic labeling and pool size data obtained by GC-MS [5–7]. Specifically, kinetic parameters are obtained by network decomposition and nonlinear least squares optimization [7, 8]. Complex metabolic networks often have a large number of unknown parameters, which leads to poor differential equation stability and multiple optima when attempting to fit all empirical equations simultaneously to the data. One way to overcome this challenge is by network decomposition, which decentralizes the complex structure of the model through creation of "subnetworks" [8]. Ultimately, one set of kinetic parameters is derived that must be able to describe network dynamics under different physiological conditions. We illustrate this by performing dynamic isotopic labeling experiments at three concentrations of exogenously supplied substrate (25, 75, and 150 mM d5-Phe) and test the estimated parameters against data from each of these experiments.

Before a kinetic model can be used for metabolic control analysis or to simulate network response to perturbations it must be validated against experimental data that was not used for the construction of the model and possibly refined based on the outcome. Although beyond the scope of this chapter, an example of model validation is detailed in Marshall-Colon et al. [7] for a kinetic model of the *Petunia* benzenoid network. In brief, the model was used to predict the redistribution of metabolite pool size and labeling patterns in response to a specific genetic modification which was then statistically evaluated against experimentally derived data from transgenic plants of the same genetic modification. Linear regression analysis of simulated and experimental data revealed an R^2 of 0.89 and a slope of 0.93, indicating the model predictions were in close agreement with the observed data.

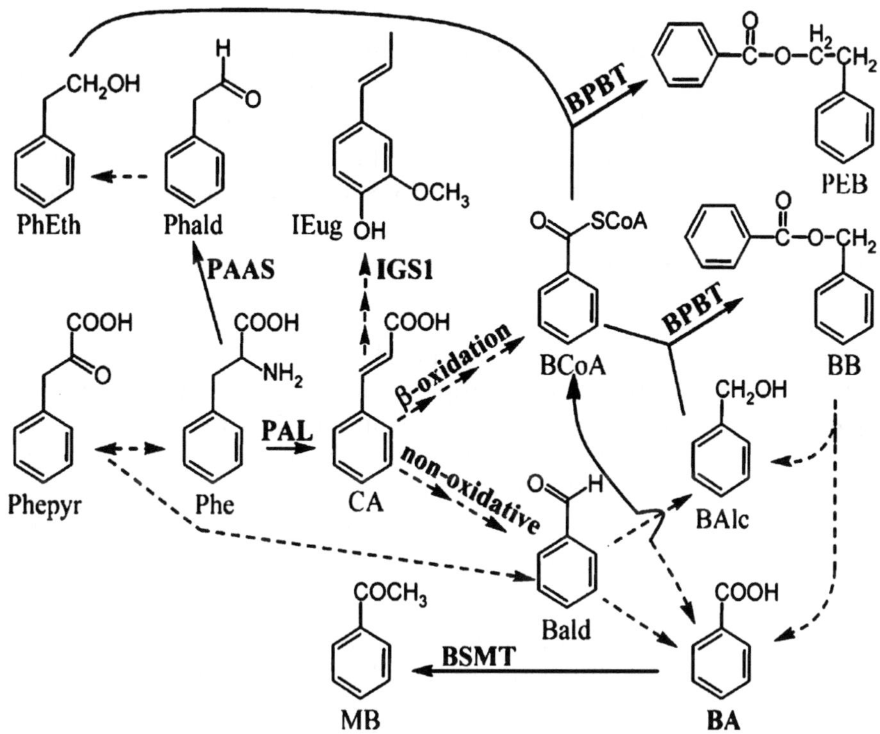

Fig. 1 The benzenoid network of *Petunia hybrida* [6]. Copyright American Society of Plant Biologists; www.plantcell.org

Once a robust set of kinetic parameters have been estimated for each reaction in the model, Metabolic Control Analysis (MCA) can be performed to pinpoint flux control within a network. MCA is based on the principle that changes in enzyme amounts and metabolite concentrations will have a direct effect on fluxes in a metabolic network [9] and can provide a quantitative understanding of metabolic networks [3] through the calculation of flux control coefficients (FCCs). FCCs are the relative change in the steady-state flux resulting from an infinitesimal change in the activity of an enzyme of the pathway divided by the relative change of the enzymatic activity [10].

In this chapter we describe a kinetic modeling strategy that combines (1) dynamic isotope labeling through feeding pathway precursors, (2) top-down kinetic parameter estimation, and (3) metabolic control analysis. This approach is applied to the benzenoid network of *Petunia hybrida* [7] as an example (Fig. 1).

2 Materials

1. L-Phenylalanine-ring-$^2H_5$, 99 % (Cambridge Isotopes).
2. A closed-loop system with an adsorbent trap consisting of a 120×7 mm glass tube containing 100 mg of polymer

Porapak Type Q (80/100 mesh; Alltech, Deerfield, IL) held in place with plugs of silanized glass wool and an additional adsorbent trap of identical construction placed at the inlet of the container to purify incoming air.

3. Agilent 6890 GC-MS with a capillary column (Agilent 19091S-433; HP-5MS 5 % Phenyl Methyl Siloxane).

3 Methods

3.1 Transient Labeling and Quantification of Metabolite Pools and Isotopic Abundance

Transient stable isotope labeling and the quantification of metabolite pools and isotopic abundance from time-course experiments are thoroughly discussed in Part I of this book. To illustrate the top-down kinetic modeling approach we use the benzenoid network of *P. hybrida*, which is the primary pathway for the biosynthesis of scent compounds in this species [7], where phenylalanine is the pathway substrate (Fig. 1). The labeling time-course was as follows:

1. Place the cut edge of excised corollas (total ten corollas per each experiment; cut just above the sepals) from 2-day-old flowers on filter paper (Whatman #1) wetted with 2 ml of stable isotope labeled substrate dissolved in water inside of a closed-loop system [5] (*see* **Note 1**). For dynamic labeling experiments select a range of substrate concentrations such as 25, 75, and 150 mM deuterium-labeled phenylalanine (L-phenylalanine-ring-$^2H_5$, 99 %; Cambridge Isotopes).
2. Collect volatile traps and tissue at timed intervals (i.e., 0, 30, 60, 120, and 240 min); sealing traps with plastic wrap and immediately flash freezing tissue in liquid N_2.
3. Quantify endogenous and exogenous pool sizes and labeling patterns of as many pathway metabolites as possible with either LC- or GC-MS as described in Chapters 7–9 of this book.

3.2 Equations for Simulation of Labeling Patterns and Metabolite Pools Over Time

1. Determine the most appropriate rate equation to model each reaction in a network. The following equations were used to model the benzenoid network in *P. hybrida* [7]:
2. Michaelis-Menten type reaction mechanisms can be assumed for most enzyme kinetic reactions:

$$v = \frac{V_{max}[S]}{K_m + [S]} \tag{1}$$

where v is the velocity of the enzyme catalyzed reaction, S is the substrate concentration in nmol/gFW, V_{max} is the maximum reaction velocity (nmol/gFW/min), and K_m is the substrate concentration at ½ V_{max}.

3. More complicated, two substrate kinetic reactions can be modeled with:

$$v = \frac{V_m [A][B]}{K_m + [A][B]} \tag{2}$$

where A is the concentration of one substrate and B is the concentration of the other substrate, both in nmol/gFW.

4. The emission rates of scent compounds are modeled as first-order diffusion reactions:

$$v_{scent} = K[A] \tag{3}$$

where K is the coefficient (min^{-1}) and A is the concentration of the emitted compound (nmol/gFW).

5. If an inverse correlation is observed between the uptake rate and internal concentration of the unlabeled compound then the uptake rate of labeled substrate can be assumed as a diffusion reaction, i.e., if external substrate uptake rate decreases as internal substrate concentration increases:

$$v1 = K_{1a}(\text{Sub}_{ex} - \text{Sub}_{in}\Phi) \tag{4}$$

where K_{1a} is the diffusion coefficient, Sub_{ex} is the external substrate concentration (mM), Sub_{int} is the intracellular substrate concentration (nmol/gFW), and Φ is the conversion factor that converts the Sub_{int} into concentration units of mM.

3.3 Solving for Kinetic Parameters Using Network Decomposition

1. Network Decomposition: Divide the network into subnetworks, as illustrated in Fig. 2, where for simple linear networks the most downstream reaction is the first subnetwork (Box 1). The second subnetwork would include the second to last and the last downstream reaction (Box 2), this is repeated until the largest network is achieved (Box 4). This reduces the number of parameters to be estimated simultaneously by testing the model in pieces [8] and entails solving for unknown kinetic parameters by the fitting of a series of empirical equations to

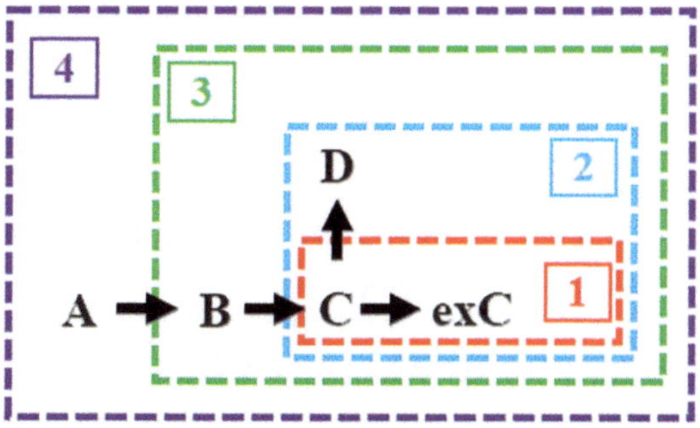

Fig. 2 A simplified example of network decomposition

experimentally determined pool size and labeling patterns for all species in a network.

For a more complex network with branchpoints emanating from a single node, the general strategy would be that each branch represents a large subnetwork consisting of the smaller linear subnetworks, described above. Once all the branches are solved for, the large subnetworks are combined at the branch-point node to solve for the reaction upstream of the node.

2. Simulation of Pools and Labeling Patterns: Solve for kinetic parameters by simultaneously curve fitting simulated pool size and labeling patterns to experimental data for each measured species from all labeling experiments (i.e. 25, 75, and 150 mM d5-Phe), using empirical equations for mass balance and isotope enrichment balances for the smallest subnetwork (Box 1 in Fig. 2) (*see* **Note 2**). The general form of the differential equations used by the optimizer is

$$\frac{\mathrm{d}C_i}{\mathrm{d}t} = \sum_{k=1}^{n} v_k \tag{5}$$

where C_i is the concentration of *i*th species (nmol/gFW) and v are the rate of reactions (nmol/gFW/min) involved in the production and consumption of a particular species.

The general isotopic labeling balance is given by

$$\frac{\mathrm{d}f_i C_i}{\mathrm{d}t} = \sum_{k=1}^{n} v_k f_k \tag{6}$$

where f_i represents the fractional labeling of the species (*see* **Note 3**).

3. Estimate Initial Kinetic Parameters: Obtaining the kinetic parameters is an optimization problem and is dependent on the initial guesses of these parameters; however, there can be more than one optimum solution. The inclusion of pool size and isotopic labeling data from different ranges to identify kinetic parameters that simultaneously satisfy these ranges increases the likelihood of the estimation of global optimum. Generally the initial starting guesses for the first subnetwork to be solved are educated guesses within the concentration ranges observed experimentally, but different guesses can be tried to find the solution that provides the minimum residue (described in the next section). Interestingly, we have found that we arrive at the same solution irrespective of the starting guess values when using the subnetwork strategy combined with data derived from experiments with multiple substrate concentrations.

4. Parameter Optimization: Perform optimization function to minimize the sum of the square of the difference between the simulated and experimental pool size and isotopic abundance data in the subnetwork.

$$\mathrm{Obj} = \sum_{i}^{N} (\mathrm{ZCal} - \mathrm{ZExp})_i^2 \tag{7}$$

where N is total number of species in the subnetwork (that is pool sizes and fractional labeling), ZCal is the calculated values, and ZExp is the experimental value of a particular experimental species. Thus the objective was to minimize the difference between the simulated and experimental data (*see* **Note 4**).

5. Once estimated parameters are obtained for subnetwork 1 (Box 1, Fig. 2), increase the network size (Box 2, Fig. 2) and repeat **steps 2** and **3**. This will result in estimated parameters for the increased subnetwork (Box 2) and refined values from the first subnetwork (Box 1). Repeat this process until the largest network (Box 4) and a final set of kinetic parameters (i.e., K_m and V_{max}) are obtained for the network. The kinetic parameters estimated for the benzenoid network are listed in Table 1 and can be visualized in Fig. 3.

Table 1
Model-optimized kinetic parameters (V_{max}, K_m, and K_i) for reactions in the benzenoid network of petunia based on pool size and labeling data from two feeding experiments with control petunia flowers [7]

K_m and K_i values (nmol/gFW)		V_{max} (V_m) and diffusion rates (K) (nmol/gFW/min)	
K_m2	39.9	V_m2	5.5
K_m3	11,006.0	V_m3	8.9
K_m4	391.8	V_m4	2.7
K_m5	60.4	V_m5	1.4E-08
K_m8	4.7	V_m8	0.6
K_m13	133.1	V_m13	1.3
K_m14	0.7	V_m14	1.3
K_m16	6.9	V_m16	5.3
K_m17	26.4	V_m17	0.008
K_m18	42.9	V_m18	17.2
K_m20	203.6	V_m20	26.5
K_m25	6.3	V_m25	0.2
K_m27	57.4	V_m27	9.0
K_m32	309.2	V_m32	1.6
K_m35	0.3	V_m35	0.5

(continued)

Table 1 (continued)

K_m and K_i values (nmol/gFW)		V_{max} (V_m) and diffusion rates (K) (nmol/gFW/min)	
K_m36	2,944.2	V_m36	14.9
K_m37	14.8	V_m37	4.9
K_m49	1.0	V_m49	0.002
K_m50	1.0	V_m50	0.001
K_m51	1.0	V_m51	0.003
K_m54	$2.1E^{-7}$	V_m54	0.8
K_m57	0.006	V_m57	0.2
K_m58	1.0	V_m58	0.001
K_m63	3.9	V_m63	0.9
K_m6	2,833.4	V_m6	20.0
K_i6	4.8	K_{1a}	4.6
		K_{1b}	0.001
		K_{19}	0.1
		K_{26}	0.0
		K_{30}	0.2
		K_{34}	0.010
		K_{62}	0.005
		K_{64}	3.3

Reactions are as indicated in Fig. 3

6. Sensitivity Analysis: To statistically test the significance of the estimated kinetic parameters, sensitivity analysis is performed on the model simulations by determining the 95 % confidence intervals for the estimated parameters. Sensitivity coefficients are the partial derivatives of the simulated values at the optimum solution with respect to the estimated parameters:

$$J_{ij} = \frac{\partial x_i^{\text{opt}}}{\partial p_j} \tag{8}$$

where J_{ij} is the sensitivity of simulated measurement (x_i) at the optimum with respect to estimated parameter p_j (*see* **Note 5**).

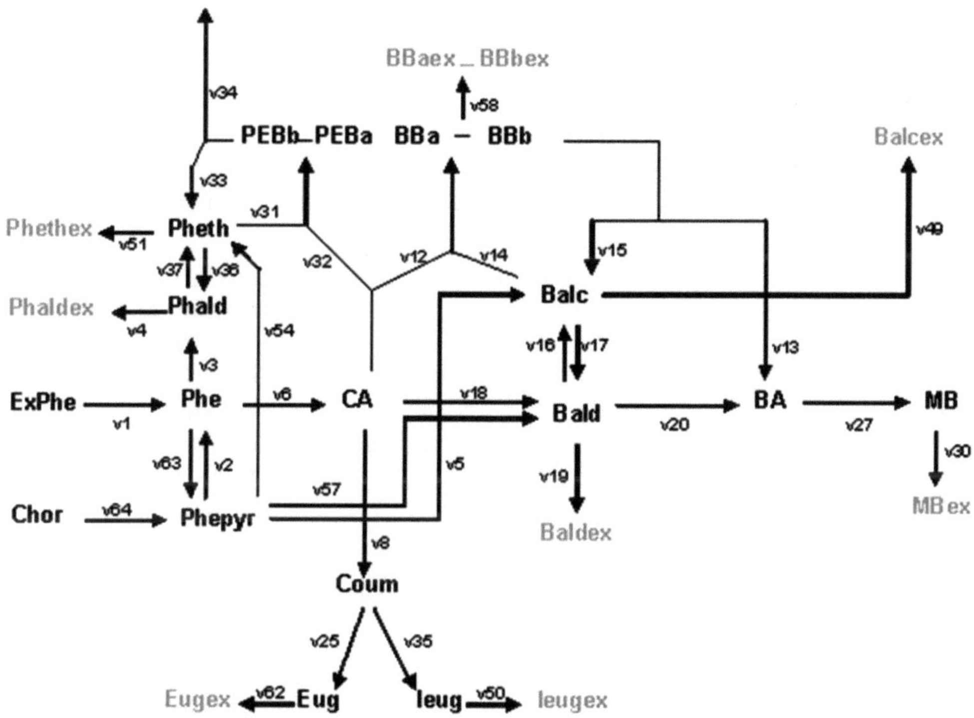

Fig. 3 Schematic of the benzenoid network reactions in petunia. *Arrows* represent flux (*v*) and *grey font* and the abbreviation "ex" indicate emitted volatiles [7]

7. Next the Hessian matrix was calculated, which is used in calculation of the covariance of estimated parameters and subsequently the confidence intervals and is defined as:

$$H = J^{\mathrm{T}} J \tag{9}$$

The covariance matrix of the estimated parameters can be calculated as:

$$\sum\nolimits_{\text{parameters}} = \frac{\Phi(H^{-1})}{n - p} \tag{10}$$

where Φ is the value of objective function at the optimum, H^{-1} is the inverse of the Hessian matrix, n are the number of observations, and p are the number of parameters. The diagonal elements of the $\Sigma_{\text{parameters}}$ are the variances of the estimated parameters, from which the approximate 95 % confidence intervals can be calculated as:

$$p_i = p_{i\text{opt}} \pm 2\sqrt{\sum\nolimits_{ii\ \text{parameters}}} \tag{11}$$

3.4 Metabolic Control Analysis

1. Calculation of the flux control coefficients (FCCs):

$$\mathrm{FCC}_{ij} = \frac{\partial \ln J_i}{\partial \ln E_j} \tag{12}$$

where E_j is the enzyme activity or parameter (determined by the kinetic model in Subheading 3.3) that affects the particular flux J_i [9].

2. An MCA Example: For the benzenoid network, the external Phe (exogenous substrate) concentration and each of the enzyme activities (V_m) were perturbed in range of ±10 %, and the fluxes were evaluated for each perturbation at the steady state [10]. To determine the amount of control of the enzyme phenylacetaldehyde synthase (*PAAS*) has over the network, its estimated V_{max} value (V_m3, Table 1) was changed in steps over a range from 8.9 to 9.1. The step size should be such that it generates 5–10 data points or more. For each increment of V_m3, starting from the original value ($V_m3 = 8.9$, Table 1), the entire network is solved to a distant point in time such that differential equations reach a steady state while keeping all the other parameters constant. The newly determined steady-state flux values are recorded, and the process is repeated changing the value of V_m3 stepwise until the range maximum is reached ($V_m3 = 9.1$).

3. The flux values (5–10 for each reaction) generated by the V_{max} perturbation in **step 2** are used to calculate FCCs [10]. The FCC is the slope of the log–log plot of the flux and enzyme activity pair and is determined by curve fitting using LSQNONLIN (MATLAB 2007). The FCC can be positive or negative depending on the direction of the slope. This was implemented by fitting a straight line through the data which is of the form:

$$\ln(J) = \ln(J_o) + m \times \ln(V_{max}) \tag{13}$$

where $\ln(J_o)$ is intercept and m is the slope as determined by curve fitting. The slope (m) is the FCC. For LSQNONLIN in MATLAB, which implements least square optimization, Eq. 13 was formulated as follows:

$$\mathrm{Obj} = \sum_{i}^{N} (\ln(J_{cal}) - \ln(J_{obs}))_i^2 \tag{14}$$

where J_{obs} are the steady flux values for a particular reaction from the perturbation analysis and J_{cal} are the flux values calculated as a function of $\ln(J_o)$, m (constants) and $\ln(V_{max})$ (perturbation values) from Eq. 13. At minimum, the values of ln (J_o) and m are obtained. Thus for each flux an FCC can be calculated as a measurement of the effect of the enzyme perturbation (Fig. 4).

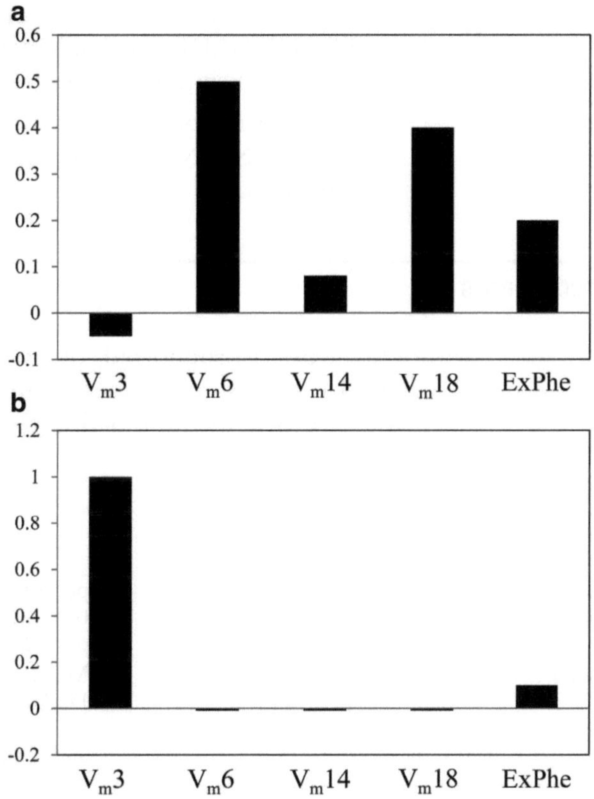

Fig. 4 Flux control coefficients calculated for fluxes around the Phe branch point. FCCs calculated for flux toward formation of CA (**a**) and Phald (**b**), in which ExPhe is the exogenous concentration of supplied Phe (150 mM), and V_{max} corresponds to the reaction in the benzenoid network (Fig. 1) exerting control on the particular emission flux [7]

4 Notes

1. Closed-loop system should have a constant stream of air flowing at 450 ml/min drawn through the container with a vacuum pump.
2. We used the CURVEFIT tool of MATLAB (2007) to fit the empirical equations to the data over a time-course, which has no biological significance but provides a smooth fit representation of the experimental data.
3. Either ODE45 in MATLAB that implements an explicit Runge-Kutta-(4,5) formula or stiff solver ODE15s were used to solve the ODEs for the subnetworks depending upon the stiffness of the differential equations.

4. The optimization was performed in MATLAB (2007) using LSQNONLIN that implemented nonlinear least square optimization using the sub-space trust region algorithm.
5. The sensitivity coefficients were obtained directly from the LSQNONLIN routine in MATLAB, calculated at the optimum.

Acknowledgments

This work is supported by the National Institutes of Health (NRSA fellowship number GM095273 for AMC) and by the US National Science Foundation (grant number MCB-0615700 for JAM and DR).

References

1. Rohwer JM (2012) Kinetic modelling of plant metabolic pathways. J Exp Bot 63(6): 2275–2292
2. Morgan JA, Rhodes D (2002) Mathematical modeling of plant metabolic pathways. Metab Eng 4:80–89
3. Rios-Estepa R, Lange BM (2007) Experimental and mathematical approaches to modeling plant metabolic networks. Phytochemistry 68: 2351–2374
4. Schomburg I, Chang A, Ebeling C et al (2004) BRENDA, the enzyme database: updates and major new developments. Nucleic Acids Res 32:D431–D433
5. Boatright J, Negre F, Chen X et al (2004) Understanding in vivo benzenoid metabolism in petunia petal tissue. Plant Physiol 135:1993–2011
6. Orlova I, Marshall Colon A, Schnepp J et al (2006) Reduction of benzenoid synthesis in petunia flowers reveals multiple pathways to benzoic acid and enhancement in auxin transport. Plant Cell 18:3458–3475
7. Marshall-Colón AM, Sengupta N, Rhodes D, Dudareva N, Morgan J (2010) A kinetic model describes metabolic response to perturbations and distribution of flux control in the benzenoid network of Petunia hybrida. Plant J 62: 64–76
8. Rizzi M, Baltes M, Theobald U, Reuss M (1997) In vivo analysis of metabolic dynamics in Saccharomyces cerevisiae: II Mathematical model. Biotechnol Bioeng 55: 592–608
9. Kacser H, Burns J (1973) The control of flux. Symp Soc Exp Biol 27:65–104
10. Stephanopoulos GN, Aristidou AA, Nielsen J (1998) Metabolic engineering: principles and methodologies. Academic, San Diego

Chapter 14

High-Throughput Data Pipelines for Metabolic Flux Analysis in Plants

C. Hart Poskar, Jan Huege, Christian Krach, Yair Shachar-Hill, and Björn H. Junker

Abstract

In this chapter we illustrate the methodology for high-throughput metabolic flux analysis. Central to this is developing an end to end data pipeline, crucial for integrating the wet lab experiments and analytics, combining hardware and software automation, and standardizing data representation providing importers and exporters to support third party tools. The use of existing software at the start, data extraction from the chromatogram, and MFA analysis at the end allows for the most flexibility in this workflow. Developing iMS2Flux provided a standard, extensible, platform independent tool to act as the "glue" between these end points. Most importantly this tool can be easily adapted to support different data formats, data verification and data correction steps allowing it to be central to managing the data necessary for high-throughput MFA. An additional tool was needed to automate the MFA software and in particular to take advantage of the course grained parallel nature of high-throughput analysis and available high performance computing facilities.

In combination these methods show the development of high-throughput pipelines that allow metabolic flux analysis to join as a full member of the omics family.

Key words Metabolic flux analysis, High-throughput, Automated workflow, Workflow management

1 Introduction

Plant functional genomics currently relies on transcriptomics, proteomics, and metabolomics. Taken together these research fields are utilized in a high-throughput manner to systematically analyze the effect of different genetic backgrounds and alterations. The best description of the phenotype of an organism is given by characterizing its metabolism [1]. Two complementary approaches for this characterization are (a) determining the metabolic composition of parts of the organism [2, 3] and (b) the determination of the flow of matter in a metabolic network through metabolic flux analysis (MFA) [4, 5].

Martine Dieuaide-Noubhani and Ana Paula Alonso (eds.), *Plant Metabolic Flux Analysis: Methods and Protocols*, Methods in Molecular Biology, vol. 1090, DOI 10.1007/978-1-62703-688-7_14, © Springer Science+Business Media New York 2014

In metabolic flux analysis (MFA), fluxes are defined as the flows of molecules between different metabolite pools catalyzed by the corresponding enzymes and/or transporters. To determine the in vivo fluxes of a given metabolic network, MFA combines a stoichiometric model, as the mathematical representation of the metabolic network, and measurement data from isotope labeling experiments [6, 7]. Taken together the model and the metabolite labeling information facilitate the calculation of in vivo fluxes not accessible by direct techniques.

MFA is usually applied to systems in metabolic and isotopic steady state, but nevertheless can be applied to non-stationary isotopic states, which increases experimental and computational effort but opens up the possibility for investigating autotrophic metabolism and pathways without carbon rearrangement. The same automation that can be applied to high-throughput steady-state MFA can also support the additional processing and analysis overhead of non-stationary experiments resulting in a series of measurements at different time points.

The interest in developing high-throughput MFA techniques is evidenced by recent labeling studies in microbes and mammalian cells. Steady-state MFA has been performed on 135 strains of *Bacillus subtilis* [8] and 92 strains of *Escherichia coli* [9]. High-throughput techniques are also required for smaller experiments involving time course measurements such as labeling experiments involving human cells reported in [10] or *Clostridium acetobutylicum* [11]. However these examples of high-throughput experiments are still the exception and have not been reached in plant-related studies, which typically investigate from one to three different genotypes or involving two different treatments [12–16]. Presumably this is due to the high demands of this method concerning tissue culture, analytics, and computation. Additionally the number of analytes for which labeling can accurately be detected in a single run is now potentially much larger [17–21] and the value of multiple biological replicates using different label combinations has been demonstrated [13, 15, 22].

In other "omic" technologies higher throughput rates have evolved through the development of more efficient workflows. This involves automating and integrating different steps of the analytical process. The example of metabolomics is instructive where sample preparation, spectrometric analysis, and data processing are now routinely integrated [23–26]. Although the importance of metabolic flux as a component of plant functional genomics has been recognized [27, 28], it is not yet an integral part of the omics platforms, since MFA currently is lacking the capability for high throughput. Therefore, in order to move towards fluxomics it is required to find means to increase the throughput of the flux analysis methodology. As parallel processing of samples and automated instrumental analyses have become

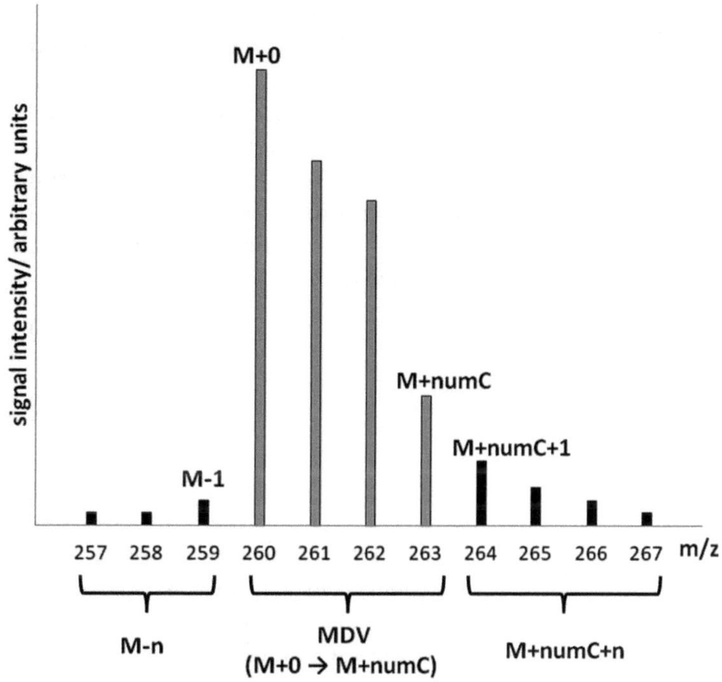

Fig. 1 Illustration of the mass isotopomer distribution vector (MDV) of a three carbon compound, e.g., alanine, the signals in *grey*. This MDV comprises the fractions of completely unlabeled (M+0), singly labeled (M+1), doubly labeled (M+2), and completely labeled (M+numC in general, i.e., M+3 in this example) analyte. Also indicated are the preceding (M−*n*) and following (M+numC+*n*) masses forming a boundary around the analyte masses. Adapted from ref. [41]

common, accurate processing of labeling data can limit the throughput of flux analysis or profiling. A typical isotope tracer experiment can result in multiple chromatograms, each containing mass spectrometric information on dozens of analytes, each of which can yield mass isotopomer patterns of multiple fragments. The entirety of mass isotopomers of a fragment ion is called the mass isotopomer distribution vector (MDV) [29], *see* Fig. 1.

1.1 Overview of MFA Workflow

Metabolic flux analysis combines isotopic labeling experiments with stoichiometric-based models to elucidate metabolic fluxes or enzymatic activity. Most commonly MFA labeling experiments use substrates with one or more carbons labeled by ^{13}C; however, it should be noted that none of the methods for high-throughput analysis explicitly rely on this.

The general form of the stoichiometric model is described by the following equation,

$$S \cdot v = \frac{dX}{dt} \quad (1)$$

where the stoichiometric matrix, S, multiplied by the vector of metabolic fluxes, v, is equal to the rate of change of the metabolite concentrations, X.

MFA may then be classified based on whether the experiment is performed in metabolic and isotopic stationary or instationary states [30, 31]. Equation 2 illustrates this classification through the general form of the isotopomer balance for the rate of change of metabolite concentrations X with fractional labeling x_i for isotopomer i,

$$\begin{aligned}
&(a)\frac{d(X \cdot x_i)}{dt} = 0 \\
&(b)X \cdot \frac{d(x_i)}{dt} \neq 0 \\
&(c)\frac{d(X \cdot x_i)}{dt} \neq 0
\end{aligned} \tag{2}$$

Three classifications emerge: (Eq. 2a) metabolic and isotopic stationary (steady-state MFA), (Eq. 2b) metabolic stationary and isotopic instationary (dynamic MFA), and (Eq. 2c) metabolic and isotopic instationary (kinetic MFA).

Not surprisingly the workflow in each of these cases is similar. The main differences arise in the labeling experiment and the computational analysis. For the labeling experiment the number of samples to be taken is no longer just a single time point sample as at isotopic stead state. Instead a series of time points has to be sampled which is dense enough to adequately describe the continuous incorporation and thus the increase in isotope label in the metabolites under investigation. Since for each time point a respective number of biological replicates are required, the number of necessary samples and thus the experimental effort for the labeling experiments and all subsequent steps including sample processing and measurement can increase drastically. For the computational analysis the mathematical representation of the model changes from algebraic equations in steady state to differential equations in dynamic and kinetic. The fluxes may be treated as constants or parameterized representations of enzyme kinetics (such as Michaelis–Menten kinetics). Furthermore kinetic modeling may require the separation of variables for complete solutions.

As illustrated in Fig. 2, the MFA workflow can be divided into three distinct parts, the development of (1) biological (actual) measurements, (2) computational (simulated) measurements, and (3) flux values. The development of biological measurements begins with the cultivation of a heterotrophic or mixotrophic cell type, tissue or organ of a plant (e.g., embryo, cell suspension culture) which is fed an organic substrate with a predetermined labeling pattern in the carbon chain, with stable isotopes (^{13}C). As the substrate progresses through the metabolic network the carbon

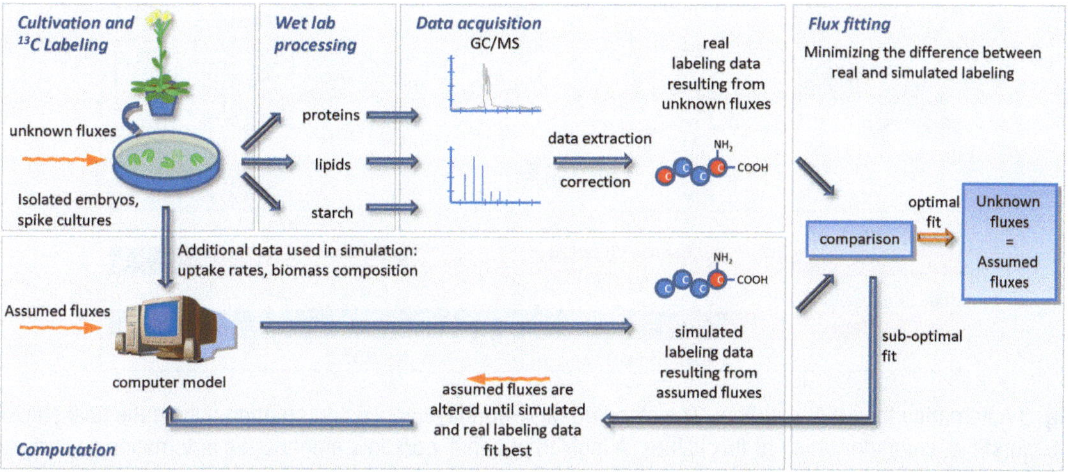

Fig. 2 Illustration of the workflow of steady-state ^{13}C metabolic flux analysis in plants, adapted from ref. [66]

chain is rearranged in the formation of new metabolites. The resulting isotope labeling patterns can be partially detected by mass spectrometric (MS) or nuclear magnetic resonance (NMR) techniques in biomass components such as starch, proteins, or lipids after hydrolysis to their monomers [12, 32, 33]. In the second part, the computational measurements seek to reproduce the isotope labeling pattern by solving the system of equations that make up the model for a specific set of flux values. The solution itself may make use of additional biological information such as uptake rates or metabolite concentrations. Finally the flux values themselves are developed using an iterative optimizer or parameter fitting algorithm. In this step the algorithm evaluates the current state of flux values by comparing the measured values X_M to the calculated values X_C using a predefined metric such as the weighted sum of squared errors, typically given by the residual R

$$R = \sum \frac{(X_M - X_C)^2}{W} \tag{3}$$

where X_M is the mean of the experimental replicates and the weight W is the standard deviation. This weighting controls the effect of the individual measurements on the flux value so as to be relative to the accuracy of the respective measurement.

1.2 High-Throughput MFA

In metabolic flux analysis the standard workflow (Fig. 3) is mostly manual. Automation mainly occurs at the final step, the computational analysis for determining the optimized flux values. Even then this automation is typically for a single experiment at a time. However automation software is available over the entire range of the workflow including MS data extraction, e.g., [34, 35], data correction [36, 37], model development, and analysis [38, 39].

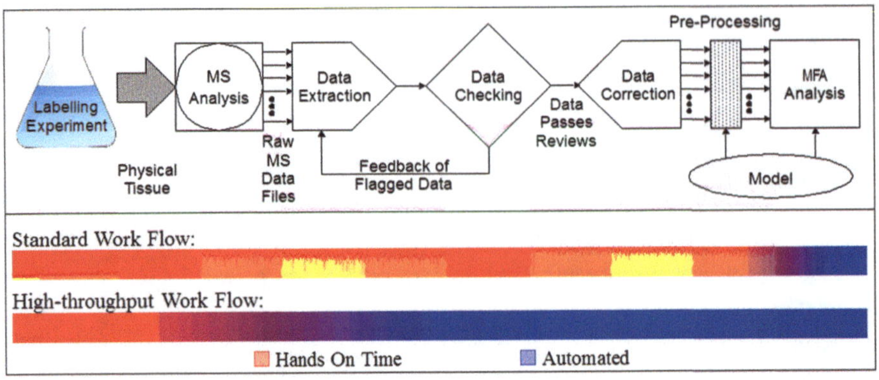

Fig. 3 Automating the MFA workflow. The standard MFA workflow applies automation only at the final step of the workflow, computer fitting of flux values. A high-throughput workflow emphasizes automation as early as possible in the first step, developing biological measurements

There is only one software platform which attempts to combine these parts, FiatFlux [40], however in doing so it is severely restricted and not suitable for plants [41].

In this chapter we focus on developing a more efficient workflow supporting high-throughput methodology by pushing automation to as early as possible in the first step, the development of biological measurements. In doing so, we will combine first and third party tools to promote efficiency and flexibility in the workflow. The workflow must be flexible in order to be robust, allowing for new laboratory and computational techniques to be added as necessary.

Throughout this chapter the automation of the MFA workflow will be presented with an example workflow in steady-state MFA using carbon labeling experiments (CLEs) and measured by GC-MS. To maximize flexibility a combination of proprietary and open software are utilized. There are many options available for working with chromatograms and performing the analysis, from these we have selected Waters QuanLynx™ (*see* **Note 1**) for the extraction of MS measurement data and 13CFLUX (*see* **Note 2**) for MFA. Missing is the automation of steps to process large quantities of measurement data for inclusion in the analysis, generate the models and a automatically manage the analysis. For this we have developed to publicly available high-throughput automation programs, iMS2Flux (*see* **Note 3**) for processing the data and autoFLUX (*see* **Note 4**) for automating the MFA process.

2 Materials

The equipment section describes the required hardware/software for automation starting from the point where experimental data has been extracted from the MS chromatograms.

2.1 Equipment

1. The minimum requirement is a single personal computer running a Linux operating system or Windows with a Linux virtual machine installed.

   For running VANTED a screen resolution of at least 1024 × 768, mouse, 2 GB RAM (32 bit) or 4 GB RAM (64 bit) are recommended.
2. Perl v.5 or higher: http://www.perl.org/get.html

   PERL comes installed as standard with most Linux installs.
3. iMS2Flux.pl: http://sourceforge.net/projects/ims2flux/
4. 13CFLUX: http://www.13cflux.net/13cflux/index.jsp

   autoFlux.pl: Available on request.
5. VANTED: http://vanted.ipk-gatersleben.de/

   Java runtime v. 6 or higher:
   https://www.java.com/en/download/manual.jsp
   FluxMap add-on for VANTED:
   http://vanted.ipk-gatersleben.de/addons/fluxmap/

2.2 Equipment Setup

The software packages may be installed in all or in part as you desire. Both iMS2Flux and VANTED are platform independent (they may be run on Windows or MacOS) requiring only PERL or the Java runtime, respectively. Each piece of software comes with its own installation guide and/or built in installer.

The installation of 13CFLUX is straight forward but requires basic knowledge of Linux. The software comes ready to run, compressed in a tar archive. It comes compiled and statically linked so as to run on any linux system on an x86-based computer. The software is run from the command line allowing for easy automation and requires the use of the Bash shell. After extraction it is always a good idea to check that the binaries (i.e., the programs) are set to be executable.

3 Method

When performing MFA on plants, there is little direct automation available during the initial phase throughout the labeling experiment and sample processing. For high-throughput it is necessary to streamline the process through simplification and organization: Simplification by minimizing steps and organization by developing a pipeline for experimental data. For the latter imperative, the proper management of a high-throughput experiment should allow for multiple steps to be performed in parallel but on different individuals. Implementing an experimental pipeline overcomes the first obstacle to high-throughput MFA, maximal data flows through the system as it becomes available.

3.1 Cultivation and Isotope Labeling

The first step is key to high-throughput metabolic flux analysis. There is little that can be done to increase the speed of biological experimentation from the growing of plants to achieving metabolic and isotopic steady state in labeling experiments. The cultivation procedures for MFA experiments using plants are typically in vitro, although in the area of kinetic MFA *in planta* experiments using ^{13}C labeled CO_2 are emerging. These experiments may use one or more organic substrates of different isotopic composition and require times ranging from several days to several weeks to reach both metabolic and isotopic steady state [12, 13, 15, 42].

For high-throughput MFA it is desirable to simultaneously minimize the number of cultures and simplify the processing pipeline. When constructing the labeling experiment the number of media compositions should be minimized. If a single labeling experiment can produce sufficiently informative data, then only a single labeling media should be used. Note, even a single labeling media may be composed of multiple labeling substrates.

3.2 Sample Processing

Before performing the measurements it is necessary to fraction the samples into the desired biomass components, such as proteins, and each of these must then be further broken down into their respective monomers. To support high-throughput MFA it is desirable to minimize both the number of these biopolymers and the number of processing steps.

It is possible to restrict the biomass components to those which contain the majority of information for the pathways of interest. For example the central metabolism of seeds can be sufficiently described by proteins, starch, lipids, and cell wall [12, 15, 43]. Robust protocols already exist for the extraction and fractionation of the plant material into these biopolymers [12, 14, 15, 42, 44]. Protein may then be further hydrolyzed into amino acids, providing a broad coverage of the central carbon metabolism [45]. Lipids may be either trans-esterified to fatty acid methyl esters (FAMEs) or derivatized to butylamides, which may provide insight into compartmentation of lipids synthesis [12, 42].

Further processing may be used depending on the type of measurement, in particular derivatization for measurement with GC-MS.

3.3 Sample Measurement

A variety of analytical platforms are available for the determination of isotope labeled compounds. The most common platforms are gas chromatography (GC) or liquid chromatography (LC) in combination with mass spectrometry (MS), and nuclear magnetic resonance (NMR). The choice of platforms may be dictated by a number of factors including preference, requirements, and availability. For example LC-MS/MS may be more suitable for analyzing cell wall compounds [46], whereas GC-MS allows for fragmentation yielding more data points for each compound.

Since the sensitivity of MS exceeds that of NMR it is often preferred, allowing for the analysis of smaller samples and less abundant metabolites [44].

For high-throughput experiments it is less likely that different measurement platforms may be combined, as such the use of a single platform allows for the most automated throughput of data. While each platform has advantages, as a single solution GC-MS is the most versatile. In GC-MS derivatized compounds fragment during the ionization step yielding both additional data points and redundancy. In the case of a contamination which may co-elute with a particular fragment a redundant fragment (containing the same labeling information) may be used for [47, 48]. Furthermore, with GC-MS the majority of biomass components (starch as glucose, proteins as amino acids, and lipids as FAMEs or butylamides and glycerol) can be measured in multiple runs on a single system with the same configuration.

A unified derivatization scheme should be employed. A simplified chemistry in line with metabolic profiling is sufficient for most compounds [2], specifically silylation and methoxamination. Thus with basic add-ons the analytical process may be fully automated from derivatization to chromatogram. Autosamplers, pipetting robots equipped with two different syringe volumes, and shaking incubators are used to perform the chemical derivatization of the analytes. The stages of derivatization, including pipetting of chemicals and sample incubation, and the actual measurements are performed in parallel, allowing for the continuous generation of data and fitting with the earlier established pipeline.

3.4 Data Processing

With the isotope labeling measurements in an electronic form (mass chromatograms) the remainder of the workflow deals with extracting, processing and then using the measurement data. The traditional MFA workflow approaches these tasks in a mostly manual way, applying copy and paste techniques to manually move data through the pipeline. In addition to being monotonous even with small data sets it is also error prone, resulting at best in unnecessary and time consuming repetition and at worst unreliable results. The frequency of these errors scales relative to the amount of data in the pipeline making a manual approach unsuitable for high-throughput MFA.

The next two sections provide an overview of the steps and the use of two software solutions to automating these tasks.

3.4.1 Extraction of Measurement Data

The first step in processing the labeling measurement data is to quantify the chromatograms from each MS experiment. This requires first preprocessing the chromatograms, identifying the retention times of the peaks of interest, and then quantifying each of the masses that make up that peak. On a per chromatogram basis most software (proprietary and public) provide algorithms for automatically performing the necessary preprocessing steps.

These include applying a baseline to the chromatogram, smoothing the spectrum, determining the retention times from standard libraries, and adjusting for spread in retention time across related chromatograms. Once preprocessed, the peaks are quantified via integration. However what is integrated is often manually selected, be it peak area, half height area or simply the peak height itself. Strangely one feature often missing is the ability to export this data into a simple, text based format for further use, again resulting in copy and paste workflows. To support high-throughput it is necessary to be able to standardize all of these configurations, apply them automatically to the chromatograms and generate a usable output. There are numerous software solutions, both proprietary manufacturer software and freely available third party software (often based on the generic ANDI/netCDF MS Data Interchange format or CDF), in particular those dedicated to the alignment of multiple MS chromatograms and extracting the relevant MS data, e.g., TagFinder [34], MZmine 2 [35], or CalSpec [49]. One example is the manufacturer supplied QuanLynx™ tool, a standard part of the MassLynx™ software from Waters. This tool is used both in the extraction and again during the data checking to verify identified data errors.

For a detailed description of working through this format see the getting started guide; MSto13C with QuanLynx™ and iMS2Flux.pdf (*see* **Notes 1** and **3** for details). Regardless of the choice of extraction tool, further automation requires standardized representation of data. It is common to provide MS data in tabular form, such as a tab separated values (TSV) or comma separated values (CSV), for easy manipulation and visualization in spreadsheets and other third party software.

3.4.2 Processing Measurement Date

As illustrated in Fig. 3 there is a significant amount of processing between the extraction of the measurement data and the MFA analysis. Isotope labeling experiments are unique in that there are both naturally occurring stable isotopes (NOIs) present in the biomass and the experimental label, requiring the measurement data to be corrected. In addition to this and other corrections, there are also necessary data verification steps. To support high-throughput MFA, it was critical to develop a piece of software that could oversee all of the processing tasks as well as remain generic and easily extendible. To this end the iMS2Flux software was developed [41] and is freely available (*see* **Note 3**). This project was executed in an interpreted language (PERL, the Practical Extraction and Reporting Language) so that the source code is immediately available and runs on all major computing platforms. The program is run from the command line utilizing a configuration file; additionally a graphical user interface is available for Microsoft Windows™. The software is modular in design and can easily be extended to support new biological compounds, import and export formats, data checks, and data corrections.

iMS2Flux is organized into four main sections: input, data verification, data correction, and output. The data verification provides feedback that may be used either for targeted manual review of the extracted MDVs of the original chromatogram or to apply automatic pruning via a rule base.

3.4.3 Input

The software developed for processing and managing the extracted MS data supports both a standard generic format (TSV) in addition to multiple proprietary formats including the report format used by QuanLynx. Similar importer/converters may be easily implemented. Although this software is command line oriented, most of the options are located in the configuration file, with only a minimal number of command line switches to override the default operation. Together, this configuration file with one or more MS data files, form the required input to the program. Depending on the configuration options it is also possible to provide additional optional data, each in their own file. For complete information on the configuration file see the iMS2Flux website (*see* **Note 3**).

As part of the modular design the spreadsheet compatible TSV file can be used to provide all necessary data. This is necessary for processing options that require additional information. Although it is a flat file (containing only a single type of data, e.g., MS measurements), multiple files of the same format can be used to provide complementary data (e.g., retention times). The format is column oriented, with the first two columns used to identify the analyte by name (first column) and mass (second column). Similarly the first two rows identify the experiment, with an optional title in the first element of the first row and the second row containing identifiers for each chromatogram. Two examples of the TSV file are shown in Fig. 4, one containing retention time data (Fig. 4b), the second containing the results of the data check after measurement extraction (Fig. 4c). Please note the coloring has been automatically added via macros in the spreadsheet view of this data, the TSV files themselves do not contain any formatting. For additional examples of the TSV file format with MS data see the iMS2Flux website (*see* **Note 3**).

The MS data consists of one or more experiments each containing a set of MDVs (Fig. 1) for each measured analyte or fragment. Biological replicates can be treated individually or automatically concatenated into means and standard deviations. In order to be automatically processed the software requires information about the biochemistry of different compound types, such as the elemental composition. This information is maintained in a series of data classes, one per compound, and the supporting derivatization agent classes. Currently iMS2Flux supports the basic compounds required for high-throughput MFA, with compound classes for amino acids, fatty acid methyl esters, glucose (starch), glycerol, cell wall compounds, and soluble metabolites (a generic class

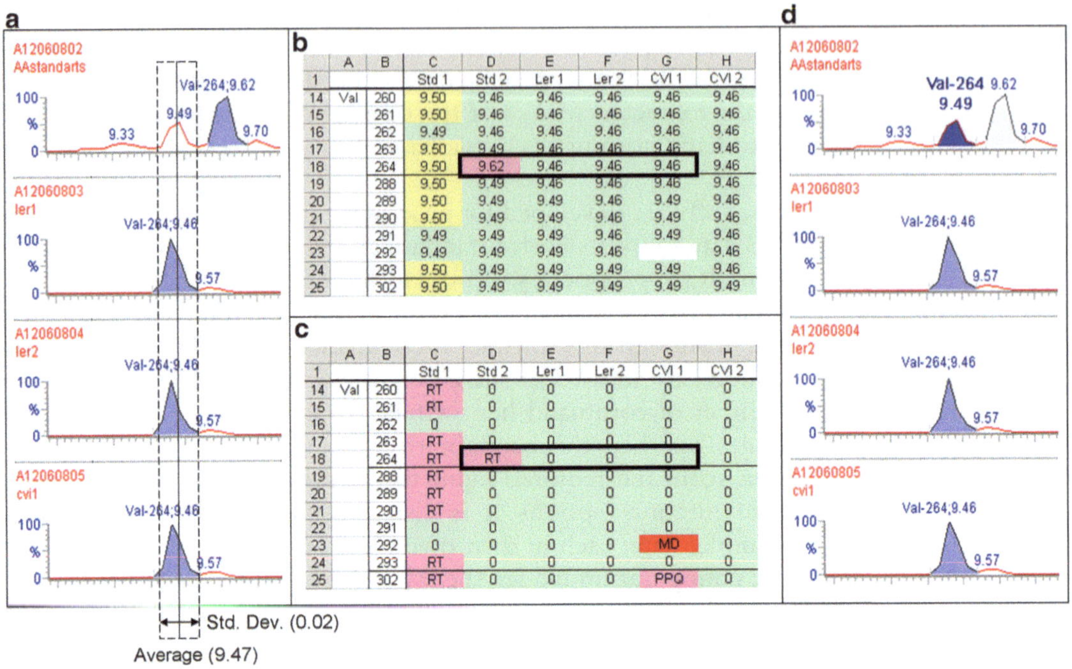

Fig. 4 Data representation and verification. (**a**) The original set of measurement peaks as automatically extracted by QuanLynx™ for valine mass 264 from the four chromatograms corresponding to the highlighted cells in (**b**) and (**c**). (**b**) The matrix of retention time data in TSV format (generated by iMS2Flux after importing from the QuanLynx™ report format). The color is an indication of severity based on retention time error, *green*-no problem, *yellow*-flagged but reasonable, *pink*-requires manual review. (**c**) The data check report summary generated by iMS2Flux. A zero indicates all checks were passed, otherwise a flag is set to indicate failed check, RT—failed retention time check, PPQ—has poor peak quality and MD is missing data (note the corresponding empty cell in (**b**)). (**d**) After manual review the erroneous valine 264 peak has been identified and corrected

seeking to provide the maximum data from a single sample process). Using a class-based modular approach additional compounds, processed in different ways, can be easily added using the existing compounds as a template and only minor interaction with the main program (to register the new compound class).

3.4.4 Data Checking

For any experiment the reliability of the underlying data is most crucial. While the use of automated data extraction eliminates many of the sources of error, it does not guarantee that the underlying data is valid and may even introduce its own errors. As Fig. 3 shows, the step after data extraction is user guided data checking. The most basic data check is for one of the most common problems: missing data values. Missing data occurs for a myriad of reasons including misidentified peaks and high noise levels. Secondary data checks apply threshold limits, identifying measurement intensities below a desired minimum (noise) and outside the linear detector range (maximum). Experience with using QuanLynx™ for the data extraction identified an occasional error that would misidentify peaks. To counter this, the regular use of retention time analysis was added to the standard data checks. Unlike the other data checks which are performed individually on each MDV from each chromatogram, the retention time check is applied to the entire analyte (which may include multiple MDVs if fragmented) and over all chromatograms. This data check is of particular benefit in a high-throughput environment with a sufficient number of chromatograms for the retention time statistics to be meaningful. Figure 4 provides an example illustrating the use of retention time analysis as an integral data check. This is an actual example that arose using QuanLynx™ for automated data extraction. No cause was found for the mistaken selection of the peak at 9.62 rather than the one at 9.49.

The example in Fig. 4 illustrates the general results from performing a data check. Given user provided constraints in the configuration file if any one or more of the MDVs fails a data check feedback is provided via a standard formatted TSV file. Figure 4c illustrates the partial results from this feedback file. The coloring has been applied via an automated macro in the spreadsheet. Each cell corresponds to a single mass reading from a single chromatogram. If all data checks are passed the value of the cell is zero (green), otherwise the cell contains an abbreviation for each data check that was failed. In this case valine mass 292 for the chromatogram CVI1 is missing the measurement value (MD, Red) while below it the valine fragment 302 has not met the minimum threshold, and is identified as having poor peak quality (PPQ, pink). Although only the 302 mass is shown, this data check, and consequently its failure, is applied to the entire MDV. Finally a number of measurements appear to fail the retention time analysis (RT, pink) for chromatograms STD 1 and STD 2. To better evaluate this the corresponding fragment of the retention time data (Fig. 4b) shows that most of these flagged values just exceed one standard deviation (0.02) from the mean retention time (9.47), and is actually consistent with the overall retention times. Only the single retention time for valine 264 of chromatograms STD 2 is then flagged for manual review. Figures 4a and 4d show this chromatogram before and after correction.

Thus the combination of automated data extraction and data checking minimizes the need for manual interaction with chromatograms allowing high-throughput MFA.

3.4.5 Post Correction Checks

While most data checking is performed before any further processing occurs, with isotope labeling data there is an additional verification that can be performed after correction. The average carbon labeling (ACL) check is particularly useful for high-throughput experiments; however it also requires a set of additional unlabeled

Avg. C Labeling

		Unlabeled	Exp. 1	Exp. 2	Exp. 3	Exp. 4
ala	158	10.45%	19.33%	30.91%	19.34%	24.14%
ala	232	0.00%	11.95%	23.94%	11.19%	17.45%
ala	260	0.04%	11.02%	16.45%	11.06%	16.99%
...	...	...	...	...	...	...
tyr	302	0.09%	11.04%	13.52%	10.43%	13.69%
tyr	364	0.51%	10.99%	13.61%	9.90%	13.19%
tyr	438	2.73%	13.44%	18.52%	12.33%	15.73%
tyr	466	0.10%	10.78%	13.26%	9.95%	13.01%
tyr	508	0.27%	10.71%	13.13%	10.01%	12.97%
		3.02%	10.07%	16.83%	10.47%	15.16%

Fig. 5 Results from an average carbon labeling check. This result illustrates the average carbon labeling for the fragments of the first and last amino acids from a set involving 20 chromatograms. The chromatograms represent the concatenated results from four experimental replicates each of four different lines (denoted Exp. 1–Exp. 4) and one unlabeled experiment from the wild type used to identify contamination. From these partial results contaminations (highlighted in *red*) are found for alanine (ala) 158 and tyrosine (tyr) 438

experiments for comparison. ACL can be used to identify contaminations (e.g., from known or unknown coeluting substances) in the measurement values. Simply put, after correcting an unlabeled sample for natural abundance (see next section), the MDV should have 100 % M+0. Of course this is an ideal result; in practice the amount of remaining label should be very small, typically less than 1 %.

Figure 5 illustrates a partial example from running the ACL. An unlabeled data set is processed in parallel with each group of labeling data in the pipeline. After measurement and correction the ACL is used to determine which analyte or fragment measurement values include contamination. When the average carbon labeling of an analyte or fragment in the unlabeled experiment is greater than the threshold (1 %), then that fragment/analyte is determined to be contaminated. In the example two fragments are contaminated, alanine (ala) 158 and tyrosine (tyr) 438. Once identified, these fragments can be removed prior to inclusion with the model for MFA analysis or other use.

3.4.6 Data Correction

For the isotope labeling experiments used in MFA it is necessary to correct the data before proceeding with the analysis. In particular correcting for natural abundance, the presence of NOIs in the measurements, is required. As the later analysis is based on the redistribution or change over time of the label during the experiment it is crucial to remove the affect of NOIs [50]. Additionally, depending on the experiments and type of processing involved other corrections may be required. In the MS process some analytes may gain or lose an individual hydrogen atom, requiring a correction for proton loss [51] (Fig. 6b) or proton

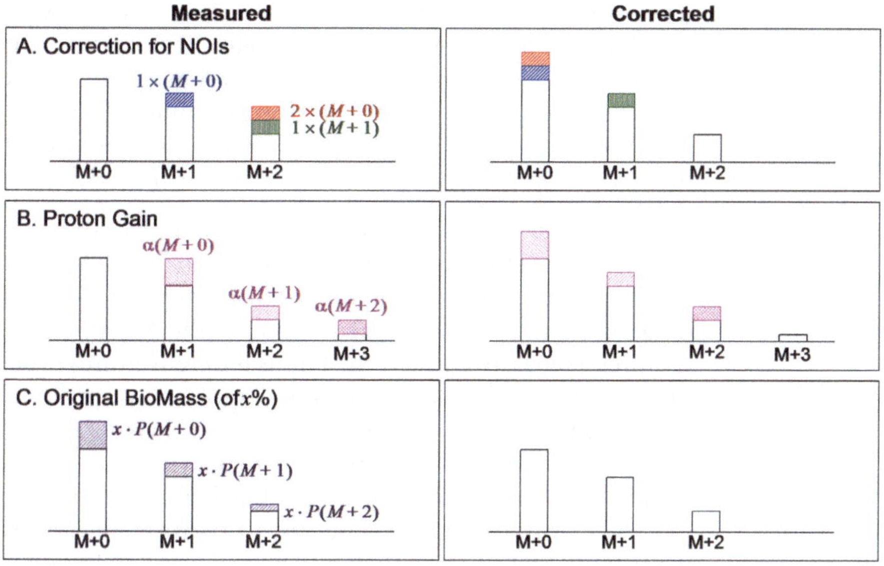

Fig. 6 Overview of MS data correction methods. Each of the three illustrated corrections show the mapping of an MDV with two carbons from their measurement values (on the *left*) to their corrected values (on the *right*), including amounts of each correction. These corrections are (**a**) correction for NOIs, (**b**) correcting for the gain of a proton, which requires at least one extra measurement after M+numC (M+2 in this example), similarly correction for losing a proton follows the same principles, but with the additional measurement before the first one in the MDV, i.e., M+0, and (**c**) the influence of original biomass. Adapted from ref. [41]

gain [12, 42]. For labeling experiments that generate relatively small amounts of biomass, the original (unlabeled) biomass (OBM) may make up a significant percentage of the final biomass, requiring a similar correction as with NOIs but relative to the percentage of OBM [50].

The selected corrections are applied to each MDV individually, and the scaled MDV_{cor} normalized to itself. When working with previously corrected or processed MS data, it is possible to select "no corrections", in which case the data is simply scaled.

Correcting for natural abundance (NA). The NA correction is central to labeling experiments. Figure 6a gives an overview to the correction. On the left hand side the measured abundances show a component of M+1 and M+2 come from the previous masses with either an additional 1 or 2 mass units added via NOIs (with either 1 or 2 additional neutrons). These NOIs can occur in any of the atoms that make up the analyte/fragment as well as any derivatizing agent (if used). The actual correction is based on the probability of natural abundance in the analyte [32],

$$\text{Natural Abundance} = N! \cdot \prod_{i=1}^{n} \frac{p(I_i)^{f(I_i)}}{f(I_i)!} \tag{4}$$

where for a given element, the natural abundance of N atoms with n stable isotopes I_i, each with probability $P(I_i)$ and occurrance $f(I_i)$ times in the analyte. The implementation of this correction in iMS2Flux considers the labeling probabilities of the M+0, M+1, and M+2 isotopes, due to the rapidly diminishing probabilities of heavier isotopes of the constituent elements [29, 52, 53].

Correcting for proton gain or loss. The gain or loss of one (or more) protons is characteristic of the type of analytic (MS device) and analyte. If this gain or loss is limited to a percentage of the molecules of the analyte it causes a partial shift in the MDV, Fig. 6b, which requires a separate correction. This is different from a process with a predictable gain or loss that affects all molecules equally and may be accounted for in the biochemistry of the analyte. For MFA, fatty acid methyl esters are an important analyte that may experience proton gain in the McLafferty ion when measured by GC-MS [12, 42]. As illustrated in Fig. 6b, to perform this correction the probability that an individual molecule is affected is independent of the presence of labeling. Thus the correction in Fig. 6b must determine this constant probability, called the scaling factor, α,

$$\begin{aligned} M_0^{\text{meas}} &= (1-\alpha)M_0^{\text{corr}} \\ M_1^{\text{meas}} &= (1-\alpha)M_1^{\text{corr}} + \alpha M_0^{\text{corr}} \\ M_2^{\text{meas}} &= (1-\alpha)M_2^{\text{corr}} + \alpha M_1^{\text{corr}} \\ M_3^{\text{meas}} &= (1-\alpha)M_3^{\text{corr}} + \alpha M_2^{\text{corr}} \end{aligned} \tag{5}$$

Where M_i^{meas} is the ith measured mass (on the left side) and M_i^{corr} is the ith corrected mass (on the right side).

Correcting for original biomass. The original biomass correction may be required when the starting biomass is relatively large or when the time frame of the experiment is relatively short [50] (such as some time course experiments). In these cases the original unlabeled biomass may bias the measured labeling data. The correction (*see* Fig. 1c) is similar to correcting for NOIs. The main difference is that this correction applies to the NOI of the original unprocessed sample, and as such does not include NOIs from the derivatizing agent(s). Thus carbon remains as the only element having a significant affect on the probability equations, and the OBM is performed considering carbon alone. To perform this correction requires the percentage of original biomass, one value per chromatogram and is provided through an auxiliary file (*see* **Note 3** for iMS2Flux website).

3.4.7 Targeted Output

In the final step, iMS2Flux generates the requested output. This can be as simple as generating TSV file(s) for the corrected measurement data (individually and/or averaged over replicate chromatograms) or the automatic generation of model files for analysis with the desired MFA tools. Standard output is made in the previously described TSV format, with one type of data per file.

If additional boundary measurements are provided with each MDV they are only saved with the uncorrected data, while only the central MDV is saved with processed data.

To complete the data pipeline for high-throughput MFA, iMS2Flux can automatically integrate the measurement data with one or more model files. When generating model files the average and standard deviation of replicate sets are always used. For each provided model file the software will generate a new model file for each experimental replicate. One caveat to this automation is the use of consistent naming of the analytes in the model file. iMS2Flux currently supports the generation of MS data for use with three third party MFA tools, the FTBL format for use with 13CFLUX [38], the FML format of 13CFLUX2, and the CSV format of OpenFLUX [39].

3.5 Analysis

The actual MFA is performed by joining the measurement data to a model representing the stoichiometry and atom transition of the system. MFA is an optimization or parameter fitting problem to determine the set of constant fluxes or enzyme kinetic parameters. Finally the resulting set of fluxes/parameters must be evaluated.

3.5.1 Model Development

Model development remains a mostly manual endeavor, encompassing up to 94 individual steps for a genome scale reconstruction [54]. Initial attempts to automate this process have led to a semi-automated model generation tool for microbes [55]. The main limitation to fully automating this type of tool is a dearth of high quality, reliable, and digitized data. Despite this, manual model generation is not a hindrance to high-throughput MFA, as typically only a very small number of different models are used. With steady-state ^{13}C MFA the model specifies the included reactions together with the underlying atom transitions, and any existing domain knowledge such as known flux values or flux ranges. For high-throughput MFA the analysis of a large number of lines for a given species may all be performed using the same model. The model is then easily adapted to variants which may be used to explore alternate pathways [56] or targeted mutations.

3.5.2 Optimization

When performing metabolic flux analysis the type of MFA and desired results often dictate whether available software tools are sufficient. For steady-state ^{13}C MFA there are a plethora of choices available, based on different labeling descriptions, optimization methods, platforms, and even the target (e.g., fluxes, flux ratios and fluxomers). Wiechert et al. [57] introduced 13CFLUX a proprietary linux-based tool utilizing the cumomer labeling description, supporting multiple optimization strategies and statistical analysis. 13CFLUX uses a simple text based column oriented tabular model file, the Flux TaBLe (FTBL) format which can be easily automated, as well as a converter to a more programming oriented text based file (FLX). A completely different approach was taken by FiatFlux [40], a proprietary MATLAB™ based tool calculating flux ratios. Additionally FiatFlux provides a single solution platform from chromatogram to results, although at the expense of flexibility. In general these tradeoffs limit the use of FiatFlux in plant-based MFA. Antoniewicz et al. [58] proposed an alternative to cumomers based on elementary metabolite units (EMU), which has spawned two software tools, Metran [59] and OpenFLUX [39]. Metran is proprietary and MATLAB™ based. It features a GUI interface for model design and a profile likelihood based approach to calculating confidence intervals. OpenFLUX is an open source MATLAB™ based tool that supports a simple FTBL like model file. By default OpenFLUX relies on the MATLAB™ *fmincom* optimizer, however as it is open source this allows for alternative optimizers to be implemented. Using the fluxomer iterative algorithm (FIA), the Fluxomer [60] tool is also open source based on Python and utilizes a composite variable composed of both isotopomer abundances and metabolic fluxes. Finally an update to 13CFLUX, 13CFLUX2 [61], has been made available. This version uses a completely new interface and file formats, but greatly extends the abilities of 13CFLUX, but remains Linux based, command line oriented and provides backwards support to the FTBL file format. The software extends support to larger models, supporting both EMU and cumomer based solutions and automatic determination of feasibility for initial flux values. The software has been written as a tool set which may be easily automated in part or in whole. Despite this there are issues that must be overcome, in particular the installation has become quite complicated and limited to only the most recent versions of certain linux platforms. In addition the academic license now includes provisions for uploading each model for license verification.

To support high-throughput using an available MFA tool limits the choice of tools. For ease of automation we required that the software must support a command line interface and have a text-based model format that could be manipulated automatically. As the nature of high-throughput MFA allows for course grained parallelism, e.g., each measurement set may be run independently of the others (*see* Fig. 7), it was desirable that the software could be run in a Linux environment to make use of existing high performance computing resources. Thus 13CFLUX, OpenFLUX (assuming the availability of MATLAB™), and 13CFLUX2 were selected, and 13CFLUX fully implemented (support for the others is ongoing).

3.5.3 autoFLUX/clusterFLUX

An automation tool, autoFLUX, was developed to implement the high-throughput analysis using the existing MFA software as described above. autoFLUX is designed to run on a single user computer but supports parallelization over multiple processing

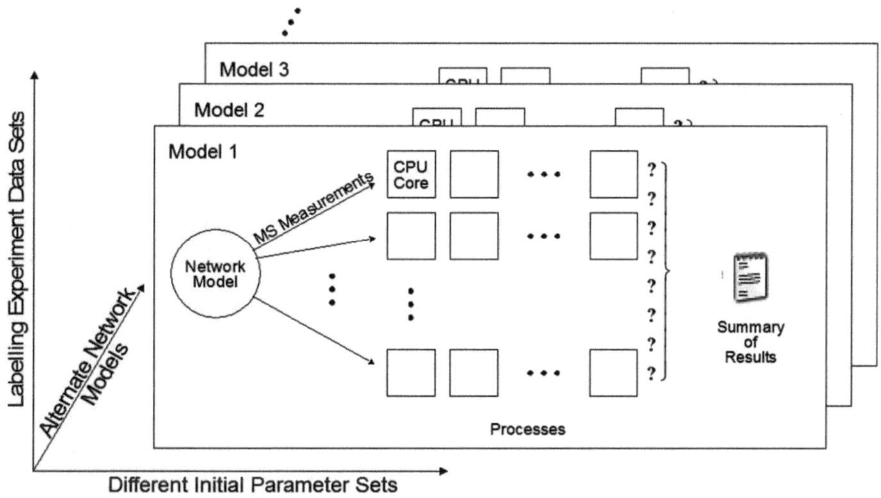

Fig. 7 Course-grained view of MFA parallelization. High-throughput MFA can be viewed as having three dimensions of parallelism. The *x*-axis represents the different sets of initial values for the fluxes or parameters, which provide a targeted or random search of the flux space. The *y*-represents the different experimental data sets (one for each labeling experiment). The *z*-axis represents the number of biological models (sets of biochemical pathways). Together the number of processes is the product of these three quantities, each of which may be executed independently

cores. This was then extended to support MFA in a cluster environment as a separate tool, clusterFLUX.

The basic form of each tool automates the search of the input space for a single model with a single measurement set (*see* Fig. 7, the *x*-axis). There is support for four search types, a grid search (good for small numbers of free fluxes), a random search (based on user defined parameter ranges), a targeted or list based search and a confidence interval search based on the profile likelihood approach of Metran [59]. For high-throughput the use of these tools is extended to support searching the input space of multiple model files which may consist of both different models and different measurement data (*see* Fig. 7, the *y*- and *z*-axes). The software summarizes the results for each model with a degree of detail specified in the configuration file.

3.5.4 Statistical Analysis, Results, and Visualization

Most of the steady-state ^{13}C MFA analysis platforms include some form of statistical analysis, typically the asymptotic confidence interval based on the curvature of the likelihood (derived from the Hessian matrix) and finite sample confidence interval, where the confidence region is based on the profile likelihood [62].

Whether performing a Monte Carlo simulation or some other search of the input space, the result may be an overwhelming number of solutions. One set of flux values must be extracted from each combination of model and measurement set. In purely

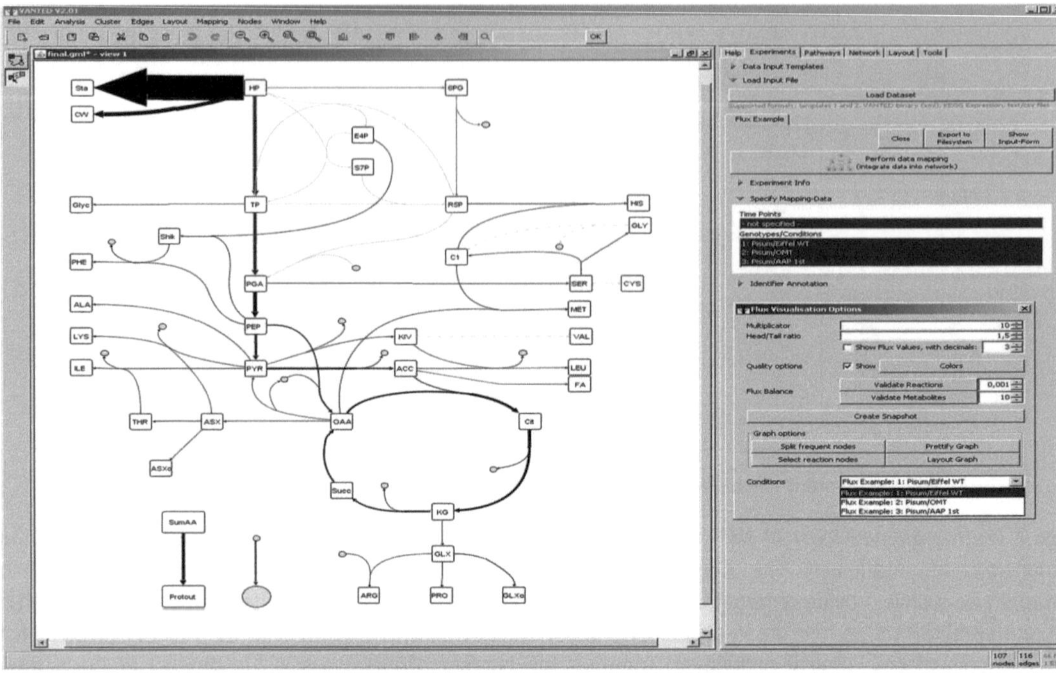

Fig. 8 Automatically generated flux map using VANTED [63] with the FluxMap plugin [65]. The visualization is automatically generated based on the selected set of flux values. The thickness of each *arrow* is a direct indication of carbon flow based on the net fluxes, while the darkness (*grey scale*) of the *lines* indicates the confidence interval

numerical terms the solution with the lowest residual (*see* Eq. 3) is the optimum; however it is crucial to ensure that this solution is also biologically relevant. More typically a clustering of the subset of the best results from the input space search reveals a finite number of flux solutions which can be more easily verified for biological significance.

Typically the final step in MFA is visualization of these results. To support high-throughput MFA there are both general biological visualization tools such as VANTED [63] and those which specifically target MFA, such as OMIX [64], that are capable of automating both the design and integration of data into the visualization. Figure 8 shows one such visualization from VANTED using the FluxMap plugin [65], given a single network layout the software automatically applies the net fluxes and standard deviations visually, as line width and color gradient respectively. The flux data is provided in a simple spreadsheet format that can be generated as a text-based TSV file, and containing the results from multiple MFA experiments. Each experimental result may be selected individually or an animation type slideshow may be used.

4 Notes

1. QuanLynx™ is a standard component of the MassLynx™ mass spectrometry software from Waters used for automated quantification of MS chromatograms.

   A users guide for QuanLynxTM as part of this workflow is provided with the iMS2Flux software in the *Getting_Started* folder.

2. 13CFLUX is a closed source metabolic flux analysis toolkit that runs on Linux. The software contains several tools including four optimization engines and likelihood-based statistical analysis. This tool is widely used and documented for both plant and microbial MFA. The software is available as a free trial version and a licensed full version (free for academic use), http://www.13cflux.net/13cflux/

3. iMS2Flux is a PERL-based program for managing MS data from isotope labeling experiments, specifically tailored to support metabolic flux analysis. The distribution is fully open and may be extended as desired to meet individual requirements. This tool requires a PERL interpreter be installed and has been tested on Windows, OSX, Linux, and Unix; however, it should run on any platform with PERL (version 5 and up).

   The iMS2Flux software and all auxiliary files including a full example with extracted chromatogram files and instructions can be downloaded from the SourceForge project website: http://sourceforge.net/projects/iMS2Flux

4. autoFLUX/clusterFLUX is an automation tool for the 13CFLUX MFA software. autoFLUX is a generic tool for a single user computer supporting course grain parallel support over multiple processors/processing cores. clusterFlux is a port of autoFlux to the Michigan State High Performance Computing Center (ICER). These are both PERL-based programs and also require the accompanying 13CFLUX software which requires the Linux environment. The autoFLUX software is available upon request. The clusterFLUX software may also be obtained from the authors; however, it may require modification for different cluster environments.

Acknowledgments and Funding

Work at Michigan State University by CHP and YSH was funded by the DOE Great Lakes Bioenergy Research Center.

This work was supported by the German Federal Ministry of Education and Research (BMBF) through grants 0315295 and 0315426A.

References

1. Allen DK, Libourel IGL, Shachar-Hill Y (2009) Metabolic flux analysis in plants: coping with complexity. Plant Cell Environ 32: 1241–1257
2. Roessner U, Wagner C, Kopka J, Trethewey RN, Willmitzer L (2000) Simultaneous analysis of metabolites in potato tuber by gas chromatography–mass spectrometry. Plant J 23:131–142
3. Dunn WB (2008) Current trends and future requirements for the mass spectrometric investigation of microbial, mammalian and plant metabolomes. Phys Biol 5:011001
4. Sauer U (2004) High-throughput phenomics: experimental methods for mapping fluxomes. Curr Opin Biotechnol 15:58–63
5. Ratcliffe RG, Shachar-Hill Y (2006) Measuring multiple fluxes through plant metabolic networks. Plant J 45:490–511
6. Blum JJ, Stein RB (1982) On the analysis of metabolic networks. In: Goldberger RF (ed) Biological regulation and development. Plenum, New York, pp 99–124
7. Wiechert W, de Graaf AA (1996) In vivo stationary flux analysis by ^{13}C labeling experiments. In: Sahm H, Wandrey C (eds) Metabolic engineering. Springer, Berlin, pp 109–154
8. Fischer E, Sauer U (2003) Large-scale in vivo flux analysis shows rigidity and suboptimal performance of Bacillus subtilis metabolism. Nat Genet 37:636–640
9. Haverkorn van Rijsewijk BRB, Nanchen A, Nallet S, Kleijn RJ, Sauer U (2011) Large-scale 13C-flux analysis reveals distinct transcriptional control of respiratory and fermentative metabolism in Escherichia coli. Mol Syst Biol 7:477
10. Munger J, Bennett BD, Parikh A et al (2008) Systems-level metabolic flux profiling identifies fatty acid synthesis as a target for antiviral therapy. Nat Biotechnol 26:1179–1186
11. Amador-Noguez D, Feng X-J, Fan J et al (2010) Systems-level metabolic flux profiling elucidates a complete, bifurcated tricarboxylic acid cycle in Clostridium acetobutylicum. J Bacteriol 192:4452–4461
12. Schwender J, Ohlrogge JB (2002) Probing in vivo metabolism by stable isotope labeling of storage lipids and proteins in developing Brassica napus embryos. Plant Physiol 130: 347–361
13. Junker BH, Lonien J, Heady LE, Rogers A, Schwender J (2007) Parallel determination of enzyme activities and in vivo fluxes in Brassica napus embryos grown on organic or inorganic nitrogen source. Phytochemistry 68: 2232–2242
14. Lonien J, Schwender J (2009) Analysis of metabolic flux phenotypes for two Arabidopsis mutants with severe impairment in seed storage lipid synthesis. Plant Physiol 151:1617–1634
15. Alonso AP, Val DL, Shachar-Hill Y (2011) Central metabolic fluxes in the endosperm of developing maize seeds and their implications for metabolic engineering. Metab Eng 13:96–107
16. Williams TCR, Miguet L, Masakapalli SK et al (2008) Metabolic network fluxes in heterotrophic Arabidopsis cells: stability of the flux distribution under different oxygenation conditions. Plant Physiol 148:704–718
17. Halket JM, Waterman D, Przyborowska AM et al (2005) Chemical derivatization and mass spectral libraries in metabolic profiling by GC/MS and LC/MS/MS. J Exp Bot 56:219–243
18. Kopka J, Schauer N, Krueger S et al (2005) GMD@CSB.DB: the Golm metabolome database. Bioinformatics 21:1635–1638
19. Beckonert O, Keun HC, Ebbels TMD et al (2007) Metabolic profiling, metabolomic and metabonomic procedures for NMR spectroscopy of urine, plasma, serum and tissue extracts. Nat Protoc 2:2692–2703
20. Brown M, Wedge DC, Goodacre R et al (2011) Automated workflows for accurate mass-based putative metabolite identification in LC/MS-derived metabolomic datasets. Bioinformatics 27:1108–1112
21. Chan ECY, Pasikanti KK, Nicholson JK (2011) Global urinary metabolic profiling procedures using gas chromatography-mass spectrometry. Nat Protoc 6:1483–1499
22. Libourel IGL, Gehan JP, Shachar-Hill Y (2007) Design of substrate label for steady state flux measurements in plant systems using the metabolic network of Brassica napus embryos. Phytochemistry 68:2211–2221
23. Schauer N, Steinhauser D, Strelkov S et al (2005) GC-MS libraries for the rapid identification of metabolites in complex biological samples. FEBS Lett 579:1332–1337
24. Lisec J, Schauer N, Kopka J, Willmitzer L, Fernie AR (2006) Gas chromatography mass spectrometry-based metabolite profiling in plants. Nat Protoc 1:387–396
25. Erban A, Schauer N, Fernie AR, Kopka J (2007) Nonsupervised construction and application of mass spectral and retention time index libraries from time-of-flight gas

chromatography-mass spectrometry metabolite profiles. Humana Press, Totowa, NJ

26. Allwood JW, De Vos RC, Moing A et al (2011) Plant metabolomics and its potential for systems biology research background concepts, technology, and methodology. Methods Enzymol 500:299–336
27. Fernie AR, Geigenberger P, Stitt M (2005) Flux an important, but neglected, component of functional genomics. Curr Opin Plant Biol 8:174–182
28. Moxley JF, Jewett MC, Antoniewicz MR et al (2009) Linking high-resolution metabolic flux phenotypes and transcriptional regulation in yeast modulated by the global regulator Gcn4p. Proc Natl Acad Sci USA 106:6477–6482
29. Wittmann C, Heinzle E (1999) Mass spectrometry for metabolic flux analysis. Biotechnol Bioeng 62:739–750
30. Nöh K, Wiechert W (2006) Experimental design principles for isotopically instationary 13C labeling experiments. Biotechnol Bioeng 94:234–251
31. Wahl SA, Noh K, Wiechert W (2008) ^{13}C labeling experiments at metabolic nonstationary conditions: an exploratory study. BMC Bioinformatics 9:18
32. Allen DK, Ratcliffe RG (2009) Quantification of isotope label. In: Schwender J (ed) Plant metabolic networks. Springer, New York, pp 105–149
33. Masakapalli SK, Le Lay P, JHuddleston E et al (2010) Subcellular flux analysis of central metabolism in a heterotrophic Arabidopsis cell suspension using steady-state stable isotope labeling. Plant Physiol 152:602–619
34. Luedemann A, Strassburg K, Erban A, Kopka J (2008) TagFinder for the quantitative analysis of gas chromatography–mass spectrometry (GC-MS)-based metabolite profiling experiments. Bioinformatics 24:732–737
35. Pluskal T, Castillo S, Villar-Briones A, Orešič M (2010) MZmine 2: modular framework for processing, visualizing, and analyzing mass spectrometry-based molecular profile data. BMC Bioinformatics 11:1–11
36. Wahl SA, Dauner M, Wiechert W (2004) New tools for mass isotopomer data evaluation in 13C flux analysis: mass isotope correction, data consistency checking, and precursor relationships. Biotechnol Bioeng 85:259–268
37. Huege J, Goetze J, Schwarz D et al (2011) Modulation of the major paths of carbon in photorespiratory mutants of *Synechocystis*. PLoS One 6:e16278
38. Wiechert W, Mollney M, Petersen S, de Graaf AA (2001) A universal framework for 13C metabolic flux analysis. Metab Eng 3:265–283
39. Quek LE, Wittmann C, Nielsen LK, Krömer JO (2009) OpenFLUX: efficient modelling software for 13C-based metabolic flux analysis. Microb Cell Fact 8:25
40. Zamboni N, Fischer E, Sauer U (2005) FiatFlux—a software for metabolic flux analysis from 13C-glucose experiments. BMC Bioinformatics 6:209
41. Poskar CH, Huege J, Krach C et al (2012) iMS2Flux—a high-throughput processing tool for stable isotope labeled mass spectrometric data used for metabolic flux analysis. BMC Bioinformatics 13:295
42. Allen DK, Shachar-Hill Y, Ohlrogge JB (2007) Compartment-specific labeling information in ^{13}C metabolic flux analysis of plants. Phytochemistry 68:2197–2210
43. Allen DK, Ohlrogge JB, Shachar-Hill Y (2009) The role of light in soybean seed-filling metabolism. Plant J 58:220–234. doi:10.1111/j.1365-313X.2008.03771.x
44. Kruger NJ, Masakapalli SK, Ratcliffe RG (2011) Strategies for investigating the plant metabolic network with steady-state metabolic flux analysis: lessons from an Arabidopsis cell culture and other systems. J Exp Bot 6: 2309–2323
45. Szyperski T (1995) Biosynthetically directed fractional 13C-labeling of proteinogenic amino acids. Eur J Biochem 232:433–448
46. Alonso AP, Piasecki RJ, Wang Y, LaClair RW, Shachar-Hill Y (2010) Quantifying the labeling and the levels of plant cell wall precursors using ion chromatography tandem mass spectrometry. Plant Physiol 153:915–924
47. Mawhinney TP, Robinett RSR, Atalay A, Madson MA (1986) Analysis of amino acids as their tert.-butyldimethylsilyl derivatives by gas–liquid chromatography and mass spectrometry. J Chromatogr A 358:231–242
48. Huege J, Sulpice R, Gibon Y et al (2007) GC-EI-TOF-MS analysis of in vivo carbon-partitioning into soluble metabolite pools of higher plants by monitoring isotope dilution after $^{13}CO_2$ labelling. Phytochemistry 68: 2258–2272
49. Talwar P, Wittmann C, Lengauer T, Heinzle E (2003) Software tool for automated processing of 13C labeling data from mass spectrometric spectra, vol 35. Eaton, Natick, MA, p 2
50. Fischer E, Sauer U (2003) Metabolic flux profiling of Escherichia coli mutants in central carbon metabolism using GC-MS. Eur J Biochem 270:880–891
51. Klapa MI, Aon JC, Stephanopoulos G (2003) Ion-trap mass spectrometry used in combination with gas chromatography for

high-resolution metabolic flux determination. Biotechniques 34:832–836, 838, 840
52. Lee WNP, Byerley LO, Bergner EA, Edmond J (1991) Mass isotopomer analysis: theoretical and practical considerations. Biol Mass Spectrom 20:451–458
53. Van Winden WA, Wittmann C, Heinzle E, Heijnen JJ (2002) Correcting mass isotopomer distributions for naturally occurring isotopes. Biotechnol Bioeng 80:477–479
54. Thiele I, Palsson BO (2010) A protocol for generating a high-quality genome-scale metabolic reconstruction. Nat Protoc 5:93–121
55. Feng X, Xu Y, Chen Y, Tang YJ (2012) MicrobesFlux: a web platform for drafting metabolic models from the KEGG database. BMC Syst Biol 6:94
56. Schwender J, Goffman F, Ohlrogge JB, Shachar-Hill Y (2004) Rubisco without the Calvin cycle improves the carbon efficiency of developing green seeds. Nature 432:779–782
57. Wiechert W, de Graaf AA (1997) Bidirectional reaction steps in metabolic networks: I. Modeling and simulation of carbon isotope labeling experiments. Biotechnol Bioeng 55:101–117
58. Antoniewicz MR, Kelleher JK, Stephanopoulos G (2007) Elementary metabolite units (EMU): a novel framework for modeling isotopic distributions. Metab Eng 9:68–86
59. Yoo H, Antoniewicz MR, Stephanopoulos G, Kelleher JK (2008) Quantifying reductive carboxylation flux of glutamine to lipid in a brown adipocyte cell line. J Biol Chem 283: 20621–20627
60. Srour O, Young J, Eldar Y (2011) Fluxomers: a new approach for 13C metabolic flux analysis. BMC Syst Biol 5:129
61. Available from: http://www.13cflux.net/13cflux2/
62. Raue A, Kreutz C, Maiwald T et al (2009) Structural and practical identifiability analysis of partially observed dynamical models by exploiting the profile likelihood. Bioinformatics 25:1923–1929
63. Junker BH, Klukas C, Schreiber F (2006) VANTED: a system for advanced data analysis and visualization in the context of biological networks. BMC Bioinformatics 7:109
64. Droste P, Weitzel M, Wiechert W (2008) Visual exploration of isotope labeling networks in 3D. Bioprocess Biosyst Eng 31:227–239
65. Rohn H, Hartmann A, Junker A, Junker BH, Schreiber F (2012) FluxMap: a VANTED add-on for the visual exploration of flux distributions in biological networks. BMC Syst Biol 6:33
66. Junker B (2011) Steady state-13C-Flussanalysen in Pflanzen. BIOspektrum 17:230–232

Part III

Predicting Fluxes in Plants

Chapter 15

Analysis of Enzyme Activities

Guillaume Ménard, Benoit Biais, Duyên Prodhomme, Patricia Ballias, and Yves Gibon

Abstract

The evaluation of enzyme activities, especially their capacities, represents an important step towards the modelling of biochemical pathways in living organisms. The implementation of microplate technology enables the determination of up to >50 enzymes in relatively large numbers of samples and in various biological materials. Most of these enzymes are involved in central metabolism and several pathways are entirely covered. Direct or indirect assays can be used, as well as highly sensitive assays, depending on the abundance of the enzymes under study. To exemplify such methods, protocols for UDP-glucose pyrophosphorylase (E.C. 2.7.7.9) operating in real time and for pyrophosphate:fructose-6-phosphate 1-phosphotransferase (E.C. 2.7.1.90) are presented.

Key words Enzyme activity, Microplate, Direct assay, Stopped assay, Cycling assay

1 Introduction

Enzymes, the players of metabolic networks, have been discovered more than 100 years ago [1] and intensively investigated since. Many of them have been extracted, purified, and assessed in vitro for their properties and classified. These efforts have been conducted in a wide range of species, organs, tissues or cell types as well as growth conditions; they led to a very good knowledge of the topology of metabolism when combined to the exploration of metabolite diversity. Plenty of mutants with altered enzyme properties have been isolated and characterized and genes encoding wild type or mutated enzymes got cloned and sequenced, which led to a growing understanding of structure–function relationships when combined to data obtained from crystallization. Molecular biology opened up the possibility to increase or decrease the activities of enzymes "à la carte" in model organisms. Although such genetic

Conflict of Interest The authors declare that they have no conflict of interest.

Martine Dieuaide-Noubhani and Ana Paula Alonso (eds.), *Plant Metabolic Flux Analysis: Methods and Protocols*, Methods in Molecular Biology, vol. 1090, DOI 10.1007/978-1-62703-688-7_15, © Springer Science+Business Media New York 2014

Table 1
Examples of activities found in leaves and/or fruits of several plant species

	Organ/tissue	Pyrophosphate:fructose-6-phosphate 1-phosphotransferase (nmol.min^{-1}.g^{-1}FW)	UDP-glucose pyrophosphorylase (nmol^{1}.min^{-1}.g^{-1}FW)
Solanum lycopersicum	Leaf	74	2,810
	Green fruit	596	3,309
	Red fruit	491	2,972
Nicotiana tabaccum	Leaf	203	3,388
Arabidopsis thaliana	Leaf	190	3,300
Zea mays	Ear	448	2,541
	Leaf	37	986
Pinus sylvestris	Leaf	17	3,650

For each enzyme, activity has been measured in one sample of a "biological standard" used on the enzyme platform. Technical error is usually below 8 % (coefficient of variation, not shown)

manipulations provided great experimental evidence [2–4] validating the metabolic control theory [5], and accessorily generated plenty of interesting phenotypes, it must be recognized that attempts to improve organisms by engineering enzyme properties haven't been that successful [6]. Indeed, altering the in vivo activity of enzymes does not necessarily affect fluxes or metabolite concentrations. For example, the overexpression of an enzyme that has only very little control over a flux would probably have no noticeable effect. In other words, it became obvious that studying each enzyme separately would not lead to a comprehensive understanding of metabolic networks. Furthermore, such understanding will be needed for a better selection of targets for genetic improvement of organisms. Clearly, the huge accumulation of knowledge about enzymes that can be found in laboratories, literature, and databases such as BRENDA [7] now needs to be assembled into a global understanding of metabolic networks.

The collection of data about enzyme "capacities," which are defined as activities of extracted enzymes measured in vitro under optimal and standardized conditions, as close as possible to in vivo conditions (pH, temperature, ion concentrations), represents a critical step for the parameterization of models describing metabolic networks, especially kinetic models [8]. Indeed, capacities, which also reflect the amount of the corresponding active protein, vary dramatically (Table 1) between species, organs, tissues, and cell types, but also throughout development or in response to the environment. In contrast, kinetic parameters such as K_m or K_i are usually considered as less variable for a given enzymatic isoform and can be gathered from literature or databases [9]. Additionally, efforts have been undertaken in order to standardize kinetic data in terms of, e.g., pH, temperature, and ion concentrations (10, *STRENDA initiative*). However, variations in kinetic properties of enzymes also occur in vivo as a consequence of posttranslational modifications such as phosphorylation or oxido-reduction that occur in response to environmental or developmental cues. Ideally, such modifications would also be assayed, at least for enzymes such as nitrate reductase [11] or sucrose phosphate synthase [12] that are often assumed to exert a strong control over their respective pathways.

Paradoxically, while enzymes have been studied for decades and protocols made available for almost every known enzyme (e.g., 13), to obtain quantitative activity profiles for an entire metabolic pathway remains a complicated task for modellers. The determination of enzyme activities recently became increasingly rare and only a few laboratories have developed platforms dedicated to such profiling. Thus, in plant research there is one in Germany (*Stitt lab at the Max Planck Institute of Molecular Plant Physiology, Potsdam, Germany*, 14), one in the USA (*Rogers lab, Environmental Sciences Department, Brookhaven National Laboratory, Upton, USA*, 15) and the one we have recently established in Bordeaux, France. These platforms use microplate technology including robotics and offer the possibility to quantify the activities of up to 60 enzymes involved in central carbon [16] and nitrogen metabolism [17] as well as redox metabolism [18]. On these platforms, preexisting assays have been optimized for microplates and new assay principles developed. Still, they can be used in any lab equipped with a basic microplate reader as most of them are performed in the UV-visible range. In the present chapter we describe two exemplary assays, one for the usually highly active UDP-glucose pyrophosphorylase (E.C. 2.7.7.9) operating in real time and one for pyrophosphate:fructose-6-phosphate 1-phosphotransferase (E.C. 2.7.1.90), which is a stopped assay followed by a determination step involving a highly sensitive "cycling assay" (*see* Table 1 for examples of activities found in plant species for each activities). Protocols for other enzymes can be obtained by contacting the authors.

Activity of a given enzyme is expressed as the amount of product formed or substrate consumed per unit of time and per unit of, e.g., protein, fresh weight, dry weight, or leaf surface. The SI unit for enzyme activity is the katal (mol/s), but units such as μmol/min or nmol/min are very often found in literature. Measurements are performed under known conditions of temperature, pH, and substrate concentration. As such conditions may vary from batch to batch, especially temperature, it is recommended to use biological standards in addition to technical blanks (extraction buffer alone). A biological standard can be defined as material that is representative for the samples under study and in which measured activities are known (e.g., they have been previously determined using chemical standards). Activity measurements usually consist in calculating the initial rate, which is defined as the slope of the tangent to the progress curve at time 0.

Briefly the process can be summarized as follows:

1. Sampling the plant tissue, homogenizing, and exact weighing of sample aliquots.
2. Preparing sample plates by randomizing samples and adding blanks and biological standards.
3. Extracting enzymes concomitant with protease inactivation.
4. Diluting samples.
5. Preparing assay plates.
6. Mixing extracts and assay mixes.
7. Reading of microplates, directly or indirectly after having stopped the reaction.
8. Calculations.

2 Materials and Reagents

2.1 General

1. Centrifuge with microplate adapter (capable of 2, 900 × *g*), (e.g., Eppendorf 5810).
2. Micronic tubes 1.1 ml or similar.
3. Flat bottom polystyrene microplates.
4. MilliQ water.
5. Hand pipettes (single and 8-channel), 0.5–10 μl, 5–100 μl, 50–1,200 μl ranges (e.g., Eppendorf research Pro).
6. Thermoblock (capable of heating to up to 95 °C).
7. Liquid nitrogen supply.
8. Vortex and microplate shaker.
9. Scalpel blades, aluminum foil, 96-well plates, spatula, scales ($d = 0.1$ mg).

2.2 Extraction Buffer for Enzymes

2.2.1 Stock Solutions

1. Pre-buffer—10×: 500 mM HEPES, 100 mM $MgCl_2$, 10 mM EDTA, 10 mM EGTA, 10 mM Benzamidine, 10 mM ε-aminocapronic acid, 2.5 % (w/v) Bovine serum albumine (protease free)

   When preparing the buffer, bring volume to about 40 ml, adjust pH to 7.5 with 10 M KOH then adjust the volume to 50 ml. Store at −20 °C.
2. 2 mM Leupeptin—freeze in liquid nitrogen and then store at—80 °C.
3. 500 mM DTT—freeze in liquid nitrogen and then store at—80 °C.

4. 100 mM Phenylmethylsulfonyl fluoride (PMSF) in isopropanol—store at 4 °C, protect from light, stable for 1 month.
5. 10 % (v/v) Triton X100—store at room temperature.
6. Glycerol.

2.2.2 Preparing the Buffer for One Batch of 48 Extracts

Pipette 2.5 ml of pre-buffer into a 50 ml plastic tube; add 2.5 ml glycerol, 1.25 ml of Triton X100, 250 μl of leupeptin and PMSF and 25 μl DTT. Adjust with milliQ™ water at 25 ml. The buffer is stable for a few hours when stored on ice. Note that if dilutions of extracts are planned, the necessary supplemental volume of extraction buffer needs to be prepared.

2.3 Assay Mix for UDP-Glucose Pyrophosphorylase

2.3.1 Stock Solutions

1. Assay buffer 5× (store at −20 °C): 50 mM $MgCl_2$, 0.5 M Tricine–KOH, pH 8, 10 mM EDTA, 0.25 % Triton X100.
2. 20 mM $NADP^+$ [−80 °C].
3. Glucose-6P dehydrogenase (E.C. 1.1.1.49) 50 U/ml in 200 mM Tricine–KOH pH 8.0, 10 mM $MgCl_2$ [−80 °C].
4. Phosphoglucomutase (E.C. 5.4.2.2) 50 U/ml in 200 mM Tricine–KOH pH 8.0, 10 mM $MgCl_2$ [−80 °C].
5. PPi 5 mM [Fresh].
6. 1 M Tricine–KOH pH 8.0 [4 °C].
7. 50 mM UDPG [−80 °C].

2.3.2 Assay Buffer for One Microplate

In a 15 ml plastic tube kept on ice, pipette as following: 2.16 ml H_2O, 2.4 ml Assay buffer 5×, 720 μl $NADP^+$, 240 μl UDPG, 240 μl G6PDH, 240 μl PGM, 5.4 ml H_2O (V_{blank}) or PPi (V_{sat}).

Store the assay mix at 4 °C before use (no longer than an hour as UDP-glucose is very unstable).

2.4 Assay Mix for Pyrophosphate: Fructose-6P Phosphotransferase

2.4.1 Stock Solutions

1. Assay buffer 5× (stored at −20 °C): 0.25 M HEPES–KOH pH 7.5, 10 mM $MgCl_2$, 0.25 % Triton X100.
2. 0.5 M HCl mixed with 100 mM Tricine–KOH, pH 8 [RT].
3. 0.5 M NaOH [RT].
4. 1 M $MgCl_2$ [RT].
5. 1 M Tricine–KOH, pH 8 [4 °C].
6. 750 mM Fructose-6P [−80 °C].
7. Fructose-1.6-bisP: 0–50–100–200 μM in the extraction buffer (for calibration) [−80 °C].
8. 1 mM Fructose-2.6-bisP [−80 °C].
9. 25 mM NADH [−80 °C].
10. 66 mM NADH [−80 °C].
11. 2.5 mM PPi [−80 °C].

12. Aldolase 100 U/ml in 50 mM HEPES–KOH, pH 7.5, 10 mM $MgCl_2$ [−80 °C].
13. Glycerol-3P dehydrogenase 200 U/ml in 100 mM Tricine–KOH pH 8, 10 mM $MgCl_2$ [−80 °C].
14. Glycerol-3P oxidase 500 U/ml in 100 mM Tricine–KOH pH 8, 10 mM $MgCl_2$ [−80 °C].
15. Triose-P-isomerase 100 U/ml in 50 mM HEPES-KOH, pH 7.5, 10 mM $MgCl_2$ [−80 °C].

2.4.2 Assay Buffer for One Microplate

In a 15 ml plastic tube kept on ice, pipette as following: 1,296 μl H_2O, 480 μl assay buffer 5×, 24 μl fructose-6P, 24 μl Fructose-2,6-bisP, 24 μl aldolase, 24 μl triose-P isomerase, 24 μl glycerol-3P dehydrogenase, 24 μ NADH 25 mM, 240 μl H_2O (V_{blank}) or PPi (V_{sat}).

Store the assay mix at 4 °C before use (no longer than a day).

2.4.3 Determination Mix for One Microplate

In a 15 ml plastic tube kept on ice, pipette as following: 4,536 μl H_2O, 1,200 μl Tricine–KOH pH 8 buffer, 120 μl 66 mM NADH, 60 μl glycerol-3P dehydrogenase, 60 μl glycerol-3P oxidase, 60 μl $MgCl_2$.

Store the determination mix at 4 °C before use (no longer than a day).

3 Methods

3.1 Sampling and Sample Preparation

The sampling procedure may vary significantly depending on the species or the organ under study (*see* **Note 1**). However, whatever is being studied, samples should be plunged as quickly as possible into liquid nitrogen or dry ice, so that freezing would occur almost instantaneously. Furthermore, it should take only a few seconds between sampling and freezing. To achieve this, samples can be quickly put into aluminum bags before dropping them into liquid nitrogen or dry ice, or directly plunged into polyethylene vials containing liquid nitrogen. Note that liquid nitrogen will strongly boil as long as a sample's temperature has not equilibrated. When samples are too big, freezing may last for minutes. In that case, it is recommended to cut samples into small pieces before freezing them. Once frozen (*see* **Note 2**), samples can be stored at −80 °C [19].

Before extracting enzymes, samples have to be ground. Depending on sample's size, grinding in a mixer mill can be performed with metal balls, after precooling with liquid nitrogen, directly in small tubes (e.g., small leaf discs), in 20 ml polyethylene vials (up to 2–3 g fresh weight), or in steel cylinders. Shaking at 30 Hz has proven optimal for most sample types. However, duration may vary from one cycle of 1–2 min to several cycles. Sample containers need to be cooled down in liquid nitrogen between each grinding cycle, but should never be closed with liquid nitrogen inside because of explosion risk.

For each sample, weigh out approximately 20 mg of powder twice into precooled storage tubes and keep in liquid nitrogen or store at −80 °C until use.

3.2 Extraction

1. For each tube—stored on liquid nitrogen or dry ice—add 500 μl of extraction buffer.
2. Vortex for 10 s, then transfer to ice.
3. Centrifuge for 7 min at 2,500–3,000 × *g*, keep supernatants.
4. If necessary, dilute supernatants.
5. For backup (*see* **Note 3**), pipette an aliquot into a *V*-shaped polyethylene microplate and shock freeze immediately by placing the microplate on liquid nitrogen. Store at −80 °C.

3.3 Assay of UDP-Glucose Pyrophosphorylase

1. For each extract, pipette 5 μl in two separate wells of a polystyrene flat bottom microplate (*see* **Note 4**) suitable for measurements at 340 nm and maintained at 4 °C. Aliquots are needed for V_{blank} and V_{sat} measurements. For convenience, and if a full plate is used, V_{blank} measurements can be performed in rows 1–4 and V_{sat} ones in rows 5–8.
2. Pipette 95 μl of both V_{blank} and V_{sat} mixes. Because UDP-glucose is highly unstable, the mixes should be brought at 25 °C just before use.
3. Mix the microplate for 10 s on a plate shaker.
4. Read the microplate at 340 nm (1 reading every 30 s, temperature set at 25 °C). Wait until rate stabilizes in every well.
5. Use the microplate reader's software to estimate the time interval at which all rates are constant then calculate the rate for each well, by using the linear regression function. Transfer the output data expressed in OD/min/well into a spreadsheet.
6. Calculations. For each well, retrieve the averaged technical blanks (do it for V_{blank} and V_{sat}, respectively). Retrieve V_{blank} from V_{sat} for each sample, to obtain activity expressed as OD/min/well. Convert the latter into nmol/min/g FW by using the following equation (*see* **Note 5**).

$$\text{Activity (nmol/gFW/min)} = \frac{\text{OD/well/min}}{2.85 \times 6.22 \times 10^{-3}} \times \frac{500}{5} \times \frac{1}{\text{gFW}} \qquad (1)$$

Where 2.85 corresponds to the path length (cm/l) (*see* **Note 6**), 6.22×10^{-3} is the absorption coefficient of NADPH at 340 nm (l/nmol/cm), 500 is the volume of the extract (μl), 5 the volume of extract used (μl), 1 the dilution factor of the extract (not diluted here), and gFW the amount of fresh weight that has been extracted.

3.4 Assay of Pyrophosphate: Fructose-6-Phosphate 1-Phosphotransferase

1. For each extract, pipette 2 μl for both V_{blank} and V_{sat} wells of a into a polystyrene flat bottom microplate[5] suitable for measurements at 340 nm and maintained at 4 °C.
2. Pipette 18 μl of both V_{blank} and V_{sat} mixes brought at 25 °C just before use.
3. Mix the microplate for 10 s on a plate shaker.
4. Incubate the plate for, e.g., 20 min at 25 °C (a preliminary experiment performed with an adequate biological standard aiming at checking linearity with time and fresh weight dilution is recommended).
5. Stop the enzyme reaction by adding 20 μl of HCl 0.5 M.
6. Neutralize by adding 20 μl of NaOH after 10 min.
7. Add 50 μl of determination mix.
8. Read the microplate at 340 nm (1 reading every 30 s, temperature set at 25 °C). Wait until rate stabilizes in every well.
9. Use the microplate reader's software to estimate the time interval at which all rates are constant then calculate the rate for each well, by using the linear regression function. Transfer the output data expressed in OD/min/well into a spreadsheet.
10. Calculations. For each well, retrieve the averaged technical blanks (do it for V_{blank} and V_{sat}, expressed as mOD/min/well, respectively). In each well this corrected rate is proportional to the amount of intermediates (fructose-1,6-bisP but also glycerol-3P or glycerone-3P) that were present in the extract and generated during the first step of the assay.
11. When fructose-1,6-bisP standards have been used, perform a linear regression analysis, for both V_{blank} and V_{sat} measurements, to obtain the respective slopes and the intercepts (Fig. 1), which can then be used to calculate the amount of fructose-1,6-bisP accumulated during step 1 in every well (nmol V_{blank} and nmol V_{sat}, respectively). To calculate the activity, use the following equation (*see* **Note 5**):

$$\text{Activity (nmol/gFW/ min)} = (\text{nmol } V_{\text{sel}} - \text{nmol } V_{sat}) \times \frac{500}{2} \times \frac{1}{\text{gFW}} \quad (2)$$

Where 500 is the volume of the extract (μl), 2 the volume of extract used (μl), 1 the dilution factor of the extract (not diluted here), and gFW the amount of fresh weight that has been extracted.

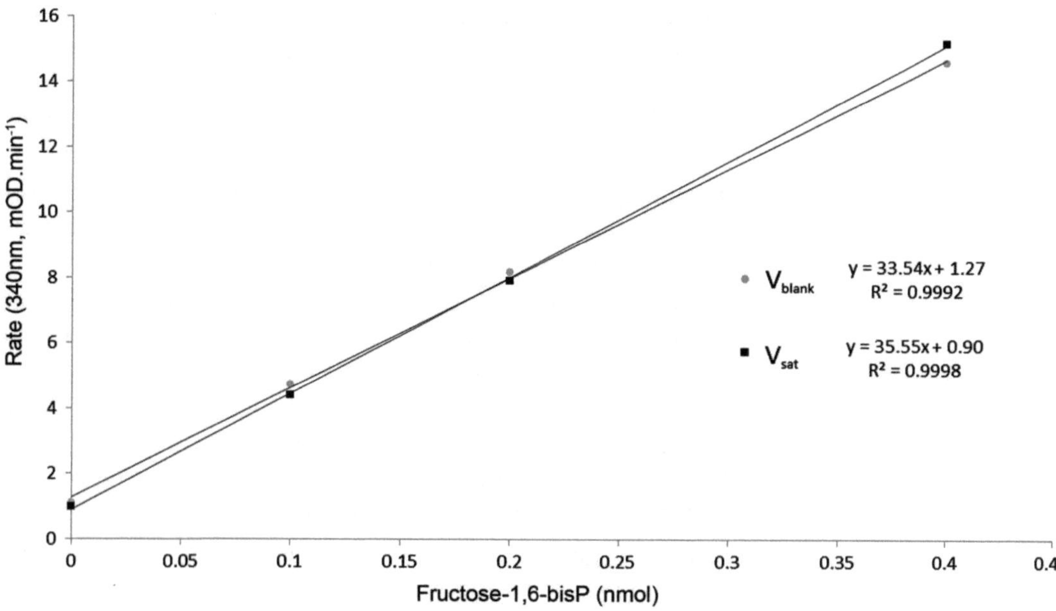

Fig. 1 Standard curves for the assay of pyrophosphate:fructose-6-phosphate 1-phosphotransferase. Fructose-1,6-bisP, which is a product of the catalyzed reaction, was used as the standard

12. When biological standards have been used, a ratio between slopes (*see* **Note 7**) obtained above has to be calculated:

$$\text{SlopeRatio} = \frac{\text{slope } V_{\text{blank}}}{\text{slope } V_{\text{sat}}} \tag{3}$$

Then, for each extract use the following equation:

$$\text{Activity (nmol/gFW/ min)} = \frac{\dfrac{\dfrac{\text{corr } V_{\text{sat}}(\text{extract})}{\text{SlopeRatio}} - \text{corr } V_{\text{blank}}(\text{extract})}{\text{gFW (extract)}}}{\dfrac{\dfrac{\text{corr } V_{sat}(\text{bio SD})}{\text{SlopeRatio}} - \text{corr } V_{\text{blank}}(\text{bio SD})}{\text{gFW (bio SD)}}} \times \text{Activity (bio SD)} \tag{4}$$

Where corrV is the corrected rate obtained by retrieving the respective averaged technical blanks and bio SD the biological standard.

4 Notes

1. Although most enzymes are probably quasi stable throughout a day and night cycle, sampling time needs to be carefully evaluated, as there are exceptions. A classic example is nitrate

reductase; its total activity rises at night and drops in the second part of the day, whereas it is activated during the day and inactivated at night [14]. Another well-known example is ADP-glucose pyrophosphorylase, which increases at night and drops during the first hours of each photoperiod [14].

2. Once samples have been frozen, they should not thaw out even before extraction. All components used for sample handling should be precooled in liquid nitrogen before use.
3. We noticed that activities of most enzymes tested so far could be recovered after a freeze/thawing cycle of the extracts. The use of 20 % glycerol is nevertheless essential.
4. Protocols given here are suitable for hand pipetting. Thus extracts are pipetted first into wells because this step is much slower than adding the assay mixes. In case a pipetting robot is used, it is more convenient to load assay mixes before extracts.
5. Equations given for activity calculations do not take into account possible dilutions of the extract.
6. In a microplate reader, OD is proportional to the number of moles of the absorbing molecules but is independent of the light path; this is why the light path is expressed in $cm \cdot l^{-1}$, *see* also ref. [20]. It is recommended to determine the light path.
7. The "SlopeRatio" parameter of Eq. 3 is needed when slopes—obtained by performing the regression analysis with chemical standards—are different for V_{blank} and V_{sat} conditions.

Acknowledgments

Financial support from INRA, Région Aquitaine, Eranet KBBE SAFQIM, Eranet Erasysbio+ FRIM, FP7 KBBE DROPS and SFR BIE is gratefully acknowledged.

References

1. Buchner E (1897) Alkoholische Gärung ohne Hefezellen. Berichte der deutschen chemischen Gesellschaft 30:1110–1113
2. Kruckeber AL, Neuhaus HE, Feil R et al (1989) Decreased-activity mutants of phosphoglucose isomerase in the cytosol and chloroplast of *Clarkia xantiana*—Impact on mass-action ratios and fluxes to sucrose and starch, and estimation of flux control coefficients and elasticity coefficients. Biochem J 261:457–467
3. Stitt M, Schulze D (1994) Does Rubisco control the rate of photosynthesis and plant-growth—an exercise in molecular ecophysiology. Plant Cell Environ 17:465–487
4. Poolman MG, Olcer H, Lloyd JC et al (2001) Computer modelling and experimental evidence for two steady states in the photosynthetic Calvin cycle. Eur J Biochem 268:2810–2816
5. Kacser HJ, Burns A (1973) The control of flux. Symp Soc Exp Biol 27:65–104
6. Jenner HL (2003) Transgenesis and yield: what are our targets? Trends Biotechnol 21:190–192
7. Schomburg I, Chang A, Ebeling C et al (2004) BRENDA, the enzyme database: updates and major new developments. Nucleic Acids Res 32:D431–D433

8. Rohwer JM, Botha FC (2001) Analysis of sucrose accumulation in the sugar cane culm on the basis of in vitro kinetic data. Biochem J 358:437–445
9. Rowher JM (2012) Kinetic modelling of plant metabolic pathways. J Exp Bot 63:2275–2292
10. Gardossi L, Poulsen PB, Ballesteros A et al (2010) Guidelines for reporting of biocatalytic reactions. Trends Biotechnol 28:171–180
11. Kaiser WM, Huber SC (2001) Post-translational regulation of nitrate reductase: mechanism, physiological relevance and environmental triggers. J Exp Bot 52:1981–1989
12. Winter H, Huber SC (2000) Regulation of sucrose metabolism in higher plants: localization and regulation of activity of key enzymes. Crit Rev Plant Sci 19:31–67
13. Bergmeyer HU (1987) Methods of enzymatic analysis. VCH, Weinheim, Germany
14. Gibon Y, Blaesing OE, Hannemann J et al (2004) A robot-based platform to measure multiple enzyme activities in Arabidopsis using a set of cycling assays: Comparison of changes of enzyme activities and transcript levels during diurnal cycles and in prolonged darkness. Plant Cell 16:3304–3325
15. Junker BH, Lonien J, Heady LE et al (2007) Parallel determination of enzyme activities and in vivo fluxes in Brassica napus embryos grown on organic or inorganic nitrogen source. Phytochem 68:2232–2242
16. Gibon Y, Pyl E-T, Sulpice R et al (2009) Adjustment of growth, starch turnover, protein content and central metabolism to a decrease of the carbon supply when Arabidopsis is grown in very short photoperiods. Plant Cell Environ 32:859–874
17. Tschoep H, Gibon Y, Carillo P et al (2009) Adjustment of growth and central metabolism to a mild but sustained nitrogen-limitation in Arabidopsis. Plant Cell Environ 32:300–318
18. Gillespie KM, Rogers A, Ainsworth EA et al (2011) Growth at elevated ozone or elevated carbon dioxide concentration alters antioxidant capacity and response to acute oxidative stress in soybean (Glycine max). J Exp Bot 62:2667–2678
19. Biais B, Bernillon S, Deborde C et al (2012) Precautions for harvest, sampling, storage, and transport of crop plant metabolomics samples. Methods Mol Biol 860:51–63, Clifton NJ ed
20. Rogers A, Gibon Y (2009) Enzyme kinetics: theory and practice. In: Svhwender J (ed) Plant metabolic networks. Springer, New York, pp 71–103. ISBN 978-0-38-778744-2

Chapter 16

Analytical Kinetic Modeling: A Practical Procedure

Gilles Curien, María L. Cárdenas, and Athel Cornish-Bowden

Abstract

This chapter describes a practical procedure to dissect metabolic systems, simplify them, and use or derive enzyme rate equations in order to build a mathematical model of a metabolic system and run simulations. We first deal with a simple example, modeling a single enzyme that follows Michaelis–Menten kinetics and operates in the middle of an unbranched metabolic pathway. Next we describe the rules that can be followed to isolate sub-systems from their environment to simulate their behavior. Finally we use examples to show how to derive suitable rate equations, simpler than those needed for mechanistic studies, though adequate to describe the behavior over the physiological range of conditions.

Many of the general characteristics of kinetic models will be obvious to readers familiar with the theory of metabolic control analysis (Cornish-Bowden, Fundamentals of Enzyme Kinetics, Wiley-Blackwell, Weinheim, 327–380, 2012), but here we shall not assume such knowledge, as the chapter is directed toward practical application rather than theory.

Key words Enzyme, Kinetic modeling, Metabolic pathway, Simulation, Rate equation

1 Introduction

Analytical kinetic modeling consists of building computer models of biological systems on the basis of detailed knowledge of the elements of the system and their properties. A kinetic model must be regarded as a tool for achieving a specific goal. This goal may be to check the reliability of biochemical data [2], identify regulatory steps [3], understand the function of allosteric interactions [4, 5], explain specific observations about the system [6], help designing novel metabolic pathways, or improve existing ones for biotechnological applications [7]. Any interested biologist can now build a kinetic model and carry out simulations without needing any specific mathematical skills, as user-friendly bioinformatic tools have become readily available, including BerkeleyMadonna, CoPaSI, CellDesigner [8], Systems Biology Workbench, and others. However, anyone who has tried to build a mathematical model of a metabolic system has been immediately confronted with

Martine Dieuaide-Noubhani and Ana Paula Alonso (eds.), *Plant Metabolic Flux Analysis: Methods and Protocols*,
Methods in Molecular Biology, vol. 1090, DOI 10.1007/978-1-62703-688-7_16, © Springer Science+Business Media New York 2014

methodological issues to dissect the metabolic system, simplify it, and use or derive enzyme rate equations, and it is on these that we shall concentrate in this chapter.

2 Kinetic Modeling of a Single Enzyme-Catalyzed Step

As noted in the introduction, any modeling approach aims at answering a question. The simple theoretical example shown in Fig. 1, which refers to the modeling of a single enzymatic step catalyzed by enzyme E in the middle of an unbranched pathway, is relevant to a real case study. One may want to check whether some information about the enzyme is reliable. Another reason could be to model this step as a preliminary stage of a synthetic-biology approach with the aim of adding a branch in competition with enzyme E and of predicting the output in the novel branch.

2.1 Definitions

Kinetic modeling of a single kinetic step or of an entire metabolic pathway requires isolating a *sub-system* embedded in a larger system. The sub-system consists of the elements of interest for the question asked (enzymes, ions, metabolites, and other molecules, all of which have properties that are known at least approximately) and which interact with each other. The system outside the sub-system is the *external system* (or "environment") and contains everything that is directly or indirectly connected with the sub-system; it likewise consists of enzymes, metabolites, and ions, generating fluxes into and out of the sub-system and affecting its properties. In the simple example illustrated in Fig. 1 substrate S is provided to the enzyme at a certain rate J_{input}. We assume that E is inhibited by a metabolite labeled I in Fig. 1, such as AMP, for example, whose concentration does not depend on E. The output of the enzyme action is a transformation of S into P at a certain rate J_{output}. The concentration (*see* **Note 1**) s of S and J_{output} will depend on the input flux J_{input}, concentrations i and e of I and E, and properties of the enzyme (as modeled by its rate equation and the associated kinetic parameters).

In this example the *variables* are J_{input}, e, i, s, and J_{output}, but there is an important difference between two classes of variables, and these must be clearly distinguished. J_{input}, e, and i are not determined by the sub-system and are the *external variables*. This follows in the example because neither P nor S affects their values. J_{output} and s, on the other hand, are *internal variables*, because they both depend on the properties of the sub-system (the enzyme kinetic parameters and the form of the rate equation). Of course, the internal variables also depend on the external variables.

Once the external variables are identified, the objective will be to calculate (i.e., simulate) the values of the internal variables using the known values of the external variables obtained from

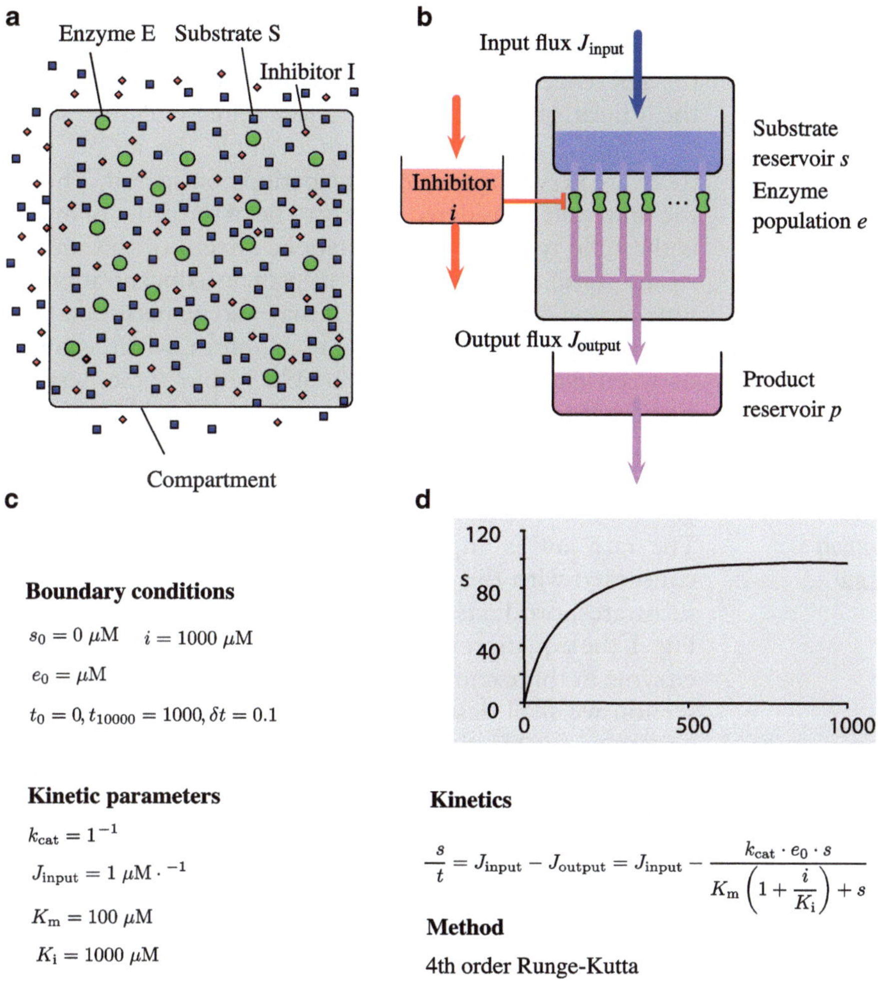

Fig. 1 Modeling of a single kinetic step in the middle of a linear branch. (**a**) A schematic representation of the sub-system to be modeled. The population of metabolites (a substrate S, a product P, and an inhibitor I) and of the enzyme E diffuse freely in a subcellular compartment. S is produced by the external system and transformed into P by the enzyme. I is produced and consumed by the external system. The concentration *i* of the inhibitor is assumed to be maintained constant by the external system. We assume that E is insensitive to P, which is consumed by the external system. (**b**) Representation using a hydraulic analogy. S is produced at a certain rate J_{input} by the external system, and this flux fills a reservoir. The height of S in the reservoir represents its concentration. The reservoir is emptied by a certain number of outlets represented by green objects that correspond to the enzyme population at concentration *e*. J_{input}, *i*, and *e* are the external variables. We assume that none of these change in the time span considered, and *s* and the rate J_{output} at which S is transformed into P are properties of the sub-system and are the internal variables

measurements carried out in the living system. Also a rate law for the enzyme is required.

This means that from the very beginning the model must be anchored in a physiological *reference state* and refer to a precise type of cell and, if pertinent, to a subcellular location, from which values of input fluxes, metabolite concentrations, and enzyme concentrations will be obtained experimentally. Also, for the subsequent validation steps the values of s and J_{output} must be obtained experimentally for the reference state considered so that they can be compared with the values obtained by simulation. Finally a time window must be defined, and in the present example the model assumes that there are no changes in e in the period considered. This means that the time scale considered is likely to be of the order of minutes, as this is the time required for a metabolic system of this kind to reach a steady state, and any changes in protein concentration during this time period can be neglected. It may also be necessary to specify the time of day, especially for plant metabolism, which is often different between day and night, or variable within these periods (as with starch accumulation during the day, for example). The location of the sub-system in the living organism and the time window chosen both determine the ***domain of validity*** of the kinetic model.

2.2 Derivation of a Rate Law

The rate law is an equation that relates the rate at which S is consumed with the concentrations of enzyme and its ligands (i.e., substrates, products, inhibitors, and activators). In the example of Fig. 1 the equation relates v to e, s, and i. As P does not inhibit the enzyme in this example, its concentration can be ignored. In a later section we shall describe the procedure for obtaining reliable rate equations for kinetic modeling. For the present, we suppose that kinetic experiments in vitro have shown that the isolated enzyme displays typical Michaelis–Menten behavior with respect to S and that I is a competitive inhibitor. The rate equation then has the following form:

$$v = e \cdot \frac{k_{\text{cat}} \cdot s}{K_{\text{M}}\left(1 + \frac{i}{K_{\text{i}}}\right) + s} \tag{1}$$

where e is the total enzyme concentration, s is the substrate concentration, i is the inhibitor concentration, k_{cat} is the catalytic constant, K_{M} is the Michaelis constant, and K_{i} is the inhibition constant for I. We also suppose that in vitro kinetic measurements have given the following values for the kinetic parameters: $k_{\text{cat}} = 1$ s^{-1}, $K_{\text{M}} = 100$ μM, and $K_{\text{i}} = 1{,}000$ μM.

2.3 Building of the Mathematical Model

The mathematical formalism that is adapted to the majority of metabolic situations, when the populations of enzyme or substrate molecules in the compartment of interest are larger than several hundreds, is a system of ordinary differential equations (often abbreviated to ODE). For modeling systems, in which the number of molecules is very small, and thus subject to sampling variation, a different approach, known as stochastic simulation, is needed. We shall consider this in Subheading 2.5. A differential equation (Eq. 2) is a way of expressing the idea that the small change δs in the concentration of S produced during a short time increment δt is the difference between the rate J_{input} at which it enters and the rate

```
METHOD RK4                     {Method used for the numerical integration, RK4 = 4th order Runge-Kutta}
STARTTIME = 0                  {start time of the simulation, in second}
STOPTIME=1000                  {stop time of the simulation, in second}
δT = 0.1                       {time increment for the integration, in second}

{Model}

{Reservoirs}
d/dt(S) =J_input-J_output      {differential equation : input of S minus output of S}
INIT S=0                       {initial value of s, μM}

{Flows}
J_output=e*k_cat*s/(K_M*(1+i/K_i)+s)   {Rate equation, competitive inhibition by I}

{Numerical values of the external variables}
J_input=1                      {in μM. s-1}
e=4                            {Enzyme E concentration, μM}
i=1000                         {Inhibitor concentration, μM}

{Numerical values of the kinetic parameters}
k_cat= 1                       {Catalytic constant of E in s-1}
K_M= 100                       {Michaelis constant, μM}
K_i=1000                       {Inhibition constant for inhibition of E by I, μM}
```

Fig. 2 Mathematical model of the system displayed in Fig. 1. The script displayed is sufficient to run a simulation using BerkeleyMadonna software. The text in {} shows comments displayed for clarity for human readers but ignored by the computer. Various numerical methods are available with most software [10]. Of these, fourth-order Runge-Kutta ("RK4") is a good popular choice

J_{output} at which it leaves. J_{output} is variable and is equal to the rate v defined by Eq. 1. In the present case a single differential equation is sufficient as there is just one reservoir to be modeled (Eq. 2):

$$\frac{ds}{dt} = J_{input} - J_{output} \tag{2}$$

Numerical integration methods then allow calculation of the evolution of the sub-system variables step by step (for time increments of δt) as a function of time.

Numerical resolution of the differential equation is the role of the modeling software and requires an initial value (starting value at t_0 of the simulation) for the internal variable s. Any value for s will do with simple systems but nonzero values of the various metabolite concentrations may be required for complicated networks, as a starting state too far from the final state will often lead to instabilities in the calculation, and some software may have difficulties with true-zero values. Numerical values also have to be given to the external variables (J_{input}, i, e).

A final thing to do before launching the simulation is to give a value to the time increment δt for the numerical resolution of the differential equation. A large value will lead to an insufficiently precise progress curve. Too small a δt value will unnecessarily increase the calculation time and the size of the file containing the results. The choice of δt depends on the dynamics of the system modeled (0.1 s was used in Figs. 2 and 3) and needs to be

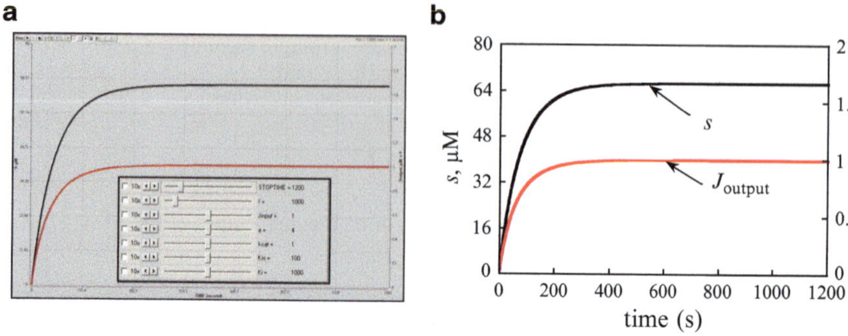

Fig. 3 Simulation results. (**a**) Screenshot of the output from BerkeleyMadonna. The table in the middle of the screen contains the values of the external variables and kinetic parameters, allowing them to be changed with ease for additional simulations. (**b**) The simulation results are reproduced for clarity reasons. Plot of *s* as a function of *t* (*black trace, left axis*) and J_{output} as a function of *t* (*red trace, right axis*) showing that *s* reaches a steady-state value (66.7 μM) in less than 1,000 s. At steady state the output flux is equal to the input flux (1 μM·s^{-1})

determined by the software. Moreover, it does not necessarily remain constant during the progress to the steady state: in the early stages a very small value may be needed to avoid spurious oscillation, but it can be much larger when the steady state is approached. This may seem a complication, but it can be ignored when modern software is used, as the step size will be determined automatically. An example of a very concise script of the model that can be used with BerkeleyMadonna is displayed in Fig. 2.

2.4 Simulation

The simulation results (Fig. 3) provide the values of the internal variables as functions of time. The important outputs of the simulation are (**a**) the existence or not of a steady state, (**b**) the time required to reach steady state with a specified degree of precision (the *characteristic time* is the time required to reach a fraction exp(−1) of the steady-state value, i.e., about 37 %), and (**c**) the steady-state values of fluxes and concentrations. A variable (flux or metabolite concentration) is considered to be in steady state when any further changes with time are negligible. For systems that approach their steady states extremely slowly, such as the one studied by [9], it is important not to mistake a slowly evolving system for one in a steady state. This can be tested computationally by measuring the eigenvalues of the Jacobian matrix in the supposed steady state. In most models of metabolic systems this complication is unlikely to arise.

Figure 3 shows the result of the simulation with the model displayed in Fig. 2. This figure shows that starting with an empty system (i.e., $s_{t=0} = 0$), s increases with time until a steady-state value is reached ($s = 66.7$ μM). J_{output} follows the same trend with a steady-state value (1 μM s^{-1}) equal to the input flux, as one would expect. The characteristic time is 24.5 s. The time required to reach steady state is thus consistent with the time window assumed for the domain of validity of the model (a matter of minutes). This is important, because if the steady state was reached only after hours the hypothesis of constant e might no longer be relevant to the physiological situation.

Simulations can be considered as virtual experiments carried out with the model. As a model may contain many parameters and external variables, the results of the simulations require a very careful examination with a laboratory notebook and reports: these are very similar to those required for experiments in the laboratory and fulfill the same function.

2.5 Stochastic Simulation

In this chapter we are mainly concerned with the simplest case of deterministic simulation, where the numbers of molecules are sufficiently large for the properties to be defined in terms of concentrations. As long as all metabolites relevant to the simulation are present in numbers of 1,000 or more, statistical fluctuations in the numbers can be ignored. As a rough guide one can assume Poisson distributions, which means that the precision of any number is about equal to its square root. So, for example, 1,000 molecules really means about 1,000 $\pm$ 31, or a precision of about 3 %. Smaller numbers of molecules imply less precision, so it is important to estimate the compartment volume before assuming that a deterministic simulation is adequate. If the compartment is an entire eukaryotic cell, or even a bacterial cell, then it is likely to be large enough, unless the model concerns a minor metabolic pathway involving metabolites present in very small quantities. However, a subcellular compartment such as a mitochondrion or a chloroplast is much smaller than the cell that contains it, and simulation of a metabolic pathway in it is accordingly more likely to require stochastic simulation, as illustrated, for example, by [11]. It is important, therefore, to estimate the numbers of molecules involved in the pathway before setting up a model. For example, 1 μM corresponds to about 120 molecules in a liver rat mitochondrion (0.2 fL) and about 20,000 molecules in a chloroplast (30 fL). Stochastic simulation requires specialized numerical methods, such as Gillespie's algorithm [12].

2.6 Validation

Validation consists in comparing the internal variables calculated with the model with the same variables measured in the organism, this organism being studied under conditions consistent with the domain of validity of the model. This can be done in various ways. In the simple example described in Fig. 1, comparison of the measured value of s in vivo with the calculated value of s obtained by simulation ("in silico") is a first validation. For example, if the calculated steady-state value of s is much higher than measured in vivo or if a steady state cannot be reached in the simulation, this is an indication that the enzyme kinetic parameters or the enzyme concentration are wrongly estimated or that some activators are missing, motivating additional in vitro kinetic experiments or measurements in vivo. The rate at which a supplementary addition of S is metabolized in vivo or the ability to reproduce phenotypes in the model are other ways to validate it. For example, the ability of the model to reproduce quantitatively the change in a metabolite that accompanies in vivo a reduction in the total enzyme concentration is a good test; an example is given by [5]. If data and model agree the model is not only validated by the phenotype, but it also offers an explanation for the observed phenotype.

Note that before validating the model with in vivo data a preliminary step might be to check the behavior of the enzyme under more physiological conditions but still in vitro. Examples of such validations are given by [4] and in the supplementary material of [5]. This kind of in vitro validation prior to in vivo validation is important, as the reasons for discrepancies between the model and the experiment are easier to identify in vitro than in vivo.

Discrepancy between model predictions and observation is a powerful way to identify missing interactions. For example, in the modeling of the aspartate pathway in *Arabidopsis* inconsistencies at an early step of the modeling process did in fact lead to the identification of activators of bifunctional aspartate kinases [13], such as alanine, which was not suspected before.

The model should be improved iteratively if the disagreement between its predictions and experimental measurements cannot be simply explained by lack of precision of the data but requires serious modifications.

2.7 Using the Model

Once agreement between the model and in vivo data is obtained in various independent situations and under conditions relevant to its domain of validity, simulations can be used to analyze the properties of the sub-system. A more intuitive understanding of the sub-system can be obtained, and novel experiments can be designed on the basis of this understanding. Metabolic control analysis [14] is a powerful tool to assess the sensitivity of fluxes and concentrations to external variables. The model can also be used to estimate the maximal capacity of the step catalyzed by an enzyme. This might be important for biotechnological applications and the model may indicate constraints that were not anticipated beforehand.

With the present example it is trivial to predict the changes without any simulations (at least qualitatively). However, for complicated networks with effectors acting simultaneously on different enzymes of the system, simulations are the only way to understand its properties using virtual experiments that would be difficult to execute experimentally or that would be inconclusive due to lack of precision of the measurements.

3 Rules for Defining Minimal Sub-systems, and Additional Definitions

3.1 Dissection of a Metabolic System Is Imposed by Its Structure

In the example described in Fig. 1b the sub-system is simple, as the enzyme communicates with the rest of the system only by its substrate and with the inhibitor I. We assumed that S does not affect the upstream enzyme and that P does not inhibit E. Thus the sub-system in Fig. 1b can be modeled autonomously and leads to (very limited) conclusions that can be considered definitive. However, in general the communication with the rest of the system is much more complicated, and information transfer [15] in the system determines the elements that must be considered as being part of the sub-system. The different sub-systems represented in Fig. 4 detailed below cannot in theory be decomposed further into smaller modules. However, we will indicate in Subheading 3.2 under which conditions simplifications can be done. In the example in Fig. 4a, the enzyme E_1 catalyzes a reversible reaction. In that case modeling the enzyme in context requires that the downstream enzyme (E_2) be part of the sub-system, as its concentration and activity potentially determine the amount of the product P_1 of E_1 and thus s_1. Neglecting the second enzyme would necessarily lead to wrong predictions in terms of system dynamics and steady-state s_1. Compared to Fig. 1 the sub-system in Fig. 4a thus contains the additional internal variables p_1, and the output flux for P_1 and the additional external variable e_2. Figure 4b shows another sub-system with inhibition of E_1 by the product P_1. For identical reasons P_1 and E_2 must be taken into account in the sub-system. Figure 4c is a slight modification of Fig. 1b where I is a by-product of the reaction catalyzed by E (rather than an independent metabolite as in Fig. 1b) but I is still present as a metabolite and it still inhibits E_1. However, whereas i was an external variable in the example in Fig. 1b it is now an internal variable as the flux at the step catalyzed by E_1 determines i. Figure 4d describes a metabolic branch-point with E_1 in kinetic competition with E_2 for the metabolite S_1. In that case the branch-point is the sub-system and the concentration of E_2, an additional external variable, and the kinetics of E_2 have to be taken into account. Finally, Fig. 4e illustrates the condensation of two metabolites S_1 and S_2 at the step catalyzed by E_1. S_2 synthesis is feedback-inhibited by itself. As the activity of E_1 determines s_2 and thus the input flux for S_2, both belong to the sub-system (they are internal variables) along with e_2 (external variable).

In summary, dissection of metabolic systems (and more generally biological systems) into sub-systems for modeling purposes is not arbitrary or subjective but is imposed by the structure of the metabolic system and by the properties of the enzymes. Both determine the minimal sub-system to model and later on to assemble to build large models. When considering these examples it might appear that it would be practically impossible to model real

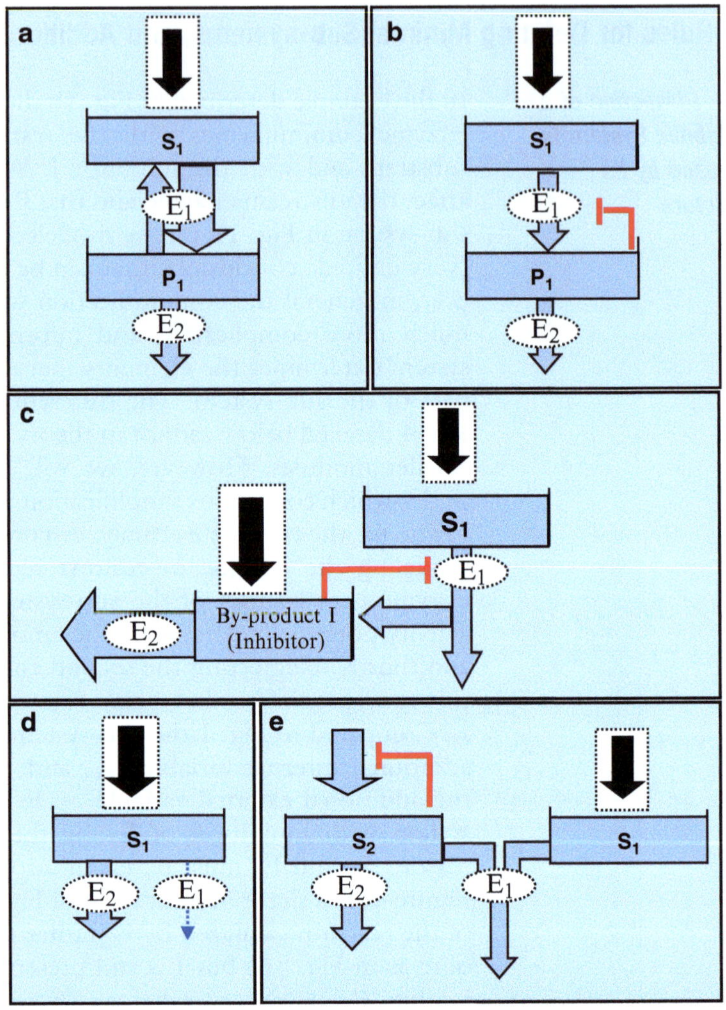

Fig. 4 Different metabolic modules defining theoretical minimal sub-systems. Fluxes symbolized by *black arrows* and enzyme concentrations symbolized by *ovals* are drawn on a *white background* to indicate that they are external variables. Internal variables (fluxes and concentrations) are in *blue*. The thickness of each *arrow* symbolizes the magnitude of the flux at steady state. In practice, the number of elements of the sub-system can be reduced further by fixing internal variable values if certain conditions are fulfilled (*see* Subheading 3.2). (**a**) Reversible reaction. (**b**) Product inhibition. (**c**) The by-product I is common to other reactions (E_2), and I inhibits E_1. (**d**) Metabolic branch-point (the relative flux has its importance for simplifications, see text). (**e**) Condensation of two metabolites at the step catalyzed by E_1. S_2 is used by another reaction. We consider here similar fluxes at the steps catalyzed by E_1 and E_2 (see text)

kinetic systems, as many components would have to be characterized before a model could be produced even for simple metabolic systems. In the following section we detail how rational simplifications can be made to define minimal autonomous sub-systems for modeling purposes. Simplification rules are also useful to derive adequate rate equations (Subheading 3.3) for enzymes with complicated rate equations, such as ones that can convert several substrates into products and are sensitive to several different inhibitors or activators.

3.2 Fixed Internal Variables

The enzymes involved in a metabolic context as simple as the one in Fig. 1 are exceptional, and they are generally not interesting for modeling purposes as they simply transmit the flux. However, most enzymes use a combination of substrates to produce a combination of products, they may catalyze reversible reactions, and they may be sensitive to many metabolites. So, for many metabolic situations, modeling an enzyme in a metabolic network requires many other enzymes to be considered or many other fluxes surrounding the step of interest, as illustrated in Fig. 4. In addition, for this kind of enzyme it is generally difficult to obtain all the kinetic parameters when more than four different ligands affect its activity. For these reasons we need to search for possible ways of simplifying the model. To rationalize these we need to define an intermediate status, at least provisionally, for some internal variables and we will call them *fixed internal variables*. This simplification is permissible if the concentrations concerned are known to vary very little over the range of the simulation or if the behavior of the sub-system is relatively independent of the changes in these variables. The first and more important case is typically illustrated by a concentration such as that of ATP, which can often be assumed to be regulated by mechanisms outside the sub-system being modeled: if it did change, the change would need to be taken into account, but if it does not change it can be treated as a constant. The other case might be illustrated by an ion such as K^+, whose concentration might change over the range of conditions considered, but could still be treated as constant if the changes had negligible effects.

Such concentrations can be set constant at their physiological values for in vitro experiments and in the mathematical model. As a fixed internal variable is fixed by the model builder its value can also be changed (if required) to estimate its quantitative importance in the behavior of the sub-system. In practice, there are four different reasons that justify fixing an internal variable, meaning then that the number of elements of the sub-system can be reduced. These reasons can be summarized as follows:

(a) The enzyme E_1 contributes quantitatively very little to the input or the output of the metabolite of interest compared to the other (outside) processes. For example if the synthesis of I by E_1 in Fig. 4c is a minute fraction of its metabolism by the cell, its concentration i, which in theory should be free to change, can be fixed. In the example of Fig. 4d, if the flux of S_1 through E_1 is several orders of magnitude lower than through

E_2, s_1 is essentially set by E_2 and it can be fixed at its measured physiological value to model the flux through E_1.

(b) The substrate or the product of the enzyme is a metabolite such as ATP with a concentration regulated by mechanisms outside the sub-system. (For example in Fig. 4e s_2, an internal variable in theory, can be so tightly controlled by allosteric inhibition that changes in the rate of S_2 consumption by E_1 may not strongly change s_2. In that case, leaving s_2 fixed at a physiological value is permissible.)

(c) Kinetic measurements show that the enzyme initial velocity in the presence of all its ligands at physiological concentration is insensitive to a change in the concentration of the metabolite considered. For example if the enzyme E_1 is saturated by its second substrate S_2 produced by another pathway, as could be the case in Fig. 4e, s_2 can be fixed.

(d) Experimental evidence exists that during the time window considered the metabolite concentration in vivo does not change.

A fixed internal variable is fixed provisionally, and its status might be changed into a real internal variable free to change if new data or expansion of the domain of validity of the model revealed that the assumed relative independence between this variable and the sub-system activity is no longer acceptable.

3.3 Hidden Variables

In an ideal world all of the variables of a model would appear explicitly in the rate equations so that their values could be changed during the simulations and their importance quantified. Unfortunately, however, for enzymes with many ligands, for which the rate equations are typically very complicated, it is not practical to derive equations that account simultaneously for the changes in the concentrations of all the ligands. However, one cannot omit a priori any known ligands of the enzyme, as otherwise its response to the other metabolites might be very different. The solution consists in measuring the enzyme activity in vitro in the presence of all the ligands [16], with their concentrations set at physiological values (values corresponding to measured concentrations in the physiological reference state), and varying the concentrations of only the ones we need to see explicitly in the equation. For example, an equation that would account explicitly for changes in the concentrations of metabolites that do not change in vivo in the time window considered is not required. The metabolites that remain constant in the assay will obviously be fixed internal variables or external variables in the model. In that case, kinetic parameters are apparent kinetic parameters that depend on these fixed concentrations. In other words the fixed variables are hidden in the apparent kinetic parameters (*hidden variables*). Although they will not appear in the equations the model should keep track of them to

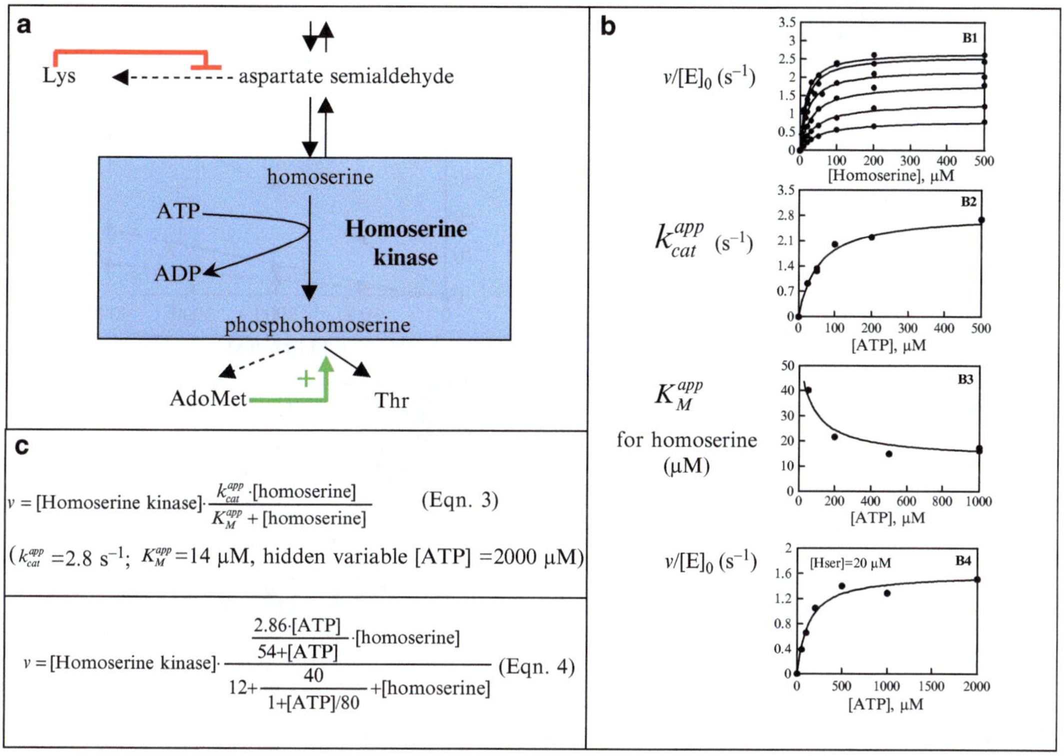

Fig. 5 Modeling the homoserine kinase reaction. (**a**) Metabolic position of homoserine kinase in the aspartate-derived amino acid pathway in plants. The enzyme catalyzes an irreversible reaction and is not inhibited by its products. (**b**) Detailed kinetic characterization of recombinant purified homoserine kinase. *B1*: Enzyme activity was measured in the presence of variable concentrations of homoserine for different [ATP]. *B2*: Apparent k_{cat} values obtained from the graph in *B1* were plotted as a function of [ATP]. *B3*: Apparent K_m values obtained from curves in the graph in *B1* were plotted as a function of [ATP]. *B4*: Enzyme activity was measured as a function of [ATP] in the presence of 20 μM homoserine. The figure shows that in the physiological range of [ATP] variations (1,000–2,000 μM) enzyme activity displays a low sensitivity to ATP. (**c**) From one of the curves in graph *B1* obtained at 2,000 μM ATP an empirical equation was derived for homoserine kinase (Eq. 3). [ATP] is a fixed internal variable hidden in the kinetic parameters ($k_{cat}^{app} = 2.8\ s^{-1}$ and $K_M^{app} = 14$ μM). A more complete equation (Eq. 4) was derived using graphs *B2* and *B3* in which K_M^{app} for homoserine and k_{cat}^{app} are functions of [ATP]

allow subsequent improvement or expansion if required (*see* for example Eqs. 3 or 5, Figs. 5 and 6, respectively).

Implicit in this discussion is that kinetic parameters obtained in the study of the mechanism of action of an enzyme will often be inappropriate for modeling purposes. In studies of mechanism it is usually desirable to study extreme conditions to amplify small differences in the predictions made by different possible mechanisms, but these differences can be ignored in a model if they become detectable only outside the physiological ranges of concentrations. For example, the differences between different types of inhibition (competitive, uncompetitive, mixed, etc.) are in principle important for distinguishing between mechanisms, but they may be very

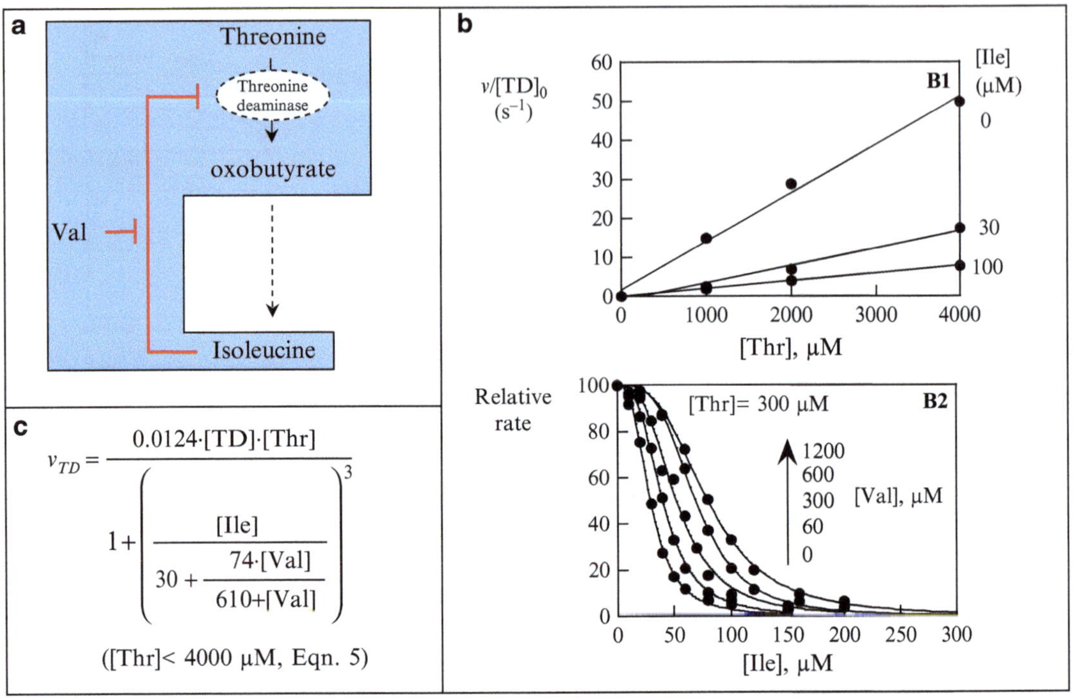

Fig. 6 Modeling threonine deaminase. (**a**) Threonine deaminase is allosterically inhibited by isoleucine and this inhibition is antagonized by valine, which thus acts as a nonobligatory activator. (**b**) Graph *B1*: Enzyme activity was measured as a function of [Thr] in the absence or the presence of isoleucine, as indicated in the graph. In the physiological range of substrate and effector concentrations, the rate is proportional to substrate concentration. The slope of the curve in the absence of threonine is 0.0124 $\mu M^{-1} \cdot s^{-1}$. This value was used as the rate constant of the uninhibited enzyme (*see* Eq. 5). Graph *B2*: Response to the inhibitor isoleucine and the activator valine at 300 μM [Thr]. (**c**) Rate equation of threonine deaminase valid for [Thr] $<$ 4,000 μM, in which [TD] is the concentration of threonine deaminase

slight if substrate concentrations are restricted to their physiological ranges. On the other hand it is also usual in mechanistic studies to choose the pH and temperature, as well as concentrations of inhibitors and activators, on the basis of experimental convenience. In particular, product concentrations are often set to zero in mechanistic studies, at least in preliminary experiments. None of this is acceptable for defining parameters for a metabolic model, which should refer as far as possible to physiological conditions, including the concentrations of all metabolites that interact with the enzyme considered.

A clear exposition of the status of the variables is important, as it shows where experimental work and additional data are required if the model needs to be improved. Its domain of validity and the conclusions drawn using the simulation results are also strongly conditioned by the decisions concerning the status of the variables. Confidence in modeling approaches will certainly improve if users of a model who did not build it can easily perceive its weaknesses and limitations.

The decision to fix internal variables and to hide external or fixed internal variables in kinetic parameters is determined by the questions to be asked, the complexity of the enzymes, their metabolic positions, the domain of validity needed for the model (cellular compartment, time window), as well as the anticipated future use of the model. For example if a modular approach is followed an external variable at a first level of integration will become an internal variable after the model is expanded. It would then be better to derive an equation in which this external variable appears explicitly in the enzyme equation, as otherwise the enzyme will have to be completely characterized again when the model is expanded.

The decisions for fixing the variables should not be subjective but need to be justified on quantitative grounds and supported by data obtained from *in vitro* kinetics or in vivo. One should try to estimate the error produced by simplifying (omitting a ligand, for example) or by degrading the status of a variable.

3.4 Kinetic Parameters and Apparent Kinetic Parameters

Mathematically all the values that are constant during the simulation can be called parameters. However, for biological systems many values that are constant at a low level of integration can become internal variables when the model is expanded to account for other reactions or if the time window is enlarged. Here we reserve the use of the term "parameter" to kinetic parameters. When modeling requires external or fixed internal variables to be hidden in kinetic parameters, it is important to specify that the kinetic parameters are apparent ones (symbolized as K_i^{app}, K_M^{app}, or k_{cat}^{app}) to make it clear to the users of the model that the enzyme kinetic parameters were obtained at fixed concentrations of the other ligands. The identities of these ligands and their concentrations should be specified along with the rate equation.

4 Deriving Rate Equations from In Vitro Measurements

4.1 General Considerations

Data available from the literature are generally inadequate for modeling purposes, and specific characterizations are required [16]. After the enzyme is obtained in its purest form from the host or as a recombinant protein, in vitro characterization has to be done under conditions that are relevant to the cellular or the subcellular compartment in which the enzyme operates in vivo in terms of pH and salt composition [17]. One should therefore characterize the metabolic environment first, prior to the kinetic characterization of the isolated enzymes, as the properties in vitro might be very different in the presence or the absence of salt or at different pH values. In addition, for enzymes with many ligands, complete equations can generally not be obtained and the allowable simplifications will depend on the information available about the metabolic environment.

A major difference between in vitro kinetics for the determination of enzyme kinetic mechanism and in vitro kinetics for kinetic modeling is that empirical equations (not mechanistic ones) are perfectly satisfactory for kinetic modeling. For example, the Hill equation is inadequate for distinguishing between different models of cooperativity, but it may well describe behavior in the physiological range just as accurately as the much more complicated equations used in mechanistic work. Of course, if complete mechanistic equations are available they can be used for modeling if they were obtained under conditions that are relevant to the metabolic context. This is only rarely the case, especially for enzymes that are important in terms of regulation. Indeed these enzymes can bind many ligands, and the equations accounting for their responses to all the ligands are practically never available. Also in most cases it is impractical to derive complete equations, because there are too many ligands. Fortunately, this is not required in the context of kinetic modeling, because the enzymes *in vivo*, in contrast with those studied in vitro, do not explore the totality of their kinetic space as fluxes and metabolite concentrations are strongly constrained by the whole network. If the range of variation of the ligand concentrations in vivo is known many simplifications can be done. We describe below how different equations describing a catalytic step can be derived without misrepresenting the quantitative behavior of the enzyme. The reader may also refer to [16] and [18] for generic equations that can be used to model different kinds of enzyme behavior.

4.2 A Simple Practical Example: Deriving the Equation of a Kinase in the Middle of a Biosynthetic Pathway

The example used is plant homoserine kinase, an enzyme involved in the synthesis of threonine, isoleucine, and methionine. It catalyzes the phosphorylation of homoserine to phosphohomoserine using ATP and forming ADP, and it is located in the chloroplast. The first thing to do is to check whether all significant interactions in this system are taken into account. Indeed, the position of the enzyme upstream of a regulated branch-point (Fig. 5) and downstream of another regulated branch-point, with a reversible catalyzed step between the two, raises the question of a possible feedback interaction that could have important regulatory functions. If phosphohomoserine inhibits homoserine kinase or if the reaction is reversible then information could be passed backward from phosphohomoserine up to the above branch-point. One thus needs to check whether the reaction is reversible or not. Even if it is not, it is still necessary to check whether phosphohomoserine inhibits the reaction. One might also be interested in the consequence of changes in the energetic status (changes in the [ATP]/[ADP] ratio, or [AMP] and [P_i]). In that case one needs to check whether changes in these concentrations have any effect on the enzyme activity.

Answers must be relevant to the metabolic context where the enzyme operates (the chloroplast). For example, there is no doubt that a 1,000-fold excess of AMP over ATP would affect the enzyme activity but that is of no interest: one needs to know whether there is any significant inhibition when all metabolites are at physiological concentrations and vary within their physiological ranges. Thus, before the kinetic characterization one needs to collect data on the concentrations of ATP, ADP, AMP, P_i, homoserine, and phosphohomoserine in the chloroplast and to express these in the same units as those used for in vitro measurements (typically µM or mM). For the sake of simplicity and to illustrate how modeling strongly depends on the question asked, we shall consider that changes in the concentrations of ATP, ADP, Pi, and AMP are not questions to be addressed. This might be justified if in the physiological states considered the concentrations of these metabolites are constant. In that case a complex mathematical description of the enzyme is not required. One simply needs to reconstitute in the test tube the metabolic context of the chloroplast (e.g., 1,500 µM ATP, 500 µM ADP, 10,000 µM Pi, 100 µM AMP (*see* **Note 2**)) and consider physiological concentrations of salts. These concentrations will not be changed, and one will examine how the enzyme activity responds to changes in homoserine concentration using HPLC analyses for example. If no effect of ADP or AMP is detected within physiological range then these metabolites can be removed from the assay and activity can be measured with a coupled enzymatic assay based on ADP detection that is much easier. A simple equation of the following form (Eq. 3) is then obtained. In this equation [ATP] is a fixed internal variable hidden in the kinetic parameters. This is specified with the concentration value indicated on the right of Eq. 3:

$$v = [\text{Homoserine kinase}] \cdot \frac{k_{\text{cat}}^{\text{app}} \cdot [\text{Homoserine}]}{K_{\text{M}}^{\text{app}} + [\text{Homoserine}]} \qquad (3)$$
$$\text{hidden variable } [\text{ATP}] = 2,000\ \mu\text{M}$$

With $k_{cat}{}^{app} = 2.8\ s^{-1}$ and $K_M{}^{app} = 14\ \mu M$.

In a more complete and detailed study, we may be interested in the effect of changing [ATP]. In that case one needs to characterize the enzyme activity in the presence of variable concentrations of ATP and homoserine (Fig. 5b). The resulting equation (Eq. 4) is an empirical one that now allows us to assess the effect on the enzyme activity of changing [ATP]:

$$v = [\text{Homoserine kinase}] \cdot \frac{\dfrac{2.86 \cdot [\text{ATP}]}{54 + [\text{ATP}]} \cdot [\text{homoserine}]}{12 + \dfrac{40}{1 + [\text{ATP}]/80} + [\text{homoserine}]} \qquad (4)$$

As [ATP] in vivo is about 2,000 μM and is essentially constant, examination of the equation shows that homoserine kinase is virtually saturated by ATP in physiological conditions and its activity is not sensitive to this concentration, Fig. 5b, graph B4. For modeling the enzyme in context it would then be legitimate to consider [ATP] as a fixed internal variable, which would play the role of an external variable. However, if flux at this step was to increase considerably (after a hypothetical biotechnological modification of the pathway) then this simplification might no longer be legitimate.

4.3 Deriving the Equation of an Allosteric Enzyme (Threonine Deaminase) with Two Effectors

Threonine deaminase (Fig. 6) catalyzes the conversion of threonine to oxobutyrate, a precursor of isoleucine, in an irreversible reaction. Threonine deaminase is inhibited by isoleucine and this inhibition is antagonized by valine, which does not have any activating effect in the absence of isoleucine. Rather than using a complex equation based on the Monod–Wyman–Changeux model [19], a simple empirical equation can be derived [5] because in vivo [Thr] is about 300 μM and is much lower than the K_M of 7,600 μM of the enzyme for threonine [20]. The threonine deaminase rate can thus be assumed to be proportional to [Thr] (Fig. 6a), in the absence or in the presence of the allosteric effectors. The response of the enzyme to the inhibitor isoleucine and the antagonist valine can then be characterized (Fig. 6b) for a single concentration of threonine (physiological concentration, 300 μM). From only about 50 experimental kinetic measurements an empirical equation, which is valid for [Thr] < 4,000 μM, was easily derived for threonine deaminase (Eq. 5, Fig. 5c):

$$v_{\mathrm{TD}} = \frac{0.0124 \cdot [\mathrm{TD}] \cdot [\mathrm{Thr}]}{1 + \left(\frac{[\mathrm{Ile}]}{30 + \frac{74 \cdot [\mathrm{Val}]}{610 + [\mathrm{Val}]}} \right)^3} \; ([\mathrm{Thr}] \qquad (5)$$

When such simplified equations are used and embedded in a larger system one will need to be cautious with the simulation results. If the simulation returns a value for the concentration of substrate that is close to $K_M{}^{\mathrm{app}}$ the simplified equation would no longer be valid.

Several other examples of procedures used to model enzymes with equations that contain many terms can be found in the supplementary material of [5]. See, for example, the equation for threonine synthase 1, which is far more complicated than Eq. 5, even after maximal simplification.

5 Conclusion: What Can One Expect from a Metabolic Simulation?

In this final section we list some of the questions that a metabolic simulation can be expected to answer:

1. How is the flux control distributed in a pathway with several enzymes?
2. What are the effects of regulatory interactions, such as cooperative feedback inhibition? How much do these affect the fluxes and metabolite concentrations?
3. Can discrepancies between experimental and calculated results be explained in terms of regulatory interactions that were not considered in the model?
4. All of these are exemplified in our study of the aspartate metabolism of *Arabidopsis thaliana* [5]. Although we have concentrated on kinetic modeling of metabolic systems in this chapter the principles are much more general and can be applied to the modeling of many other biological systems.

6 Notes

1. Concentrations are often written, for example, as [S], and this is always convenient for molecules with complicated symbols, but the equations have a simpler appearance if lower case italic letters, such as s for the concentration of S, are used. We shall use this system here.
2. As all must be expressed in the same units for the calculation it is best to express them all in the same units in the text.

Acknowledgments

We would like to thank Elisa Dell'Aglio for critical reading of the manuscript.

References

1. Cornish-Bowden A (2012) Fundamentals of enzyme kinetics, 4th edn. Wiley, Weinheim, pp 327–380
2. Teusink B, Passarge J, Reijenga CA et al (2000) Can yeast glycolysis be understood in terms of in vitro kinetics of the constituent enzymes? Testing biochemistry. Eur J Biochem 267:5313–5329
3. Eisenthal R, Cornish-Bowden A (1998) Prospects for antiparasitic drugs. The case of *Trypanosoma brucei*, the causative agent of African sleeping sickness. J Biol Chem 273:5500–5505
4. Curien G, Ravanel S, Dumas R (2003) A kinetic model of the branch-point between the methionine and threonine biosynthesis pathways in *Arabidopsis thaliana*. Eur J Biochem 270:4615–4627
5. Curien G, Bastien O, Robert-Genthon M et al (2009) Understanding the regulation of

aspartate metabolism with a model based on measured kinetic parameters. Mol Syst Biol 5:271

6. Xu YF, Amador-Noguez D, Reaves ML et al (2012) Ultrasensitive regulation of anapleurosis via allosteric activation of PEP carboxylase. Nat Chem Biol 8:562–568
7. Endler L, Rodriguez N, Juty N et al (2009) Designing and encoding models for synthetic biology. J R Soc Interface 6:S405–S417
8. Funahashi A, Matsuoka Y, Jouraku A et al (2008) Cell designer 3.5: a versatile modeling tool for biochemical networks. Proc IEEE 96:1254–1265
9. Piedrafita G, Montero F, Moran F et al (2010) A simple self-maintaining metabolic system: robustness, autocatalysis, bistability. Plos Comput Biol 6:e1000871
10. Press WH, Flannery BP, Teukolsky SA, Vetterling WT (1986) Numerical recipes. Cambridgre University Press, Cambridge, pp 547–577
11. Piedrafita G, Cornish-Bowden A, Moran F, Montero F (2012) Size matters: influence of stochasticity on the self-maintenance of a simple model of metabolic closure. J Theor Biol 300:143–151
12. Gillespie DT (1976) A general method for numerically simulating the stochastic time evolution of coupled chemical reactions. J Comput Phys 22:403–434
13. Curien G, Ravanel S, Robert M, Dumas R (2005) Identification of six novel allosteric effectors of *Arabidopsis thaliana* Aspartate kinase-Homoserine dehydrogenase isoforms: physiological context sets the specificity. J Biol Chem 280:41178–41183
14. Kacser H, Burns JA (1973) The control of flux. Symp Soc Exp Biol 27:65–104
15. Cornish-Bowden A, Cardenas ML (2001) Information transfer in metabolic pathways. Effects of irreversible steps in computer models. Eur J Biochem 268:6616–6624
16. Cornish-Bowden A, Hofmeyr JHS (2005) Enzymes in context. The Biochemist 27:11–14
17. van Eunen K, Bouwman J, Daran-Lapujade P et al (2010) Measuring enzyme activities under standardized in vivo-like conditions for systems biology. FEBS J 277:749–760
18. Rohwer JM, Hanekom AJ, Crous C, Snoep JL, Hofmeyr JHS (2006) Evaluation of a simplified generic bi-substrate rate equation for computational systems biology. Syst Biol (Stevenage) 153:338–341
19. Monod J, Wyman J, Changeux JP (1965) On the nature of allosteric transitions: a plausible model. J Mol Biol 12:88–118
20. Wessel PM, Graciet E, Douce R, Dumas R (2000) Evidence for two distinct effector-binding sites in threonine deaminase by site-directed mutagenesis, kinetic, and binding experiments. Biochemistry 39:15136–15143

Chapter 17

Flux Balance Analysis as an Alternative Method to Estimate Fluxes Without Labeling

Eva Grafahrend-Belau, Astrid Junker, Falk Schreiber, and Björn H. Junker

Abstract

The analysis of plant metabolic networks essentially contributes to the understanding of the efficiency of plant systems in terms of their biotechnological usage. Metabolic fluxes are determined by biochemical parameters such as metabolite concentrations as well as enzyme properties and activities, which in turn are the result of various regulatory events at various levels between control of transcription and posttranslational regulation of enzyme protein activity. Thus, knowledge about metabolic fluxes on a large scale provides an integrated view on the functional state of a metabolically active cell, organ, or system. In this chapter, we introduce flux balance analysis as a constraint-based method for the prediction of optimal metabolic fluxes in a given metabolic network. Furthermore, we provide a step-by-step protocol for metabolic network reconstruction and constraint-based analysis using the COBRA Toolbox.

Key words Flux balance analysis, Metabolic network reconstruction, Matlab, COBRA toolbox

1 Introduction

Plant growth and performance under varying environmental conditions substantially influence the plant's economic value as the basis for feed, food, and various biotechnological applications. Therefore, the understanding of plant metabolism, which is tightly coupled to plant fitness and development [1], is of vast importance for both science and agriculture. Plant metabolic networks are characterized by a unique complexity originating from complex plant structures being composed of multiple interacting source- and sink-organs and highly compartmentalized cells. The overall state of the whole metabolic network of certain biological system determines strongly the metabolic behavior down to single reactions [2, 3], thus requiring large-scale analysis of metabolic networks.

Martine Dieuaide-Noubhani and Ana Paula Alonso (eds.), *Plant Metabolic Flux Analysis: Methods and Protocols*, Methods in Molecular Biology, vol. 1090, DOI 10.1007/978-1-62703-688-7_17,

Several methods exist for the determination of intracellular metabolic fluxes on a larger scale, the most prominent of which are metabolic flux analysis (MFA, Chaps. 11 and 12) and flux balance analysis (FBA) [4]. Both methods rely on mass balances based on the stoichiometric coefficients of the metabolic reactions of the network. For both methods, input fluxes (usually uptake fluxes) and output fluxes (usually biomass production) have to be defined by additional direct measurements. FBA uses this information (stoichiometric matrix and input/output measurements) to determine an optimal flux distribution. As the number of unknown fluxes is usually much higher than the number of measured input/output fluxes, the mathematical problem to be solved is under-determined. Thus, an objective function is needed that is subsequently minimized or maximized, e.g., the maximization of the biomass production or minimization of energy consumption. The applied optimization algorithm will attempt to find the flux distribution for which the objective function is maximal or minimal. For MFA in contrast, the stoichiometric matrix and input/output measurements are enriched by (a) carbon transition maps, in which the fate of each carbon atom in each reaction is defined, and (b) data from stable isotope feeding experiments. By adding this surplus of data, from a mathematical point of view the under-determined problem is turned into an overdetermined one, which means that in principle no objective function is needed. Instead, a different optimization approach is used that minimizes the difference between simulated and measured isotope labeling data.

The choice of the method (FBA vs. MFA) depends on the system that is to be analyzed. While MFA allows for deriving detailed flux maps of metabolic sub-networks including the resolution of cyclic, parallel, and bidirectional fluxes, it also requires a significantly larger amount of experimental data. Thus, MFA is the method of choice for well-characterized systems. FBA in contrast is used for large metabolic networks, often all metabolic reactions of a specific cell type, but only needs few parameters to be specified by measurements. Care has to be taken in the interpretation of the resulting fluxes for both methods: while MFA due to its overdetermined nature comes closer to a *measurement* of fluxes, FBA is clearly a *prediction* of one possible flux distribution which is optimal for the objective function. Nevertheless FBA has been successfully applied to numerous photosynthesizing/plant systems over the last years, e.g., Arabidopsis [5–9], barley seeds and plants [10–12], maize leaves [13, 14], brassica seeds [15], green algae [16–18], and photoautotroph bacteria [19, 20].

In the following, we give a detailed protocol of how to reconstruct a metabolic model and how to perform constraint-based model analysis using the constraint-based reconstruction and analysis (COBRA) Toolbox, a freely available *Matlab* toolbox, in order to evaluate the metabolic behavior of the system of interest.

2 Materials

2.1 Equipment

1. A personal computer capable of running Matlab.
2. Matlab (Version 7.0 or above): http://www.mathworks.com.
3. COBRA Toolbox (Version 2.0.0 or above): http://www.cobratoolbox.org.
4. libSBML programming library (version 4.0.1 or above): http://www.sbml.org.
5. SBML Toolbox for Matlab (version 3.1.1 or above): http://www.sbml.org.
6. A linear programming solver supported by the COBRA Toolbox:
   - glpk (freeware): http://www.gnu.org/software/glpk.
   - Gurobi (commercial): Gurobi Optimization: http://www.gurobi.com.
   - CPLEX (commercial): Tomlab Optimization Inc: http://tomopt.com.

2.2 Equipment Setup

The Matlab software, the COBRA Toolbox, libSBML, the SBML Toolbox, and one of the listed solvers should be installed following the instruction of the software providers. A comprehensive documentation of the installation of the COBRA Toolbox and the requisite software is given on http://www.cobratoolbox.org and in [21–23].

3 Methods

3.1 Model Reconstruction

3.1.1 Model Definition

1. Define modeling objective. Specify scope and purpose of the model.
2. Define schematic draft model. Depending on the modeling objectives define a schematic draft model specifying (a) model content (i.e., *model system*: organism, organ, tissue or cell type; coverage: part of metabolism to be modeled; *time point*: developmental stage) and (b) model boundaries (i.e., the exchange of substances with the surrounding).

3.1.2 Model Reconstruction and Refinement

Based on the draft model, a stoichiometric network (i.e., a reaction list) of the model system is reconstructed in a stepwise manner following the procedure specified below. For each network reaction:

1. Determine reaction stoichiometry including substrate and cofactor usage.
2. Determine reaction directionality. In the absence of appropriate information, reactions should be assumed to be reversible (*see* **Note 1**).

3. Determine the chemical formula for each metabolite in the reaction.
4. Check mass balance. Verify the stoichiometric reaction equation by checking its mass balance (i.e., check if the number of chemical elements on the left side (educts) equals the one on the right side (products)) (*see* **Note 2**).
5. Verify reaction location (spatial and temporal) as specified in the draft model. In the absence of appropriate information data referring to closely related species or organism-specific data from other organs, tissues and/or plant developmental stages can be considered to substantiate the reaction.
6. Add references. To facilitate the retraceability of the experimental data supporting the data collected in **steps 1–5**, annotate each reaction with the respective references.
7. Repeat **steps 1–5** for all single-reaction steps of the pathways specified in the draft model. Collect the data based on an intensive literature and database search (*see* Table 1 for a list of potential data sources), and list the data as well as the resulting reaction list in a spreadsheet.
8. Identify and integrate add-on reactions. Reactions not assigned to a specific pathway but playing an important role for the metabolic functionality of the modeled system are called add-on reactions. Add-on reactions can be identified by considering each network reaction in its metabolic context (i.e., the connectivity of the reaction with its metabolic surrounding). Examples for add-on reactions are anaplerotic and cataplerotic reactions.
9. Add intracellular transport reactions. To reconstruct a multi-compartment-specific metabolic model, intracellular transport reactions need to be integrated allowing the exchange of substances between the cell compartments (*see* **Note 3**).
10. Add exchange reactions. The exchange of substances with the surrounding is modeled by exchange reactions representing the system boundaries. Add exchange reactions as specified in the draft model. Verify if additional exchange reactions are required.
11. Determine and add biomass reaction (for growth simulations only).
    (a) Determine the biomass composition (dry weight percent) of the modeled system taking into account all major biomass components (e.g., defined as components contributing more than 1 % to the total dry weight) such as protein, starch, cell wall components, and lipids (see, for example Table 2).
    (b) Determine the amino acid composition of the protein fraction, and calculate the percent composition of each amino acid in total biomass (dry weight percent) (see, for example Table 3).

Table 1
Data sources

	Chemical databases			Reaction/enzyme databases					Metabolic pathway databases					Protein/transporter databases		
	PubChem	ChEBI	KEGG Compound	Brenda	ExPASy-ENZYME	KEGG Enzyme	Rhea	Sabio-Rk	MetaCyc	KEGG Pathways	PlantCyc	Reactome	MetaCrop	UniProtKB	Aramemnon	Genevestigator
Reaction stoichiometry[a]				+	+	+	+	+	+	+	+	+	+			
Reaction directionality				+				+	+	+	+	+	+			
Chemical formula (metabolites)	+	+	+										+			
Reaction location (spatial, temporal)				+				+	+	(+)	(+)	(+)	+	+		+
Intracellular transport reactions									+			+	+	+	+	
Exchange reactions													+	+	+	

(+): Information on the spatial and/or temporal localization of a given reaction is only partially provided
[a]Including substrate/
cofactor usage

Table 2
Specifies the chemical composition of an (artificial) example tissue, *see* Subheading 3.1.2, step 11a

Component	% of DW
Carbohydrates	80
Starch	64
Sucrose	4
Cellulose	12
Protein	20

Table 3
Specifies the amino acid composition of the protein fraction, *see* Subheading 3.1.2, step 11b

Amino acid	% of protein
L-Alanine	10
L-Glutamate	40
L-Proline	30
L-Serin	15
L-Threonine	5

(c) Determine the neutral lipid (e.g., mono-, di-, triglycerids, free fatty acids) and phospholipid composition of the lipid fraction, and calculate the percent composition of each component in total biomass (dry weight percent).

(d) Calculate the fractional distribution of each biomass precursor to the biomass composition (mmol/g dry weight) (see, for example Table 4).

(e) Determine growth-associated maintenance energy (GAME). Use experimental data or data from the scientific literature (*see* **Note 4**).

(f) Add biomass reaction. Formulate the biomass reaction by combining all metabolic precursors and the respective stoichiometric coefficients calculated in **steps 1–3** in one single reaction. Add the GAME to the biomass reaction (*see* **Note 5**). Add the biomass reaction to the reaction list.

12. Determine and add non-growth-associated maintenance energy (NGAME) reaction. Use experimental data or data from the primary literature to estimate NGAME. Model the consumption

Table 4
Specifies the fractional distribution of each biomass precursor to the biomass composition, *see* Subheading 3.1.2, step 11d

Precursor	% of DW	g/g Cell	MW [g/mol]	mmol/g DW
Starch	64	0,64	180	3,552
Sucrose	4	0,04	342	0,117
Cellulose	12	0,12	180	0,666
L-Alanine	2	0,02	89	0,224
L-Glutamate	8	0,08	147	0,544
L-Proline	6	0,06	115	0,521
L-Serin	3	0,03	105	0,285
L-Threonine	1	0,01	119	0,084

Table 5
Software for the constraint-based model analysis

Software	Link	References
CNA	http://www.mpi-magdeburg.mpg.de/projects/cna/cna.html	[27]
COBRA Toolbox	http://opencobra.sourceforge.net/openCOBRA/Welcome.html	[21, 22]
FBA-SimVis	http://fbasimvis.ipk-gatersleben.de/	[28]
SBRT	http://www.bioc.uzh.ch/wagner/software/SBRT	[29]
SNA	http://bioinformatics.org/project/?group_id=546	[30]

of NGAME by adding an ATP drain reaction (i.e., 1 ATP + 1 H_20 => 1 ADP + 1P + H^+) to the reaction list. In the process of model initialization, the reaction rate of the ATP drain reaction will be fixed to the NGAME value.

13. Determine simulation constraints. Check the literature for uptake and excretion rates (e.g., sucrose uptake rate, specific growth rate) required for model validation (Subheading 3.1.5) and analysis (Subheading 3.2).

3.1.3 Translation into a Mathematical Model

In the following, constraint-based model validation and analysis are performed using the COBRA Toolbox [21]. Table 5 gives an overview of additional software for constraint-based model analysis. For a comprehensive overview of constraint-based software packages the reader is refered to http://cobramethods.wikidot.com. Details on initializing and using the COBRA Toolbox are given in the documentation and [21–23]. The toolbox commands described in the following only specify mandatory input parameters. For a detailed description of the command, the remaining (optional) parameters, as well as the simulation results the reader is refered to the documentation and the respective literature.

1. Translate reaction list into COBRA Toolbox model. Use the Excel template specified in the toolbox documentation (Matlab Index: cobra/io/xls2model) to convert the reaction list generated in Subheading 3.1.2 into a COBRA Toolbox-compatible Excel file (*see* **Note 6**). Alternatively, the reaction list can be manually translated into an SBML-file using an SBML editor software (http://www.sbml.org) and imported into the toolbox as described in the documentation (Matlab Index: cobra/io/readCbModel) (*see* **Note 7**).
2. Load the model into the COBRA Toolbox. Initiate Matlab following the instruction of the software provider. Navigate to the directory where the COBRA Toolbox was installed, and start the toolbox:

   > *initCOBRAToolbox*

3. Read the model into the toolbox:

   > *model = xls2model(fileName),*

   where *fileName* is the name of the Excel spreadsheet.

3.1.4 Model Initialization

1. Set objective function. The objective function can be set either via the Excel spreadsheet (column: "Objective") or by using the following toolbox command:

   > *model = changeObjective(model,rxnNameList,objectiveCoeff)*,

   where *model* is the COBRA model, *rxnNameList* is the reaction (s) to be set as objective function, and *objectiveCoeff* is the objective coefficient (i.e., *1*: maximization; *–1*: minimization).
2. Set simulation constraints. The simulation constraints can be set either via the Excel spreadsheet (columns: *"Upper bound," "Lower bound"*) or by using the following toolbox command:

   > *changeRxnBounds(model,rxnNameList,value,boundType)*,

   where *rxnNameList* is the reaction(s) for which the reaction bounds should be changed, *value* is the bound value(s), and *boundType* is the bound type ("u": upper, "l": lower, "b": both).

3.1.5 Model Validation

Model validation refers to the process of model verification, evaluation, and validation. To facilitate the process of model validation it is recommended to perform the procedure of assembling a mathematical model (i.e., a COBRA Toolbox model) in a stepwise manner: (a) select a pathway and its appropriate reactions from the reaction list generated in Subheading 3.1.2, (b) transform the selected reactions into a COBRA Toolbox model (Subheading 3.1.3), (c) validate the model by applying the validation steps specified below, (d) extend the validated model by selecting a new pathway and repeating **steps a–c**, and (e) repeat **steps a–d** until all pathways specified in the reaction list are integrated into a validated COBRA Toolbox model. To validate a given model proceed as follows:

1. Test metabolic capability of the model. Test if the model is capable of correctly predicting known quantitative properties. For example, to predict the growth rate perform an FBA using the biomass reaction as the objective function:

   $$> FBAsolution = optimizeCbModel(model, osenseStr),$$

   where *osenseStr* is the mode of optimization (*'max','min'*).

   In case the model is unable to reproduce known experimental results try to debug the model by applying further validation steps. The COBRA Toolbox offers different functionalities to evaluate a given model automatically. Some of the functions are listed below:

2. Check for blocked reactions. Use

   $$> blockedReactions = findBlockedReaction(model)$$

   to determine those reactions which cannot carry any flux for structural reasons.

3. Check for metabolic dead ends. Use

   $$> outputMets = findGaps(model)$$

   to detect dead-end metabolites.

See the documentation and [23] for additional functions of model validation. Once the model is capable of reproducing known experimental results, the validated model can be used for model analysis.

3.2 Model Analysis

In the following different functionalities of the COBRA Toolbox are listed, allowing for an in-depth analysis of plant metabolism. See the documentation for the complete list of functions provided by the COBRA Toolbox.

3.2.1 Evaluating Optimal States

1. Test metabolic capability of the model. *See* Subheading 3.1.5, **step 1**. Depending on the size of the metabolic model, the simulation constraints, and the chosen objective function, the constraint-based model analysis might result into alternative optima (i.e., multiple flux distributions with the same optimal value identified by the objective). To handle the problem of alternative optima, the COBRA Toolbox offers different functions to obtain a unique optimal solution. Use, for example,

   $$> flux = geometricFBA(model)$$

   to find a unique optimal FBA solution.

2. Check flux variability. Flux variability resulting from alternative optima might be used as an additional characteristic of the model, allowing for getting a different perspective on the model behavior. Flux variability can explicitly be analyzed using flux variability analysis:

   > *[minFlux,maxFlux] = fluxVariability(model)*

3.2.2 Varying Environmental Parameters

1. Varying one parameter: robustness analysis. The effect of varying a single parameter on the metabolic behavior of the model can be tested using robustness analysis:

   > *[controlFlux, objFlux] = robustnessAnalysis(model, controlRxn)*,

   where *controlRxn* is the reaction to be varied.

2. Varying two parameters: phenotype-phase plane (PhPP) analysis. PhPP analysis [24] is a method used to obtain the range of optimal flux distribution (i.e., optimal phenotypic behavior) in response to changes in two environmental parameters. To perform a PhPP, use

   > *[growthRates,shadowPrices1,shadowPrices2]=phenotypePhasePlane(model,controlRxn1,controlRxn2)*,

   where *controlRxn1* is the first and *controlRxn2* is the second parameter to be varied.

3.2.3 Varying Genetic Parameters

1. In silico knockout: To perform a single-reaction deletion, use

   > [grRatio,grRateKO,grRateWT] = singleRxnDeletion(model)

2. In silico reaction addition: To check the effect of introducing a specific reaction into a given model, first add the reaction into the model using

   > *model = addReaction(model,rxnName,rxnFormula)*,

   where *rxnName* is the name of the reaction to be added and *rxnFormula* is the corresponding reaction formula. For example, for adding the reaction 'newRxn1': A -> B + 2 C, use

   > *model = addReaction(model,'newRxn1','A -> B + 2 C')*

   Second, perform an optimization (*see* for example Subheading 3.1.5, **step 1**).

4 Application Example

In the following a small application example is used to illustrate the presented methodology.

4.1 Model Loading and Initialization

1. To read the example model shown in Fig. 1 into the COBRA Toolbox, fill the respective columns of the COBRA Toolbox-compatible Excel file as specified in Tables 6 and 7 (*see* **Note 6**),

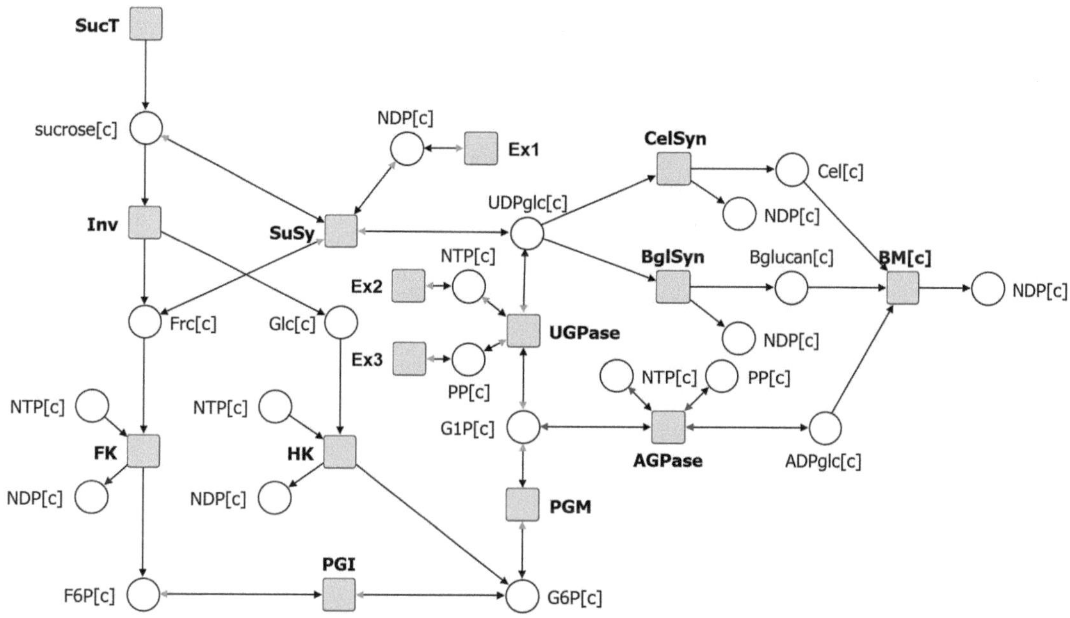

Fig. 1 The example network comprises a simplified version of the sucrose-to-starch pathway as well as cellulose and β-glucan biosynthesis pathway with ADPglucose, cellulose and β-glucan being used for the formation of biomass which is excreted from the model. Biochemical reactions and their associated catalyzing enzymes are shown as *rectangles*, and metabolites are shown as *circles*. The reversibility (resp. irreversibility) of a given reaction is shown by a single (resp. double) *arrow*. *Abbreviations*: *reactions: AGPase* ADP glucose phosphorylase, *BglSyn* β-glucan biosynthesis, *BM* biomass biosynthesis, *CelSyn* cellulose biosynthesis, *FK* fructokinase, *HK* hexokinase, *Inv* invertase, *PGI* phosphoglucoisomerase, *PGM* phosphoglucomutase, *SucT* sucrose transporter, *SuSy* sucrose synthase, *UGPase* UDP glucose phosphorylase, *metabolites: ADPglc* ADP glucose, *Bglucan* β-glucan, *Cel* cellulose, *F6P* fructose 6-phosphate, *Frc* fructose, *G1P* glucose 1-phosphate, *G6P* glucose 6-phophate, *Glc* glucose, *NDP/NTP* nucleotide di/triphosphates such as ADP/ATP and UDP/UTP, *PP* inorganic pyrophosphate, *UDPglc* UDP glucose

save the file as *example.xls*, and load the file into the Toolbox using the following command:

> *model* = *xls2model('example.xls')*

2. For model initialization, set the maximization of biomass as the objective and constrain the uptake of sucrose to a maximal value of 8 replace: mmol gDW^{-1} h^{-1}

   > *model* = *changeObjective(model,'BM',1)*;
   > *model* = *changeRxnBounds(model,'SucT',8.0,'u')*

4.2 Model Validation

To validate the model, test its metabolic capability by predicting the optimal growth rate as well as the corresponding flux distribution:

> *[solution]* = *optimizeCbModel(model)*
> *printFluxVector(model,solution.x)*

Table 6
"Reactions" tab of the example.xls spreadsheet

Abbr.	Name	Reaction	Sub-system	Reversible	Lower bound	Upper bound	Objective
AGPase	AGPase	1.0 G1P[c] + 1.0 NTP[c] ⇔ 1.0 ADPglc [c] + 1.0 PP[c]	SucBreakdown	1	−100	100	0
Inv	Inv	1.0 sucrose[c] → 1.0 Glc [c] + 1.0 Frc[c]	SucBreakdown	0	0	100	0
PGI	PGI	1.0 F6P[c] ⇔ 1.0 G6P [c]	SucBreakdown	1	−100	100	0
FK	FK	1.0 NTP[c] + 1.0 Frc [c] → 1.0 NDP [c] + 1.0 F6P[c]	SucBreakdown	0	0	100	0
PGM	PGM	1.0 G1P[c] → 1.0 G6P [c]	SucBreakdown	1	−100	100	0
UGPase	UGPase	1.0 G1P[c] + 1.0 NTP [c] ⇔ 1.0 PP[c] + 1.0 UDPGlc[c]	SucBreakdown	1	−100	100	0
HK	HK	1.0 Glc[c] + 1.0 NTP [c] → 1.0 NDP [c] + 1.0 G6P[c]	SucBreakdown	0	0	100	0
SuSy	SuSy	1.0 sucrose[c] + 1.0 NDP [c] ⇔ 1.0 Frc [c] + 1.0 UDPGlc[c]	SucBreakdown	1	−100	100	0
SucT	SucT	→1.0 sucrose[c]	Import	0	0	100	0
BM	BM	1.0 ADPglc + 1.0 Cel + 1.0 Bglucan→1.0 NDP	BMSynthesis	0	0	100	0
Ex1	Ex1	⇔1.0 NDP[c]	Im/Export	1	−100	100	0
Ex2	Ex2	⇔1.0 NTP[c]	Im/Export	1	−100	100	0
Ex3	Ex3	⇔1.0 PP[c]	Im/Export	1	−100	100	0
CelSyn	CelSyn	1.0 UDPGlc[c]→1.0 Cel [c] + 1.0 NDP[c]	BMSynthesis	0	0	100	0
BglSyn	BglSyn	1.0 UDPGlc[c] ⟶ 1.0 Bglucan[c] + 1.0 NDP[c]	BMSynthesis	0	0	100	0

Table 7
"Metabolites" tab of the example.xls spreadsheet

Abbreviation	Name	Formula (neutral)	Formula (charged)	Charge
ADPglc	ADPglc[c]	C16H25N5O15P2	C16H25N5O15P2	0
Bglucan	Bglucan[c]	C12H22O11(C6H10O5)n	C12H22O11(C6H10O5)n	0
Cel	Cel[c]	(C6H10O5)n	(C6H10O5)n	0
F6P	F6P[c]	C6H13O9P	C6H11O9P	−2
Frc	Frc[c]	C6H12O6	C6H12O6	0
G1P	G1P[c]	C6H13O9P	C6H11O9P	−2
G6P	G6P[c]	C6H13O9P	C6H11O9P	−2
Glc	Glc[c]	C6H12O6	C6H12O6	0
NDP	NDP[c]	C10H15N5O10P2	C10H12N5O10P2	−3
NTP	NTP[c]	C10H16N5O13P3	C10H12N5O13P3	−4
PP	PP[c]	H4P2O7	H4P2O7	0
UDPGlc	UDPGlc[c]	C15H24N2O17P2	C15H24N2O17P2	0
Sucrose	Sucrose[c]	C12H22O11	C12H22O11	0

As the example model does not contain any modeling mistakes (e.g., blocked reactions, dead-end metabolites) additional validation steps (*see* Subheading 3.1.5) are not required.

4.3 Model Analysis

1. To compute the range of fluxes for a given optimal solution, perform a flux variability analysis:

   > [minFlux,maxFlux] = fluxVariability(model)

   The resulting minimal/maximal flux values for each model reaction are shown in Fig. 2.

2. To study the effect of varying the flux through the enzyme AGPase on biomass production perform a robustness analysis:

   > [controlFlux, objFlux] = robustnessAnalysis(model, 'AGPase')

   The resulting plot depicting how the optimal growth rate varies as a function of flux through the AGPase reaction is shown in Fig. 3.

3. To perform a PhPP analysis based on the variation of the uptake of sucrose and NTP, set the upper bound of the sucrose uptake (SucT) to 100 mmol $gDW^{-1}h^{-1}$ and perform the analysis as follows:

   >[growthRates,shadowPrices1,shadowPrices2] = phenotypePhasePlane(model,'SucT','Ex2')

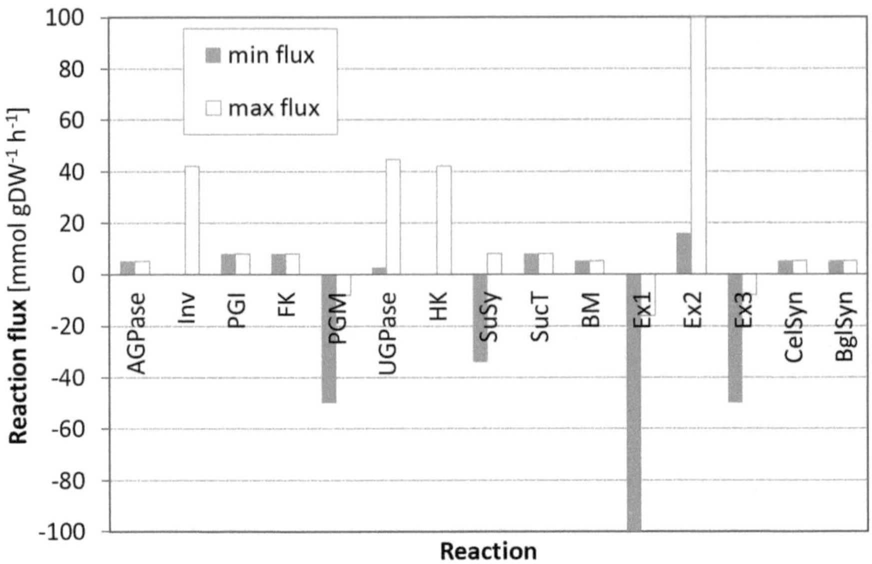

Fig. 2 Flux variability analysis performed on the application example

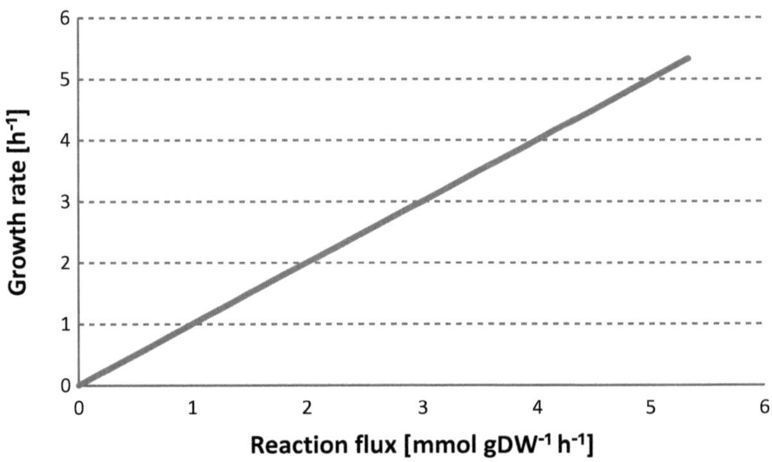

Fig. 3 Robustness analysis performed on the application example: varying the reaction flux of AGPase

The resulting 3D surface plot of the PhPP analysis is shown in Fig. 4.

4. To perform a single deletion study for each of the model reaction using standard FBA use the following command:

   > *[grRatio,grRateKO,grRateWT] = singleRxnDeletion(model)*

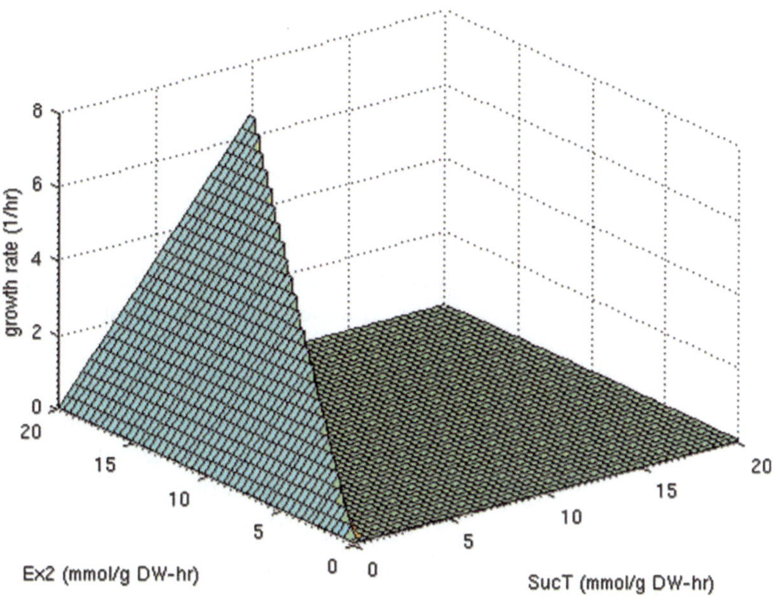

Fig. 4 Phenotype-phase plane analysis. The 3D surface plot of the PhPP depicts the maximum growth rate that is achievable at every possible combination of sucrose (exchange reaction: SucT) and NTP (exchange reaction: Ex2) uptake rates

5. To test the effect of reaction addition on the optimal growth behavior of the example model, first add the reaction (e.g., FrcT: exchange reaction of fructose) to the model, then specify its flux boundaries, and perform the growth optimization:

   > model = addReaction(model,'FrcT','-> Frc[c]')
   > model = changeRxnBounds(model,'FrcT',8.0,'u')
   > [solution] = optimizeCbModel(model)
   >printFluxVector(model,solution.x)

5 Concluding Remarks

This book chapter introduced FBA as an alternative method for the determination of flux distributions in metabolic networks using a basic set of functionalities provided by the open-access COBRA Toolbox. Simple FBA as described here will increasingly be substituted by more integrative approaches which use different kinds of -omics (e.g., transcriptomics) data for constraining the FBA. Furthermore, by model coupling with other model types FBA models will gain higher resolution in spatial and temporal dimensions.

6 Notes

1. Besides using biochemical data, the primary literature, or online database (*see* Table 1), information about the standard Gibbs free energy of formation or the presence of k_m-values might be used as a basis to make assumptions about the directionality of a given reaction.
2. Take the chemical formulas extracted in the previous step as a basis. In case your research objective just requires the model to be carbon balanced, check the mass balance by comparing the number of C-atoms only.
3. To reduce the number of alternate optimal solutions and loops, restrict the number of integrated internal exchange reactions to transporters with a definite experimental proof or well-known transporter processes reported in the respective reviews.
4. To estimate the (non-)growth-associated maintenance energy cost for different plant species, *see* for example refs. 25, 26 and related work.
5. An example for the stoichiometric equation of a biomass reaction is shown in Formula 1. Taking the biomass composition of the example tissue listed in Tables 2 to 4 as a basis,

Formula 1:

$$3,552 \text{ starch} + 0,117 \text{ sucrose} + +0,084 \text{ threonine} + x\text{ATP} + x\text{H}_2\text{O} => 1 \text{ biomass} + x\text{ADP} + x\text{P} + x\text{H}^+$$

Formula 1 specifies the biomass precursors and their respective stoichiometric coefficients as summarized in Table 4. In addition, Formula 1 includes the GAME (modeled as xATP + $x$$H_2O$ => x ADP + x P + $x$$H^+$, where x corresponds to the amount of GAME in mmol/g DW required for biomass production).

6. As described in the documentation, create an Excel spreadsheet containing one "reactions" and one "metabolites" tab. Make sure to name the tabs and required column headers as specified in the documentation, e.g., *tab names*: reactions, metabolites; *header names*: Abbreviation, Name, Reaction, etc. To perform a model analysis as described in Subheading 3.2, only the following columns need to be filled with content: "reactions" tab: col 1–3 and 7–11; "metabolites" tab: col 1–5. See the example below:

   ”reactions” tab:

Abbr.	Name	Reaction	...	Sub-system	Reversible	Lower bound	Upper bound	Objective
AGPase	ADPglc pyro-phosphory-lase	1.0 G1P[c] + 1.0 ATP[c] <==> 1.0 ADPglc[c] + 1.0 PP[c]		sucrose break-down	1	−1000	1000	0
Inv	invertase	1.0 sucrose[c] -> 1.0 Glc[c] + 1.0 Frc[c]		sucrose break-down	0	0	1000	0

In the column "Reaction" note down the *stoichiometric equation* by specifying the subcellular localization of the reaction metabolites using the suffix-notation shown in the example, i.e., put the compartment name/id in brackets behind the metabolite name. *Reversible reactions* should be specified using a double arrow (⇔) in the stoichiometric equation, entering "1" in the column "Reversible" and assigning values for the upper and lower bounds. In contrast, *irreversible reactions* are specified by using a single arrow (⟶), entering "0" in the column "Reversible" and assigning "0" for the lower bound and a positive value for the upper bound. The *objective function* can be set by specifying the objective coefficient of the respective reaction(s) (1: maximization, −1: minimization). Alternatively the objective function can be set using the toolbox command listed in Subheading 3.1.4.

"metabolites" tab:

Abbreviation	Name	Formula (neutral)	Formula (charged)	Charge
Sucrose[c]	Sucrose[c]	C12H22O11	C12H22O11	0
ADPglc[c]	ADPglc[c]	C16H25N5 O15P2	C16H25N5 O15P2	0

Column one to five are required to allow the xls spreadsheet to be read into the COBRA Toolbox. For the model analysis performed in this chapter, information about the charged formula and charge are not required; therefore these columns might be filled with dummy values (e.g., neutral formula of the metabolite and a charge value of zero) to shorten the time-consuming process of model reconstruction.

7. To validate a given SBML model use the online SBML validator (http://sbml.org/Facilities/Validator/).

References

1. Stitt M, Sulpice R, Keurentjes J (2010) Metabolic networks: how to identify key components in the regulation of metabolism and growth. Plant Physiol 152:428–444
2. Sweetlove LJ, Fell D, Fernie AR (2008) Getting to grips with the plant metabolic network. Biochem J 409:27–41
3. Kruger NJ, Ratcliffe RG (2012) Pathways and fluxes: exploring the plant metabolic network. J Exp Bot 63:2243–2246
4. Edwards JS, Ramakrishna R, Schilling CH, Palsson BO (1999) Metabolic Flux Balance Analysis. In: Lee SY, Papoutsakis ET (eds) Metabolic engineering. Marcel Dekker, New York, pp 13–57
5. Poolman MG, Miguet L, Sweetlove LJ, Fell DA (2009) A genome-scale metabolic model of Arabidopsis and some of its properties. Plant Physiol 151:1570–1581
6. Williams TC, Poolman MG, Howden AJ et al (2010) A genome-scale metabolic model accurately predicts fluxes in central carbon metabolism under stress conditions. Plant Physiol 154:311–323
7. de Oliveira Dal'Molin CG, Quek LE, Palfreyman RW et al (2010) AraGEM, a genome-scale reconstruction of the primary metabolic network in Arabidopsis. Plant Physiol 152:579–589
8. Radrich K, Tsuruoka Y, Dobson P et al (2010) Integration of metabolic databases for the reconstruction of genome-scale metabolic networks. BMC Systems Biol 4:114
9. Kleessen S, Araujo WL, Fernie AR, Nikoloski Z (2012) Model-based confirmation of alternative substrates of mitochondrial electron transport chain. J Biol Chem 287:11122–11131
10. Grafahrend-Belau E, Schreiber F, Koschutzki D, Junker BH (2009) Flux balance analysis of barley seeds: a computational approach to study systemic properties of central metabolism. Plant Physiol 149:585–598
11. Rolletschek H, Melkus G, Grafahrend-Belau E et al (2011) Combined noninvasive imaging and modeling approaches reveal metabolic compartmentation in the barley endosperm. Plant Cell 23:3041–3054
12. Grafahrend-Belau E, Junker A, Eschenröder A, Müller J, Schreiber F, Junker BH (2013) Multiscale metabolic modeling: dynamic flux balance analysis on a whole plant scale. Plant Physiol [Epub ahead of print]
13. Dal'Molin CG, Quek LE, Palfreyman RW et al (2010) C4GEM, a genome-scale metabolic model to study C4 plant metabolism. Plant Physiol 154:1871–1885
14. Saha R, Suthers PF, Maranas CD (2011) Zea mays iRS1563: a comprehensive genome-scale metabolic reconstruction of maize metabolism. PLoS One 6:e21784
15. Schwender J, Hay JO (2012) Predictive modeling of biomass component tradeoffs in Brassica napus developing oilseeds based on in silico manipulation of storage metabolism. Plant Physiol 160:1218–1236
16. Boyle NR, Morgan JA (2009) Flux balance analysis of primary metabolism in Chlamydomonas reinhardtii. BMC Syst Biol 3:4
17. Cogne G, Rugen M, Bockmayr A et al (2011) A model-based method for investigating bioenergetic processes in autotrophically growing eukaryotic microalgae: application to the green algae Chlamydomonas reinhardtii. Biotechnol Prog 27:631–640
18. Dal'Molin CG, Quek LE, Palfreyman RW, Nielsen LK (2011) AlgaGEM—a genome-scale metabolic reconstruction of algae based on the Chlamydomonas reinhardtii genome. BMC Genomics 12(Suppl 4):S5
19. Knoop H, Zilliges Y, Lockau W, Steuer R (2010) The metabolic network of Synechocystis sp. PCC 6803: systemic properties of autotrophic growth. Plant Physiol 154:410–422
20. Montagud A, Navarro E, Fernandez de Cordoba P, Urchueguia JF, Patil KR (2010) Reconstruction and analysis of genome-scale metabolic model of a photosynthetic bacterium. BMC Syst Biol 4:156
21. Becker SA, Feist AM, Mo ML et al (2007) Quantitative prediction of cellular metabolism with constraint-based models: the COBRA toolbox. Nat Protoc 2:727–738
22. Schellenberger J, Que R, Fleming RM et al (2011) Quantitative prediction of cellular metabolism with constraint-based models: the COBRA Toolbox v2.0. Nat protoc 6:1290–1307
23. Thiele I, Palsson BO (2010) A protocol for generating a high-quality genome-scale metabolic reconstruction. Nat Protoc 5:93–121
24. Edwards JS, Ramakrishna R, Palsson BO (2002) Characterizing the metabolic phenotype: a phenotype phase plane analysis. Biotechnol Bioeng 77:27–36

25. Amthor JS (2010) From sunlight to phytomass: on the potential efficiency of converting solar radiation to phyto-energy. New Phytol 188:939–959
26. Amthor JS (2000) The McCree±de Wit±Penning de Vries±Thornley respiration paradigms: 30 years later. Ann Bot 86:1–20
27. Klamt S, Stelling J, Ginkel M, Gilles ED (2003) FluxAnalyzer: exploring structure, pathways, and flux distributions in metabolic networks on interactive flux maps. Bioinformatics 19:261–269
28. Grafahrend-Belau E, Klukas C, Junker BH, Schreiber F (2009) FBA-SimVis: interactive visualization of constraint-based metabolic models. Bioinformatics 25:2755–2757
29. Wright J, Wagner A (2008) The systems biology research tool: evolvable open-source software. BMC Syst Biol 2:55
30. Urbanczik R (2006) SNA—a toolbox for the stoichiometric analysis of metabolic networks. BMC Bioinformatics 7:129

Chapter 18

Flux Variability Analysis: Application to Developing Oilseed Rape Embryos Using Toolboxes for Constraint-Based Modeling

Jordan O. Hay and Jörg Schwender

Abstract

Flux variability analysis enables comprehensive exploration of alternate optimal routes in a metabolic network. This method is especially useful with models such as *bna572* for the developing oilseed rape embryo which is highly compartmentalized. Here, we describe a protocol for carrying out flux variability analysis on reactions and network projections of *bna572* using well-established software (CellNetAnalyzer and COBRA) for constraint-based analysis of stoichiometric network reconstructions.

Key words Stoichiometric modeling, Constraint-based modeling, Flux balance analysis, Flux variability analysis, Metabolic network, Linear programming

1 Introduction

Constraint-based models of metabolism are based on the stoichiometric structure and thermodynamic reversibility of biochemical reactions in a metabolic network [1]. By using such minimal information, constraint-based models of metabolism facilitate the modeling of large and complex biochemical reaction networks. This information can be obtained at the genome scale as well as bibliomic level from a variety of sources including metabolic pathway databases [2, 3], biochemical literature, genome sequencing projects [4, 5], and transcriptional profiling studies [6]. Stoichiometric flux balance modeling is becoming increasingly important as a systems level approach for metabolism research in plants [7–17].

Metabolism is highly complex when considering the huge number of possible combinations of routes in a metabolic network [18]. The method of flux balance analysis (FBA) uses linear programming to provide a single metabolic flux distribution which optimizes a predefined cellular objective such as maximization of

Martine Dieuaide-Noubhani and Ana Paula Alonso (eds.), *Plant Metabolic Flux Analysis: Methods and Protocols*, Methods in Molecular Biology, vol. 1090, DOI 10.1007/978-1-62703-688-7_18, © Springer Science+Business Media New York 2014

growth rate (biomass formation) [19, 20]. FBA has been widely used in constraint-based modeling. For example, the growth rate of microorganisms on different substrates and the secretion of by-products can be predicted with good precision [21]. Lethality of gene knockouts was predicted with high accuracy [22, 23]. However, the optimal flux distribution obtained by FBA is not always unique. A large number of alternative flux distributions may have the same optimality, while only one of them is explored by solving the linear programming problem. If the goal is, for example, to interpret the predicted flux values of particular reactions within a network, more elaborate exploration of the numerical space of alternative solutions is required. For this reason, a bi-level optimization procedure called flux variability analysis (FVA) has been developed as an extension to FBA [24]. FVA explores alternate solutions in the optimal flux space by assigning predicted flux intervals (ranges) to each reaction [24], i.e., defines the range of flux values a reaction can assume under any possible optimal solution. Recently we extensively explored the potential of this approach with a model of developing oilseed rape embryos which have a highly compartmentalized distribution of metabolites in the cell [10, 11]. About 13 % of all reactions in our network were of variable flux value [10]. Subheading 3.1 gives a detailed introduction to the FVA method. The flux intervals determined by FVA have various useful interpretations and different variability types have been introduced (Table 1) [10]. For example, when both upper and lower flux bounds of a flux interval are zero, this means that the reaction is never used in the optimal flux space. Forcing flux through such a reaction in silico indeed causes suboptimal flux solutions. In contrast, if the flux value of a reaction is predicted to be strictly nonzero, simulation of the knockout of this reaction causes suboptimal flux states or even lethality. On the other hand, if a range of flux values is predicted that includes zero, the reaction can be considered redundantly essential (i.e., substitutable) in the sense that flux through this reaction is exchangeable with another route. Blocking one redundant pathway may cause another to become absolutely necessary. Furthermore, the existence of infinite flux bounds is related to substrate cycles without net conversion of metabolites.

The bi-level optimization process to derive optimal flux intervals can also be used for network projections. As explained in Subheading 3.1, if in the secondary optimization two or more reactions are selected, then the flux bounds of the combined or "lumped" reaction is obtained. ^{13}C-MFA networks usually contain lumped reactions, i.e., this procedure is useful if simulations of a highly compartmented stoichiometric network are to be compared to fluxes obtained by ^{13}C-Metabolic Flux Analysis (^{13}C-MFA) [11, 17].

Table 1
Frequency of variability types determined by flux variability analysis

Variability type	Frequency
+	275 (0.481)
-	13 (0.023)
0	209 (0.365)
[+ +Inf]	2 (0.003)
[+ +]	3 (0.005)
[- + Inf]	7 (0.012)
[- +]	6 (0.01)
[- -]	1 (0.002)
[- 0]	1 (0.002)
[-Inf + Inf]	32 (0.056)
[-Inf +]	5 (0.009)
[-Inf -]	3 (0.005)
[0 + Inf]	2 (0.003)
[0 +]	13 (0.023)

Instead of using FVA, the problem of alternative optima can also be addressed by optimization procedures that have the side effect of leading to unique flux solutions. Various objective functions that have been considered combine linear and quadratic programming [25]. Quadratic programming does not produce alternative optima. For example, assuming an organism makes maximal efficient use of enzyme capacity, optimal flux solutions can be sought for which the sum of squares of all fluxes is minimal. If done by combining linear and quadratic programming, the resulting flux solutions tend to have evenly distributed flux loads between alternative pathways. For instance, this approach was applied in a model for developing barley seeds [16].

Various software packages are available for predictive simulation of constraint-based models. Among those are two programs which we have adopted here, CellNetAnalyzer (CNA) [26] and COnstraint-Based Reconstruction and Analysis (COBRA) [27, 28]. CNA and COBRA are toolboxes for MATrix LABoratory (MATLAB®). Both packages support model simulation through command line scripting which enables the user to encode more complex and specialized tasks like FVA and the network projections discussed below. CNA also features a graphical user interface with many functionalities organized in drop-down menus.

COBRA offers a large number of functions for advanced procedures like OptKnock [29] and MOMA [30]. Based on reaction-gene associations COBRA has the comprehensive capability to be used for genome-scale models. COBRA might therefore be the currently most advanced computing environment for simulation of constraint-based metabolic models and is probably most widely used. User manuals for CNA and COBRA are available in the installation package and have been published [27]. These guides can be used as reference in addition to this chapter.

This chapter is intended for an audience that is at a beginner level with respect to constraint-based modeling. Basic skills in using MATLAB® and in matrix algebra are expected. We describe in detailed protocols the use of CNA and COBRA for basic simulations of FBA, FVA, and network projections based on a model we published that simulates seed storage synthesis in a developing oilseed. Oilseed embryos utilize light, sugars, amino acids, and inorganic nutrients to accumulate biomass and store seed reserves such as oil, protein, and starch [31]. To allow predictive simulation of these processes in developing *B. napus* embryos, the large scale, fully-compartmented stoichiometric model *bna572* was developed [10]. By characterizing the flux variability in *bna572* it is possible to make predictions about the role of each reaction in overall storage metabolism [10, 11].

2 Materials

We use a personal computer with internet access, Windows® 7 operating system, Microsoft Excel® 2010, and MATLAB® version 7.11 (R2010b) computing software. MATLAB® is a commercial programming environment for technical computing that is used extensively by scientists and engineers for numerous applications in metabolic engineering. In conjunction with MATLAB®, we use two free (for noncommercial use), periodically updated computer programs for metabolic pathway engineering named CellNetAnalyzer and COBRA (obtained specifically for this protocol on 6/8/2012).

2.1 Installation of CellNetAnalyzer (CNA)

1. Using your web browser, navigate to the following web site: http://www.mpi-magdeburg.mpg.de/projects/cna/download.html. Fill out and submit the webform. Using the address for software download that gets e-mailed to you, download CNA99_for_MATLAB75_or_higher.zip. Unzip the file into folder \CNA.
2. Start MATLAB®. Set the path. Choose "File" > "Set Path..." > "Add with Subfolders ...." In the "Browse for Folder" window select folder \CNA and click "OK" (*see* **Note 1**).

2.2 Obtaining the* Bna572 *Model and Making It Accessible to CNA

1. Navigate to http://onlinelibrary.wiley.com/doi/10.1111/j.1365-313X.2011.04613.x/suppinfo. Download TPJ_4613_sm_AppendixS5-S13.xls to folder \CNA. Open the workbook and access worksheet "S6 SBML," which contains an SBML encoding of the *bna572* model. Copy the contents and paste it into a Notepad text file saved as \CNA\bna572.xml (*see* **Note 2**).

2.3 Installation of Constraint-Based Reconstruction and Analysis (COBRA) Software

1. Using your web browser, navigate to the following web site: http://sourceforge.net/projects/opencobra/files/cobra/. Download cobra_2.0.4.zip and extract the compressed file to folder \COBRA.
2. Start MATLAB®. Set the path. Choose "File" > "Set Path..." > "Add with Subfolders ...." In the "Browse for Folder" window select folder \COBRA and click "OK."

2.4 Obtaining the* Bna572 *Model and Making It Accessible to COBRA

1. Download file TPJ_4613_sm_AppendixS4.mat (*see* **Note 3**) into folder \COBRA from http://onlinelibrary.wiley.com/doi/10.1111/j.1365-313X.2011.04613.x/suppinfo.

3 Methods

Perform protocols after booting MATLAB®, installing CellNetAnalyzer and COBRA, and making the *bna572* model accessible to the programs (*see* Subheading 2). Execute commands, indicated by "≫", at the command prompt in the MATLAB® Command Window.

3.1 Flux Variability Problem Formulation for Bna572

The following protocols are based on *bna572*'s photoheterotrophic organic nitrogen model (the PO model) [10]. The constraints for this simulation include knockout constraints on the inorganic nitrogen exchange rates (#124, 128), flux capacity constraints of 0.0725 and 2.79 μmol/h placed on the biomass flux (#241) and maintenance ATP consumption rate (#33), respectively, and a carbon flux ratio constraint of $R = 2.9$ on the carbon flux into oil (60.7118 mol C times flux #399) relative to the CO_2 efflux (#4) from the system.

Fluxes in *bna572* have units of μmol/h in accord with the rate of the biomass synthesis equation (reaction #368 in the stoichiometric model) which is expressed as

$$\sum_{i=1}^{n} \frac{W_i}{\mathrm{MW}_i} \text{mmol component } i \xrightarrow{\text{yields}} 1 \text{ g Biomass} \qquad (1)$$

where W_i is the dry weight fraction of biomass component i (e.g., oil or protein), MW_i is the molecular weight in g/mmol of biomass component i and n is the number of biomass components.

We constrain the biomass synthesis rate to a measured value of 0.0725 mg/h the units of which determine the flux units of μmol/h.

The end result of FVA is a predicted flux interval (i.e., minimum flux and maximum flux) for each reaction in the network. In practice, to compute the predicted flux interval for each reaction in *bna572* we formulate mathematical optimization problems, and then directly solve them using the revised simplex method of MATLAB® executable GLPK (glpkmex-2.4-precom-win-glpk-4.20) [32]. The primary optimization problem is formulated as

$$\begin{aligned} \text{minimize} \quad & p' = c^{\mathrm{T}}x \\ \text{subject to} \quad & Ax = b \qquad A \in R^{m\times n}, \\ & lb \le x \le ub \end{aligned} \tag{2}$$

where $c^{\mathrm{T}}x$ is the objective function, p' the optimal value, $Ax = b$ the m equality constraints (e.g., steady-state mass balances), and lb and ub vectors containing upper and lower bounds placed on the n fluxes of the optimization variable x (the flux distribution is denoted x). The primary objective function in Eq. 2 is negative one times the sum of exchange fluxes for all exchangeable metabolites (influx is defined negative) except for the gasses CO_2 and O_2, i.e., Eq. 2 minimizes total substrate and light uptake rates.

With secondary optimizations, each flux x_j is minimized and maximized subject to the optimal value of the primary problem in Eq. 2, resulting in a predicted flux interval $[p''_j\, p'''_j]$ (in μmol/h).

$$\begin{aligned} \text{minimize} \quad & p''_j = x_j & \text{maximize} \quad & p'''_j = x_j \\ \text{subject to} \quad & Ax = b & \text{subject to} \quad & Ax = b \\ & lb \le x \le ub & & lb \le x \le ub \\ & c^{\mathrm{T}}x = p' & & c^{\mathrm{T}}x = p' \end{aligned} \tag{3}$$

The principle of bi-level optimization used in FVA also has another useful application in defining network projections. Projection of the fully compartmentalized *bna572* model onto smaller, less compartmented models is useful for network simplification and comparison of FVA predictions to ^{13}C-MFA estimates [11]. For FVA of a projected reaction, instead of minimizing and maximizing the flux of a single reaction, we carry out the secondary optimizations on a weighted lumped sum of selected fluxes. That is, p_j'' and p'''_j in Eq. 3 are expressed as $c'^{\mathrm{T}}x$, instead of x_j, where the coefficients for the summation of interest are in c'.

In the following we demonstrate the computation of a single-projected flux interval. Phosphoenol pyruvate (PEP) carboxylase (PEPC) and PEP carboxykinase (PPCK) are reactions in *bna572* that interconvert phosphoenol pyruvate (PEP) and oxaloacetate (OxA). In a *B. napus* embryo model used for ^{13}C-MFA only one reaction (vPEPC) is present connecting the two metabolites.

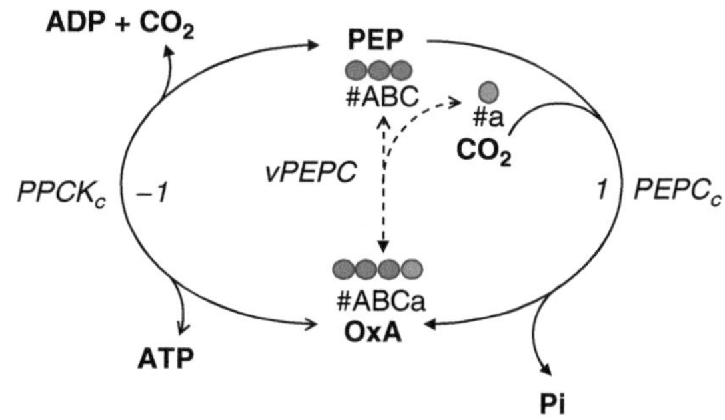

Fig. 1 Network projection. Two stoichiometric model reactions in this example, phosphoenol pyruvate carboxylase ($PEPC_c$) and carboxykinase ($PPCK_c$), are projected onto a reaction that models only carbon atom transitions. To compute the predicted flux interval of the projected reaction (vPEPC), a weighted sum of the stoichiometric model fluxes is minimized and maximized after assigning appropriate objective function coefficients to each flux (1 or −1)

In general, ^{13}C-MFA tends to use models with largely simplified network topology by lumping reactions [33]. To compare the FVA model predictions with the ^{13}C-MFA results the two *bna572* reactions can be projected onto vPEPC (Fig. 1). Since PEPC and PPCK have opposing directions in the stoichiometric model, we use "1" and "−1" as coefficients, respectively, in the objective function for secondary optimizations. The optimal values of these secondary optimizations are the flux interval bounds for "vPEPC." Examples of other projections can be found in the supplementary information of Hay and Schwender [11].

For convenience, each predicted flux interval $[p'_j\ p''_j]$ can be translated into a qualitative symbolic variability type (Fig. 2, Table 1) [10]. For example, the variability type "+" represents a reaction predicted to have a nonvariable flux in the forward reaction direction, i.e., the reaction is defined by a positive value. Square brackets denote the types associated with reactions which have variable fluxes. Reactions that have the type "[- -]," for instance, are termed finitely variable and have a reverse flux defined by two negative values. If the flux interval has a bound deemed zero, "0" is used in the variability type. For unbounded flux intervals, characteristic of reactions which belong to a biochemical loop [34], the variability type contains "Inf" for infinitely variable.

3.1.1 Flux Variability Analysis of the Bna572 Model in CellNetAnalyzer

Assuming the steps outlined in Subheadings 2.1 and 2.2 have been completed, the following protocol details how to import *bna572* from the *bna572* SBML file into CNA, and how to make basic adjustments and settings so that flux variability formerly reported for the photoheterotrophic/organic nitrogen condition, "PO" model [10], can be reproduced.

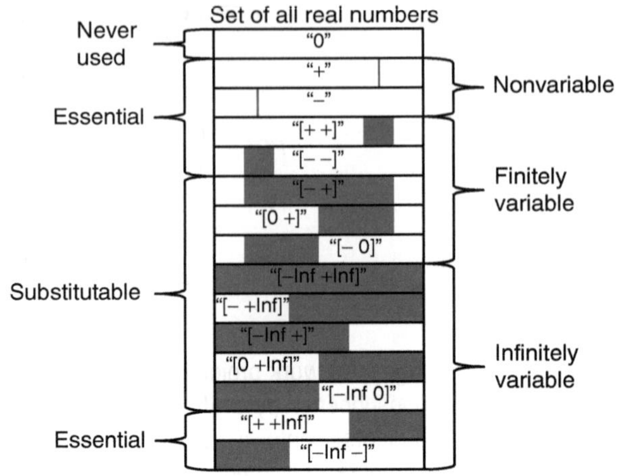

Fig. 2 Flux variability types used to categorize reactions according to flux variability analysis. Each of the 15 stacked horizontal bars represents an infinitely large real number line from negative to positive infinity. *Grey shading* is used to represent an example allowable flux interval for the variability type in *quotes*. Interpretation is provided alongside the variability types

1. Import the SBML of the *bna572* model into a CellNetAnalyzer model (*see* **Note 4**).

   ≫ bna572cna = CNAsbml2MFNetwork('bna572.xml');

   Choose "4 = ext" when prompted with "Do you think that one of the compartments listed below serves as container for external species (in the sense of stoichiometric network studies)? If so please give the number of this compartment (or choose 0 for "None")" (*see* **Note 5**).

2. Declare unbalanced water and hydrogen species as external (*see* **Note 6**).

   ≫ bna572cna.specExternal([252 253 254 255 256 265 268 269 270]) = 1;

3. Set the objective function to minimize total uptake (*see* **Note 7**).

   ≫ uptakes = [6 109 111 113 115 117 119 121 124 128 239 240];

   ≫ bna572cna.objFunc(uptakes) = 1;

4. Fix the growth rate to 0.0725 mg/h (*see* **Note 8**).

   ≫ bna572cna.reacDefault(241) = 0.0725;

5. Fix the maintenance ATP hydrolysis rate to 2.79 μmol/h (*see* **Note 9**).

   ≫ bna572cna.reacDefault(33) = 2.79;

6. Specify organic nitrogen conditions by knocking out inorganic nitrogen exchanges (*see* **Note 10**).

   ≫ bna572cna.reacDefault([124 128]) = 0;

7. Fix CO_2 exchange rate to specify photoheterotrophy (*see* **Note 11**).

   ≫ bna572cna.reacDefault(4) = 0.6041;

8. Carry out the primary optimization (*see* **Note 12**).

   ≫ fluxes_cna = CNAoptimizeFlux(bna572cna, bna572cna.reacDefault);

9. Fix uptake exchange fluxes to their optimal values (*see* **Note 13**).

   ≫ bna572cna.reacDefault(uptakes) = fluxes_cna(uptakes);

10. Substitute 1,000 for infinity in the flux lower and upper bounds (*see* **Note 14**).

    ≫ bna572cna.reacMin(isinf(bna572cna.reacMin)) = -1000;
    ≫ bna572cna.reacMax(isinf(bna572cna.reacMax)) = 1000;

11. Carry out Flux Variability Analysis (*see* **Note 15**).

    ≫ [min_fluxes_cna, max_fluxes_cna] = CNAfluxVariability (bna572cna, bna572cna.reacDefault)

3.1.2 Flux Variability Analysis of a Projected Reaction in CellNetAnalyzer

1. Set a new objective function for the model that has already been constrained by the results of the primary optimization (*see* **Note 16**).

   ≫ forwards = [12];
   ≫ backwards = [13];
   ≫ bna572cna.objFunc = zeros(572, 1);
   ≫ bna572cna.objFunc(forwards) = 1;
   ≫ bna572cna.objFunc(backwards) = -1;

2. Minimize projected vPEPC flux (*see* **Note 17**).

   ≫ fluxvector = CNAoptimizeFlux(bna572cna, bna572cna.reacDefault);

3. Compute minimal value of objective function (*see* **Note 18**).

   ≫ min_vPEPC_cna = fluxvector'*bna572cna.objFunc

4. Maximize projected vPEPC flux.

   ≫ bna572cna.objFunc = -bna572cna.objFunc;
   ≫ fluxvector = CNAoptimizeFlux(bna572cna, bna572cna.reacDefault);

5. Compute maximal value of objective function.

   ≫ max_vPEPC_cna = fluxvector'*bna572cna.objFunc*-1

3.1.3 Flux Variability Analysis of the Bna572 Model in COBRA

Assuming the steps outlined in Subheadings 2.3 and 2.4 have been completed, the following protocol details how to import *bna572* from the *bna572* MATLAB file into COBRA, and how to make basic adjustments and settings so that flux variability formerly reported for the photoheterotrophic/organic nitrogen condition, "PO" model [10], can be reproduced. Before carrying out this protocol, to avoid conflicts between the GLPK solvers packaged with COBRA and CellNetAnalyzer, make sure the subfolders added to the MATLAB path in Subheading 2.1 are removed from the path.

1. Load the *bna572* model to the MATLAB® workspace.

   ≫ load TPJ_4613_sm_AppendixS4.mat;

2. Populate reaction names of a COBRA model (*see* **Note 19**).

   ≫ bna572cobra.rxns = PO.RN;

3. Populate lower bounds (*see* **Note 14**).

   ≫ bna572cobra.lb = PO.LB; bna572cobra.lb(isinf(bna572-cobra.lb)) = -1000;

4. Populate upper bounds.

   ≫ bna572cobra.ub = PO.UB; bna572cobra.ub(isinf (bna572cobra.ub)) = 1000;

5. Populate stoichiometric matrix (*see* **Note 20**).

   ≫ bna572cobra.S = PO.A1;

6. Populate right hand side.

   ≫ bna572cobra.b = PO.b1;

7. Populate objective function (*see* **Note 21**).

   ≫ bna572cobra.c = PO.c*-1;

8. Populate reversibilities.

   ≫ bna572cobra.rev = and(isinf(PO.LB), isinf(PO.UB));

9. Populate metabolite names.

   ≫ bna572cobra.mets = PO.M;

10. Knockout inorganic nitrogen exchange fluxes (*see* **Note 22**).

    ≫ bna572cobra = changeRxnBounds(bna572cobra, {'NO3_ap_exch', 'NH4_ap_exch'}, [0 0], {'b', 'b'});

11. Constrain the growth rate.

    ≫ bna572cobra = changeRxnBounds(bna572cobra, 'Biomas s_exch', .0725, 'b');

12. Fix CO_2 exchange (i.e., carbon balance) to specify photoheterotrophy.

    ≫ bna572cobra = changeRxnBounds(bna572cobra, 'CO2_ap_exch', .6041, 'b');

13. Fix the ATP drain to 2.79 μmol/h.

    ≫ bna572cobra = changeRxnBounds(bna572cobra, 'ATP_c: H2O_c::ADP_c:Pi_c', 2.79, 'b');

14. Choose an optimization solver (*see* **Note 23**).

    ≫ changeCobraSolver('glpk');

15. Carry out Flux Balance Analysis (*see* **Note 24**).

    ≫ L = optimizeCbModel(bna572cobra, 'max');
    ≫ fluxes_cobra = L.x;

16. Perform Flux Variability Analysis (*see* **Note 25**).

    ≫ [min_fluxes_cobra, max_fluxes_cobra] = fluxVariability (bna572cobra)

3.1.4 Flux Variability Analysis of a Projected Reaction in COBRA

1. Constrain flux balance to results of the primary optimization.

   ≫ uptakes = [6 109 111 113 115 117 119 121 124 128 239 240];
   ≫ bna572cobra = changeRxnBounds(bna572cobra, bna572cobra.rxns(uptakes), fluxes_cobra(uptakes));

2. Change objective function.

   ≫ bna572cobra.c = zeros(572, 1);
   ≫ forwards = [12];
   ≫ backwards = [13];
   ≫ bna572cobra.c(forwards) = 1;
   ≫ bna572cobra.c(backwards) = -1;

3. Minimize projected vPEPC flux.

   ≫ L = optimizeCbModel(bna572cobra, 'min');
   ≫ min_vPEPC_cobra = L.f

4. Maximize projected vPEPC flux.

   ≫ L = optimizeCbModel(bna572cobra, 'max');
   ≫ max_vPEPC_cobra = L.f

3.1.5 Computation of Flux Variability Types

1. Define a numerical zero (*see* **Note 26**).

   ≫ tol = 10^-7;

2. Find unbounded intervals (*see* **Note 27**).

   ≫ UINF = abs(max_fluxes_cna) > 100;
   ≫ LINF = abs(min_fluxes_cna) > 100;

3. Find intervals bounded by a zero.

   ≫ LZ = abs(min_fluxes_cna) < tol;
   ≫ UZ = abs(max_fluxes_cna) < tol;

4. Find intervals that are nonvariable.

   ≫ fixed = abs(max_fluxes_cna-min_fluxes_cna) < tol;

5. Find intervals that are infinitely variable.

   ≫ infinitevariable = abs(max_fluxes_cna-min_fluxes_cna) > 100;

6. Find intervals that are finitely variable.

   ≫ finitevariable = and(abs(max_fluxes_cna-min_fluxes_cna) > = tol, ~infinitevariable);

7. Find intervals that have a positive upper limit.

   ≫ UP = and(max_fluxes_cna >= tol, ~UINF);

8. Find intervals that have a negative upper limit.

   ≫ UN = and(max_fluxes_cna <= -tol, ~UINF);

9. Find intervals that have a positive lower limit.

   ≫ LP = and(min_fluxes_cna >= tol, ~LINF);

10. Find intervals that have a negative lower limit.

   ≫ LN = and(min_fluxes_cna <= -tol, ~LINF);

11. Fill in symbolic variability types (*see* **Note 28**).

```
≫ temp = [UINF LINF]; VT(all(temp,2), 1) = {'[-Inf +Inf]'};
≫ temp = [LINF UZ]; VT(all(temp, 2)) = {'[-Inf 0]'};
≫ temp = [LZ UINF]; VT(all(temp, 2)) = {'[0 +Inf]'};
≫ temp = [LP UINF]; VT(all(temp, 2)) = {'[+ +Inf]'};
≫ temp = [LINF UN]; VT(all(temp, 2)) = {'[-Inf -]'};
≫ temp = [LINF UP]; VT(all(temp, 2)) = {'[-Inf +]'};
≫ temp = [LN UINF]; VT(all(temp, 2)) = {'[- +Inf]'};
≫ temp = [LZ UZ]; VT(all(temp, 2)) = {'0'};
≫ temp = [LP UP fixed]; VT(all(temp, 2)) = {'+'};
≫ temp = [LN UN fixed]; VT(all(temp, 2)) = {'-'};
≫ temp = [LZ UP finitevariable]; VT(all(temp, 2)) = {'[0 +]'};
≫ temp = [LN UZ finitevariable]; VT(all(temp, 2)) = {'[- 0]'};
≫ temp = [LN UP finitevariable]; VT(all(temp, 2)) = {'[- +]'};
≫ temp = [LP UP finitevariable]; VT(all(temp, 2)) = {'[+ +]'};
≫ temp = [LN UN finitevariable]; VT(all(temp, 2)) = {'[- -]'}
```

4 Notes

1. Adding a specific folder, in this example \CNA, to the MATLAB® path allows MATLAB® to locate all of the files included with the CNA software.
2. The systems biology markup language (SBML) [35] encodes the metabolic composition, compartmentation, and stoichiometry of the metabolic network using xml format.
3. In addition to SBML, the equations, inequalities, objective function, and related information (e.g., reaction names) were made explicitly available in the form of saved MATLAB® arrays.
4. In CNA a model is represented as a MATLAB® variable of class structure. Through reading of the xml file, fields of the structure will are filled including reacMin: [572x1 double] (lower bounds on flux variables), stoichMat: [579x572 double] (stoichiometric matrix), specExternal: [1x579 double] (boolean selection vector for external metabolites), specInternal: [1x555 double] (index vector specifying internal metabolites),

objFunc: [572x1 double] (objective function coefficient vector), reacMax: [572x1 double] (upper bounds on flux variables), and reacDefault: [572x1 double] (flux capacity constraints on flux variables).

5. External metabolites are included in the *bna572* network and denoted by the suffix "_ext". These metabolite pools, serving as sources (i.e., sugars) and sinks (e.g., Biomass), are characterized by accumulation or depletion. Hence, these external metabolites need to be declared as unbalanced.
6. In the *bna572* metabolic network reconstruction, water molecules were not strictly balanced at steady state, and not accounting for protonation states and pH, neither were many of the protons. However, protons in the lumen, cell wall, and mitochondrial intermembrane space, which are pumped in chemiosmotic processes, were forced to balance.
7. In the SBML for *bna572*, the uptake reactions were defined in the direction of influx, i.e., external species on the left hand side of the yields symbol, giving each influx a positive a positive value. CNA carries out minimization of the linear combination of fluxes specified by the objective function coefficient vector.
8. In *bna572*, the biomass synthesis reaction (#241) is a summation of biomass components in mmol to 1 g of biomass. Constraining this reaction rate to 0.0725 mg/h scales the flux units to μmol/h.
9. Flux capacity constraints are imposed on reactions in CNA through the reacDefault field.
10. Knocking out a reaction is equivalent to setting the flux capacity to zero.
11. The photoheterotrophic model of *bna572* is characterized by imposing a ratio of 2.9 for the carbon molar flux into oil relative to carbon efflux from the system. The carbon molar flux into oil is determined by the fixed growth rate. Hence, we fix the CO_2 exchange rate to enforce the flux ratio constraint.
12. The CNAoptimizeFlux function solves a flux balance problem (a linear program), and returns a solution (flux distribution) which optimizes the objective function and satisfies the constraints. The solution of the primary optimization problem (minimize total uptake) becomes a constraint in secondary optimization problems which are carried out to determine each reaction's predicted flux interval in the solution space of the primary problem.
13. By constraining the exchange reactions to their optimal fluxes we narrow down the flux space to the set of optimal flux distributions. We need to do this because CNA does not do it automatically (*see* **Notes 15** and **17**).
14. In *bna572* the lower and upper bounds imposed on the flux variables are either positive infinity or negative infinity.

When using dedicated software for constraint-based flux analysis, it is a good idea to use finite flux bounds to avoid program errors.

15. The CNAfluxVariability function determines for each reaction the range of flux values which satisfies the mass balance equations specified by fields stoichMat and specExternal as well as the flux capacity constraints specified by the second argument of the function. Carrying out this step provides a predicted flux interval (minimum flux and maximum flux) within the optimal solution space for each reaction in the network.

16. This objective function is a weighted lumped sum of two reactions in the *bna572* network. It represents a projection of reactions in *bna572* with carbon atom transitions between PEP and OxA (PEPC and PPCK, #12-13) onto the vPEPC reaction of *bna39/2006* which is a smaller model and minimally compartmentalized. When carrying out optimizations using this objective function, it is important that the feasible solutions be restricted to the optimal solution space (*see* **Note 13**).

17. As the second argument of the CNAoptimizeFlux function, we specify flux capacity constraints, including those on the exchange fluxes to restrict the network wide flux balance to the optimal set of solutions.

18. The scalar product between objective function coefficients and optimal flux distribution returns the optimal value, in this case a weighted lumped sum corresponding the projected reaction.

19. COBRA also uses the class structure in MATLAB® to encode a metabolic network model. The fields of the structure include rxns: {572x1 cell} (reaction names), lb: [572x1 double] (lower bounds), ub: [572x1 double] (upper bounds), S: [555x572 double] (stoichiometric matrix), b: [555x1 double] (right hand side values of the equality constraints), c: [572x1 double] (objective function coefficients), rev: [572x1 logical] (boolean vector defining reversible reactions), and mets: {555x1 cell} (metabolite names).

20. Internal metabolite steady-state mass balance equations are put in field S (i.e., stoichiometric matrix) of the COBRA model.

21. In TPJ_4613_sm_AppendixS4.mat, which we are using to populate the COBRA model, the exchange reaction fluxes are defined as negative in the direction of influx. With the objective being minimize total uptake rate and the optimization problem being minimization, the objective function coefficients needed to be negative. In COBRA, we make the objective function coefficients positive, and later specify explicitly to the program that we want to maximize the sum of exchange fluxes.

22. In COBRA, flux capacities can be changed through constraints placed on the lower and upper bounds of the flux variables.

23. You must specify which of the several supported mathematical optimization problem solvers you want to use. Here we chose "glpk" which is supplied together with the COBRA software.
24. In this code, "max" directs the program to perform objective function maximization. The optimizeCbmodel function returns a data structure including fields x: [572x1 double] (the optimal flux solution), f: −6.3846 (the optimal value of objective function), and y: [555x1 double] (dual solution, or shadow prices).
25. Flux Variability Analysis in COBRA automatically constrains the flux space according to optimization of the primary objective function.
26. Since variability types (Fig. 2) depend on a declaration of numerical zero, a range of values for numerical zero need to be tested before deciding on a specific value.
27. Reactions with unbounded flux intervals are members of reaction loops. The infinity threshold should be chosen carefully depending on the constraints in the model.
28. Variability types represent the predicted role of each reaction in central metabolism. Table 1 contains the distribution of variability types using the FVA output from CNA.

Acknowledgments

We would like to thank Griffin St. Clair and Dr. Zhijie Sun for helpful discussions. We gratefully acknowledge funding from the US Department of Energy (Division of Chemical Sciences, Geosciences, and Biosciences, Office of Basic Energy Sciences, Field Work Proposal BO-133).

References

1. Palsson B (2006) Systems biology: properties of reconstructed networks. Cambridge University Press, Cambridge
2. Ogata H, Goto S, Sato K et al (1999) KEGG: Kyoto encyclopedia of genes and genomes. Nucleic Acids Res 27(1):29–34
3. Mueller LA, Zhang P, Rhee SY (2003) AraCyc: a biochemical pathway database for Arabidopsis. Plant Physiol 132(2):453–460
4. Wang X, Wang H, Wang J et al (2011) The genome of the mesopolyploid crop species *Brassica rapa*. Nat Genet 43(10): 1035–1039
5. The Arabidopsis Genome Initiative (2000) Analysis of the genome sequence of the flowering plant *Arabidopsis thaliana*. Nature 408(6814):796–815
6. Troncoso-Ponce MA, Kilaru A, Cao X et al (2011) Comparative deep transcriptional profiling of four developing oilseeds. Plant J 68(6):1014–1027
7. Williams TC, Poolman MG, Howden AJ et al (2010) A genome-scale metabolic model accurately predicts fluxes in central carbon metabolism under stress conditions. Plant Physiol 154 (1):311–323
8. Poolman MG, Miguet L, Sweetlove LJ, Fell DA (2009) A genome-scale metabolic model of Arabidopsis and some of its properties. Plant Physiol 151(3):1570–1581

9. Pilalis E, Chatziioannou A, Thomasset B, Kolisis F (2011) An in silico compartmentalized metabolic model of *Brassica napus* enables the systemic study of regulatory aspects of plant central metabolism. Biotechnol Bioeng 108(7):1673–1682
10. Hay J, Schwender J (2011) Metabolic network reconstruction and flux variability analysis of storage synthesis in developing oilseed rape (*Brassica napus* L.) embryos. Plant J 67(3): 526–541
11. Hay J, Schwender J (2011) Computational analysis of storage synthesis in developing *Brassica napus* L. (oilseed rape) embryos: flux variability analysis in relation to ^{13}C metabolic flux analysis. Plant J 67(3):513–525
12. Mintz-Oron S, Meir S, Malitsky S et al (2012) Reconstruction of *Arabidopsis* metabolic network models accounting for subcellular compartmentalization and tissue-specificity. Proc Natl Acad Sci U S A 109(1):339–344
13. de Oliveira Dal'Molin CG, Quek LE, Palfreyman RW et al (2010) AraGEM, a genome-scale reconstruction of the primary metabolic network in Arabidopsis. Plant Physiol 152 (2):579–589
14. Schwender J (2008) Metabolic flux analysis as a tool in metabolic engineering of plants. Curr Opin Biotechnol 19(2):131–137
15. Saha R, Suthers PF, Maranas CD (2011) *Zea mays iRS1563*: a comprehensive genome-scale metabolic reconstruction of maize metabolism. PLoS One 6(7):e21784
16. Grafahrend-Belau E, Schreiber F, Koschützki D, Junker BH (2009) Flux balance analysis of barley seeds: a computational approach to study systemic properties of central metabolism. Plant Physiol 149(1):585–598
17. Schwender J, Hay JO (2012) Predictive modeling of biomass component tradeoffs in *Brassica napus* developing oilseeds based on *in silico* manipulation of storage metabolism. Plant Physiol 160(3):1218–1236
18. Klamt S, Stelling J (2002) Combinatorial complexity of pathway analysis in metabolic networks. Mol Biol Rep 29(1–2):233–236
19. Boyle NR, Shastri AA, Morgan JA (2009) Network stoichiometry. In: Schwender J (ed) Plant metabolic networks. Springer, New York, pp 211–243
20. Sweetlove LJ, Ratcliffe RG (2011) Flux-balance modeling of plant metabolism. Front Plant Sci 2:38
21. Edwards JS, Ibarra RU, Palsson BO (2001) In silico predictions of Escherichia coli metabolic capabilities are consistent with experimental data. Nat Biotechnol 19(2):125–130
22. Edwards JS, Palsson BO (2000) Metabolic flux balance analysis and the in silico analysis of Escherichia coli K-12 gene deletions. BMC Bioinformatics 1:1
23. Edwards JS, Palsson BO (2000) The Escherichia coli MG1655 in silico metabolic genotype: its definition, characteristics, and capabilities. Proc Natl Acad Sci U S A 97(10):5528–5533
24. Mahadevan R, Schilling CH (2003) The effects of alternate optimal solutions in constraint-based genome-scale metabolic models. Metab Eng 5(4):264–276
25. Schuetz R, Kuepfer L, Sauer U (2007) Systematic evaluation of objective functions for predicting intracellular fluxes in Escherichia coli. Mol Syst Biol 3:119
26. Klamt S, Saez-Rodriguez J, Gilles ED (2007) Structural and functional analysis of cellular networks with Cell NetAnalyzer. BMC Syst Biol 1:2
27. Schellenberger J, Que R, Fleming RM et al (2011) Quantitative prediction of cellular metabolism with constraint-based models: the COBRA Toolbox v2.0. Nat Protoc 6 (9):1290–1307
28. Becker SA, Feist AM, Mo ML et al (2007) Quantitative prediction of cellular metabolism with constraint-based models: the COBRA Toolbox. Nat Protoc 2(3):727–738
29. Burgard AP, Pharkya P, Maranas CD (2003) OptKnock: a bilevel programming framework for identifying gene knockout strategies for microbial strain optimization. Biotechnol Bioeng 84(6):647–657
30. Sègre D, Vitkup D, Church GM (2002) Analysis of optimality in natural and perturbed metabolic networks. Proc Natl Acad Sci U S A 99 (23):15112–15117
31. Baud S, Dubreucq B, Miquel M et al (2008) Storage reserve accumulation in Arabidopsis: metabolic and developmental control of seed filling. Arabidopsis Book 6:e0113
32. http://www.gnu.org/software/glpk/ GLPK (GNU Linear Programming Kit)
33. Schwender J (2011) Experimental flux measurements on a network scale. Front Plant Sci 2:63
34. Price ND, Thiele I, Palsson BO (2006) Candidate states of *Helicobacter pylori*'s genome-scale metabolic network upon application of "loop law" thermodynamic constraints. Biophys J 90(11):3919–3928
35. Hucka M, Finney A, Sauro HM et al (2003) The systems biology markup language (SBML): a medium for representation and exchange of biochemical network models. Bioinformatics 19(4):524–531

Chapter 19

Plant Genome-Scale Modeling and Implementation

Cristiana G.O. Dal'Molin, Lake-Ee Quek, Robin W. Palfreyman, and Lars K. Nielsen

Abstract

Considerable progress has been made in plant genome-scale metabolic reconstruction and modeling in recent years. Such reconstructions made it possible to explore metabolic phenotypes through appropriate model formulation and optimization methods. As a result, plant genome-scale modeling has increasingly attracted interest from the plant research community. In this chapter, the first generation of plant genome-scale metabolic reconstructions is presented, along with the important concepts behind model and constraint formulation. A brief protocol describing the use of constraint-based reconstruction and analysis (COBRA) Toolbox in flux simulation and model modification is provided. This is followed by a presentation of metabolic constraints required to generate fluxes in AraGEM using COBRA that describe photosynthesis, photorespiration, and respiration, respectively. Overall, plant genome-scale modeling is a powerful approach that is accessible and readily adopted.

Key words Plant metabolic reconstruction, Genome-scale modeling, AraGEM, Constraint-based reconstruction, COBRA Toolbox, Flux simulation

1 Introduction

1.1 First Generation of Plant Metabolic Reconstruction

Over the past decade, genome-scale metabolic models (GEMs), or more accurately genome-scale metabolic reconstructions, have played an increasingly important role in elucidating metabolic characteristics of biological systems. A common denominator in microbial systems biology, these models have proven valuable for predicting organism phenotypes from genotypes and for guiding systems-level metabolic engineering for strain improvement [1–6]. The vast majority of GEMs are for prokaryotic organisms [1, 5, 7–11] or unicellular eukaryotes [12–19], while GEMs for multicellular eukaryotes only exists for mouse [20], human [21], and recently four higher plants: *Arabidopsis thaliana* [22–24], maize (*Zea mays*) [25, 26], Sorghum (*Sorghum bi-color*), and

Martine Dieuaide-Noubhani and Ana Paula Alonso (eds.), *Plant Metabolic Flux Analysis: Methods and Protocols*, Methods in Molecular Biology, vol. 1090, DOI 10.1007/978-1-62703-688-7_19, © Springer Science+Business Media New York 2014

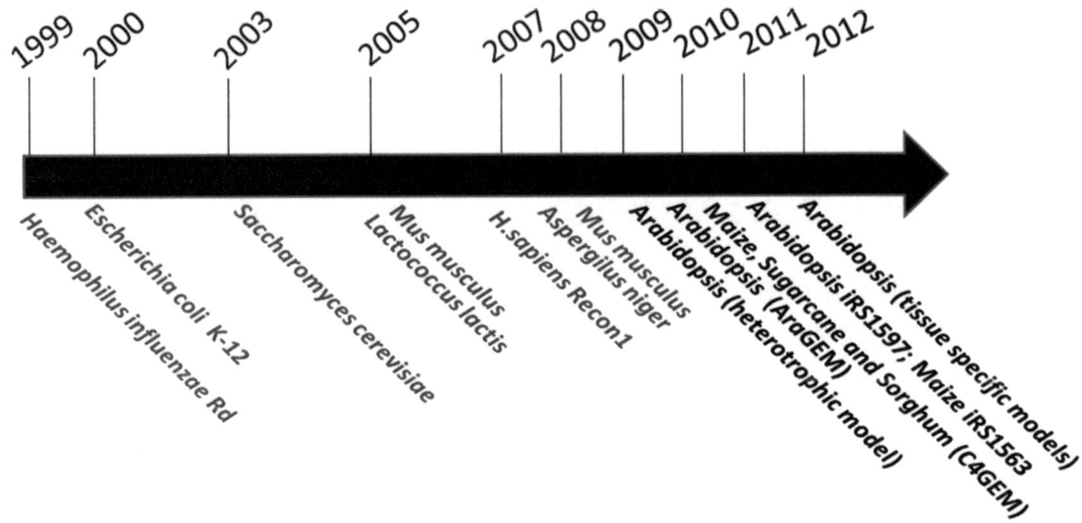

Fig. 1 Time line course for the genome-scale reconstructions of some model organisms compared to plants

sugarcane (*Saccharum officinarum*) [26]. Figure 1 shows the time line for plant metabolic reconstructions compared to other model organisms.

A detailed protocol exists for generating high-quality GEMs [27]. Two-thirds of the steps are readily automated for prokaryotes and draft GEMs have been developed for thousands of prokaryotes [28]. Compartmentation of metabolism in eukaryotes complicates matters significantly and reconstruction of plant metabolic networks requires extensive manual curation [29]. The first comprehensive plant reconstruction (AraGEM v. 1.0) [22] allocated gene products and hence reactions across five compartments (cytosol, mitochondria, plastids, peroxisomes, and vacuoles) connected via 83 inter-organelle transporters. Even for Arabidopsis, protein compartmentation and substrates for inter-organelle transporters remain open research questions. While reconstruction has already moved on to crop species, e.g., C4GEM v 1.0 [26], these models are directly or indirectly derived from homology to Arabidopsis, hence continued reconstruction efforts to close knowledge gap regarding compartmentation and transporters in Arabidopsis is essential [29].

1.2 Constraint-Based Modeling

Although reconstruction of Arabidopsis metabolism will continue for many years, one can conceive a complete model capturing all reactions encoded for in the Arabidopsis genome. In contrast, there are as many models as hypotheses that can be formulated about plant metabolism. At genome scale, it is impossible to use dynamic models with parameterized kinetics; instead constraint-based modeling approaches have been applied. In constraint-based modeling, constraints are used reduce the solution space (Fig. 2b), before the

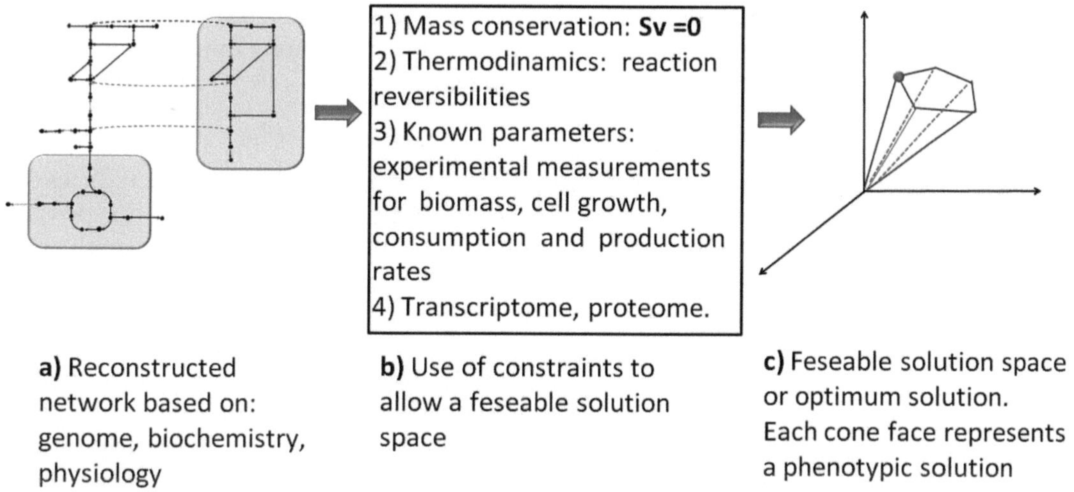

Fig. 2 Constrained-based analysis. Metabolic networks are reconstructed based on genome and other biological data sources. The reconstruction is then interrogated by imposing different constraints on the system to allow solution. The cell optimality is tested and the flux solutions are subjected to validation against biochemical evidences

feasible solution spaced is explored for either for general properties (e.g., correlations) or for solutions satisfying particular optimality criteria (Fig. 2c).

The most important constraint is the flux balance constraint

$$S \cdot \overline{v} = 0 \tag{1}$$

where S is a stoichiometric matrix with the stoichiometric coefficients for each balanced internal metabolite in the rows and each reaction in the columns, and $\overline{v}$ is the flux vector with elements corresponding to each column in S. The flux balance constraint is valid as long as the system is in pseudo-steady state, i.e., accumulation of the balanced metabolites is small compared to flux towards and the flow away from the metabolite. Rapid transients such as an abrupt heat or osmotic shock may invalidate the pseudo-steady-state assumption, but it is generally valid for *intermediate* metabolites. However, it is not valid for abundant storage metabolites such as sucrose as well as vacuolar-free amino acids and organic acids in some tissues. Sucrose can reach very high concentrations (e.g., several hundred millimolar up to more than one molar in the conducting vascular cells [30]) and maintaining these concentrations places a nontrivial load on the metabolic network. Highly abundant metabolites must be included in biomass or treated as "storage components," i.e., should be considered products rather than internal metabolites.

Additional constraints may be derived from thermodynamics (e.g., irreversibility), known capacity constraints (e.g., maximum light capture capacity), actual measurement of rates (e.g., growth rates), and expression analysis (reactions cannot occur in the absence of the relevant enzymes).

1.3 Exploring the Feasible Solution Space

The flux balance constraint combined with the additional (linear) inequality constraints reduce the feasible solution space reduces to a convex polyhedral cone called the flux cone (Fig. 2c). It is possible to explore the solution space in an unbiased manner through sampling, i.e., random collection of solutions within the feasible solution space. Statistical analysis on the sample can, for example, be used to identify novel correlations between fluxes through otherwise distant reactions. It is more common, however, to explore the solution space assuming some optimality criterion for the plant. For example, AraGEM was used to predict the flux distribution in photosynthetic leaves at fixed growth rate (measured), assuming that Arabidopsis had evolved to use captured photons efficiently [22]. This can be formulated as a linear programming problem

$$\begin{aligned} &\min_{\overline{v}} v_{\text{photons}} \\ \text{S.t.} \quad & S \cdot \overline{v} = 0 \\ & v_{\text{growth}} = \mu \\ & \overline{l} \leq \overline{v} \leq \overline{u} \end{aligned} \tag{2}$$

This is the widely used flux balance analysis (FBA) approach. There are generally many equally good solutions to Eq. 2 and the solver will only return one of these. Thus, one cannot compare and contrast two FBA solutions directly. Instead, flux variability analysis (FVA) is used to determine the range of values for each flux for which it is possible to find a solution that achieves optimal objective value, e.g., from Eq. 2 we could find the minimum and maximum value of flux v_i from

$$\begin{aligned} \min_{\overline{v}} v_i \quad &\text{or} \quad \min_{\overline{v}} v_i \\ \text{S.t.} \quad & S \cdot \overline{v} = 0 \\ & v_{\text{growth}} = \mu \\ & v_{\text{photon}} = \min v_{\text{photon}} \\ & \overline{l} \leq \overline{v} \leq \overline{u} \end{aligned} \tag{3}$$

Contrasting two FVA solutions, one can identify those fluxes for which the two flux spans do not overlap as those that significantly change between two scenarios. For example, FVA was applied to predict flux differences between tissues in C4 metabolism [26]. C4 metabolism requires two tissue types, bundle sheath and mesophyll, for its operation [31], and two instances of C4GEM were connected through transporters representing plasmodesmata [26]. Flux differences predicted using FVA were consistent with observed enrichment between the bundle sheath and the mesophyll proteomes.

Several algorithms have been developed to predict what happens when a gene is deleted, including minimum of the metabolic adjustments (MOMA) [32] and regulatory on/off minimization (ROOM) [33]. Starting with a predicted or measured wild-type flux distribution, the algorithms find the feasible flux distribution in the knockout that involves the least adjustment to the original state, "least" defined as smallest absolute difference in flux (MOMA) or minimal number of genes being activated or deactivated (ROOM). Both algorithms were used to predict the changes in flux distribution in seed following knockdown of plastidic pyruvate kinase and the predicted changes correlated strongly with changes observed using 13C MFA [34].

1.4 COBRA Toolbox

The application of constraint-based reconstruction and analysis (COBRA) methods promises to be important to the development of system level understanding of plant metabolism as well as plant metabolic engineering. For these models and tools to find common use in plant biology, however, they must be made more accessible.

The field is not settled enough for standardized, web-based modeling. Superior COBRA methods and strategies for plants are still being developed, including frameworks for multi-tissue and whole plant modeling. While this is ongoing, the best option is an integrated development environment such as Matlab with the COBRA Toolbox [35]. Matlab provides a high-level, flexible programming language for modeling, while the COBRA Toolbox provides standardized data structures for COBRA methods as well as implementation of a variety of common methods used in metabolic engineering including FBA, FVA, method of minimum metabolic adjustments (MOMA) and sampling.

COBRA Toolbox imports models written in the systems biology markup language (SBML), an XML-based format for exchange of biological models and the default format for exchanging genome scale metabolic models. In order to support, e.g., gene–protein reaction mapping information, however, the COBRA Toolbox uses its own variant of SBML. Hence, only plant genome scale models stored in the COBRA Toolbox dialect can be used in the framework.

2 Materials

The software listed below are required to perform FBA on AraGEM using COBRA Toolbox. While the following software are valid at the time of writing, new versions of the software are released over time. Users must therefore refer to the current instructions prescribed by COBRA Toolbox.

1. A computer running MATLAB version 7.0 or above (Mathworks, Inc.).

2. COBRA Toolbox 2.0.5.
3. SBMLToolbox 4.0.1 (included in COBRA Toolbox).
4. libSBML 5.1.0-b0 (*see* **Note 1**).
5. Gurobi Optimizer 5.0 (*see* **Note 2**).
6. AraGEM (*see* **Note 3**).

3 Methods

3.1 Installation

1. Install libSBML. Use the ready-to-install precompiled binaries (which are made available for most operating systems). Add directory ".\libSBML-*version*\bindings\matlab\matlab" in MATLAB's path. This folder contains the MATLAB files required to implement libSBML functions (e.g., TranslateSBML.m).
2. Install Gurobi. Follow Gurobi's instructions to activate license (*see* **Note 4**). Add directory ".\gurobi*version**OperatingSystem*\matlab" to MATLAB path. This folder contains the mex file required to implement Gurobi functions (e.g., gurobi.mexw64).
3. Download and unzip COBRA Toolbox. Add COBRA's root directory ".\cobra" to MATLAB path.
4. Run MATLAB. Type "*initCobraToolbox*" in MATLAB command line. Run *testSBML* and *testFBA* to check for errors in installation (*see* **Note 5**).
5. Download AraGEM from http://web.aibn.uq.edu.au/cssb/resources/Genomes.html

3.2 Load Model and Run Flux Simulations

1. Run MATLAB. Type "*initCobraToolbox*" in MATLAB command line.
2. Change MATLAB current directory to folder containing the metabolic model in SBML file format (e.g., AraGEM.xml).
3. Type "*model* = *readCbModel*('AraGEM.xml')" (*see* **Note 6**).
4. Type "*fluxes* = *optimizeCbModel*(*model*, 'max',*p1*,*p2*)" (*see* **Note 7**).

3.3 Modify Flux Parameters and COBRA Model

1. Load SBML model (*see* Subheading 3.2).
2. Type "*model* = *changeRxnBounds*(*model*, 'rxnID', -50, 'l')" to change the lower flux boundary value of reaction "rxnID" from existing value to −50 (*see* **Note 8**).
3. Type "*model* = *changeRxnBounds*(*model*, 'rxnID', 100, 'u')" to change the upper flux boundary value of reaction "rxnID" from existing value to 100.
4. Type "*model* = *changeRxnBounds*(*model*, 'rxnID', 0, 'b')" to change both the lower and upper flux boundary values of reaction "rxnID" from existing values to 0.

5. Type "*model* = *changeObjective*(*model*, 'rxnID', 1)" to maximize flux of reaction "rxnID" (*see* **Notes 8–10**).

3.4 Modify Reaction Equation

1. Type "*model* = *addReaction*(*model*, 'newRxnID', *metaboliteList*, [*stoic coeff*])" to add a new reaction to COBRA model (*see* **Notes 11–13**). "newRxnID" is a string that must be different from any existing reaction IDs.
2. Type "*model* = *addReaction*(*model*, 'rxnID', *metaboliteList*, [*stoic coeff*])" to modify an existing reaction "rxnID" found in *model.rxns* (*see* **Note 14**).
3. Type "*model* = *removeRxns*(*model*,*rxnRemoveList*)" to remove reactions from the COBRA model. *rxnRemoveList* is a cell array containing a list of reactions IDs found in *model.rxns* corresponding to the reactions that are to be permanently removed from the COBRA model (*see* **Note 15**).

3.5 Modify Biomass Equation

1. Identify reaction ID for biomass equation in *model.rxns* (e.g., BIO_L in AraGEM) (*see* **Note 16**).
2. Identify index of BIO_L using "rxnIndex = findRxnIDs (model,'BIO_L')".
3. Identify metabolites involved in the biomass equation and the corresponding stoichiometric coefficients using "*hitMetsIndex* = *find*(*model.S*(:,*rxnIndex*))", "*hitMets* = *model.mets*(*hitMetsIndex*)" and "*hitStoic* = full(*model.S*(*hitMetsIndex*, *rxnIndex*))".
4. Change entries in *hitMets* cell array and *hitStoic* vector in accordance to the required modifications.
5. Type "*model* = *addReaction*(*model*, 'BIO_L', *hitMets*,hitStoic)".

3.6 Save COBRA Model

1. Type "*writeCbModel*(*model*, 'sbml', fileName)" to export modified COBRA model into SBML file (*see* **Note 17**).

4 FBA Examples of AraGEM in COBRA

4.1 Flux Simulations

The section demonstrates the set up of flux constraints when performing FBA on AraGEM using COBRA. The advantaged of performing FBA is that only a few constraints and/or parameters are required to simulate vastly different metabolic phenotypes.

AraGEM is generic and has the reactions to describe photosynthetic and non-photosynthetic cells. Three different metabolic scenarios are described: photosynthesis, photorespiration, and respiration. For biological interpretation of the flux simulations for each scenario, we recommend complementary reading of the AraGEM article [22]. The parameters to run the simulations are shown in Table 2. In non-photosynthetic cells for example, photosynthetic and photorespiratory fluxes are set to zero, while sucrose

Table 1
COBRA's convention of metabolites according to cellular compartment

Tag	Compartment	Tag	Compartment
c	Cytoplasm	g	Golgi
m	Mitochondrion	r	Endoplasmic_reticulum
v	Vacuole	n	Nucleus
x	Peroxisome	p	Periplasm
e	Extracellular	l	Lysosome
t	Pool		

uptake is allowed (Table 2). Isoenzymes are also constrained according to biochemical evidences found in the literature. Figure 3 details for example the roles for isoenzymes in nitrogen assimilation in leaves and roots, assimilation of photorespiratory ammonia in leaves, and reassimilation of "recycled" nitrogen generated by amino acid deamination [36]. These isoenzymes are present in AraGEM and are further constrained to simulate nitrogen assimilation in photosynthetic or non-photosynthetic tissue. According to the literature [36], plastidic and cytosolic glutamate synthase NADH(P)-GOGAT (*GLT1*) are active in dark tissues (Fig. 3). Therefore such step reactions are constrained to zero for photosynthetic tissues, as shown in Table 2 (see reactions ID R00093_p and R00114_c).

Each metabolic example is accomplished in three general steps, which consists of (in sequence) loading the AraGEM model in MATLAB, specifying the flux constraints and objective function, and finally generating a flux solution. The exact methods involved in respective steps can be found in Subheading 3.

There are both distinct and overlapping flux constraints that are applied for AraGEM to simulate fluxes for the three different metabolic scenarios (Table 2). The rationale behind the application a given flux constraint is provided. In general, these rationales revolve around the need to (1) turn off or switch direction of an influx/efflux reaction, (2) fix growth condition to an experimental value (e.g., growth and photorespiration rate), (3) inactivate reactions that are known to be not available for the specific growth conditions, and (4) refine reaction direction based on plant-specific thermodynamic constraints. Additional, the minimum photon uptake (Ex16) objective function is used for both photosynthesis and photorespiration, while the minimum sucrose uptake (Ex17) objective function is used for respiration. While the contents in AraGEM are purely structural, physiologically relevant fluxes can be generated by providing a set flux constraints that best describe key metabolic features defining a phenotype. Rubisco favors carbon dioxide as a ligand to oxygen; however, photorespiration occurs when there is

Table 2
List of flux constraints used for photosynthesis, photorespiration, and respiration

Reaction ID	Reaction name	Rationale/motivation of constraints	Photosynthesis	Photorespiration	Respiration
Turn off influx/efflux reaction					
Ex6	Glutamate efflux	Auxotrophic	[0,0]	[0,0]	[0,0]
Ex7	Glutamine efflux	Auxotrophic	[0,0]	[0,0]	[0,0]
Ex8	Aspartate efflux	Auxotrophic	[0,0]	[0,0]	[0,0]
Ex9	Glutamate efflux	Auxotrophic	[0,0]	[0,0]	[0,0]
Ex10	Serine efflux	Auxotrophic	[0,0]	[0,0]	[0,0]
Ex13	Glucose efflux	Auxotrophic	[0,0]	[0,0]	[0,0]
Ex15	Maltose efflux	Auxotrophic	[0,0]	[0,0]	[0,0]
Switch direction of influx/efflux reaction					
Ex16	Photons influx	Photons available in light condition	[0,1000][a]	[0,1000][a]	[0,0]
Ex17	Sucrose efflux	Sucrose consumed in dark condition	[0,0]	[0,0]	[−1000,0]
Ex1	CO_2 influx	CO_2 is fixed in light condition	[0,1000]	[0,1000][a]	[−1000,0]
Experimental parameter					
BIO_L	Biosynthesis		[0.11,0.11]	[0.11,0.11]	[0.11,0.11]
R03140_p	Ribulose-bisphosphate carboxylase	Control ratio of carboxylation–decarboxylation	[0,1000][a]	[0.887,0.887]	[0,1000][a]
General inactivation					
R01221_m	Glycine synthase	Inactive during photosynthesis	[0,0]	[−1000,1000][a]	[−1000,1000][a]
R05875_p	Ferredoxin reductase (NAD)	Only active in dark condition	[0,0]	[0,0]	[−1000,1000][a]
R01195_p	Ferredoxin reductase (NADP)	Only active in light condition	[−1000,1000][a]	[−1000,1000][a]	[0,0]
R01063_p	Glyceraldehyde-3-phosphate dehydrogenase (NADP)	Only active in light condition	[−1000,1000][a]	[−1000,1000][a]	[0,0]
R00343_m	Malate dehydrogenase (NADP)	Only NAD-dependent isoenzyme is active	[0,0]	[0,0]	[0,0]

(continued)

Table 2
(continued)

Reaction ID	Reaction name	Rationale/motivation of constraints	Photosynthesis	Photorespiration	Respiration
Revise reaction direction					
R00472_x	Malate synthase	Glyoxylate shunt irreversible in the direction of glyoxylate synthesis	[0,1000]	[0,1000]	[0,1000]
R00243_c	Glutamate dehydrogenase (cytoplasmic)	NH_3 is only fixed by GS system	[0,1000]	[0,1000]	[0,1000]
R00243_m	Glutamate dehydrogenase (mitochondria)	NH_3 is only fixed by GS system	[0,1000]	[0,1000]	[0,1000]
R00253_p	Glutamine synthetase (plastid)	An irreversible ATP-consuming process	[0,1000]	[0,1000]	[0,1000]
R00253_m	Glutamine synthetase (mitochondria)	An irreversible ATP-consuming process	[0,1000]	[0,1000]	[0,1000]
R00086_c	ATP hydrolysis	Irreversible process	[0,1000]	[0,1000]	[0,1000]
Specific inactivation: turn off alternative redox and CO_2 shuttle between mitochondria and plastid					
R00093_p	Glutamate synthase (GOGAT) (NAD)	Only active in dark condition	[0,0]	[0,0]	[0,1000][a]
R00114_c	Glutamate synthase (GOGAT) (NADP)	Only active in dark condition	[0,0]	[0,0]	[0,1000][a]
R00149_p	Carbamoyl phosphate synthase	Turn off ornithine-citrulline shuttle system	[0,0]	[0,0]	[0,0]
R01398_c	Ornithine carbamoyltransferase	Turn off ornithine-citrulline shuttle system	[0,0]	[0,0]	[0,0]

[a]Default lower and upper boundary value

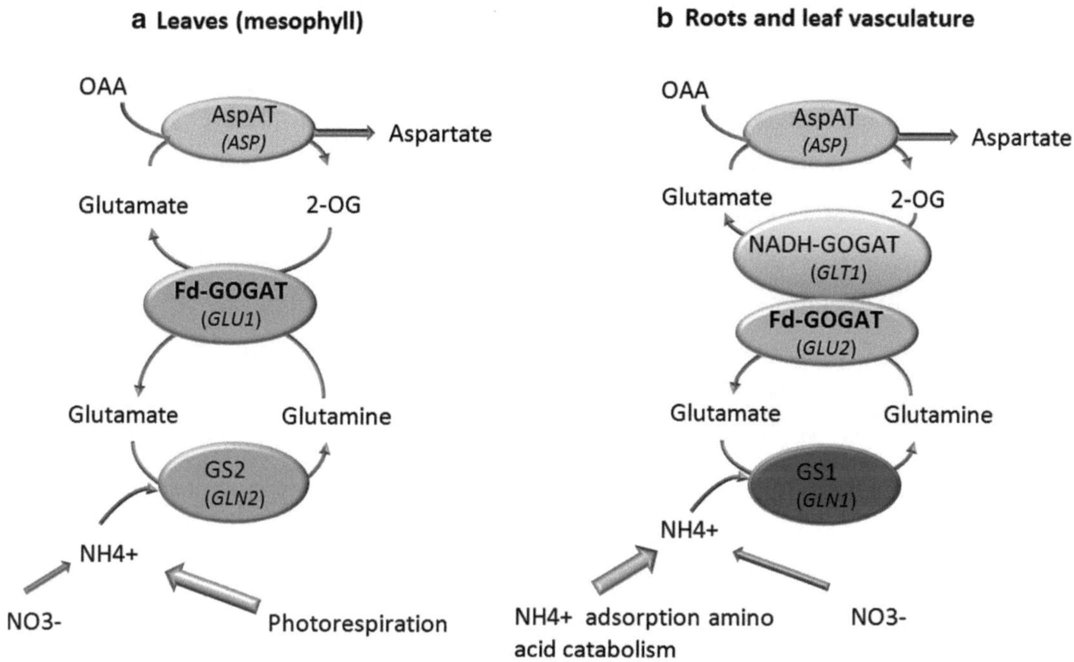

Fig. 3 Isoenzymes involved in nitrogen assimilation in photosynthetic and non-photosynthetic tissues in Arabidopsis, captured by AraGEM. Chloroplastic GS2 (GLN2) and one type of Fd-GOGAT (GLU1) seem to play a major role in recapturing the ammonia lost during photorespiration and perhaps in the primary assimilation of ammonia in leaves under light conditions (high C:N ratio). Cytosolic GS1 (GLNI), NADH-GOGAT (GLTI), and another Fd-GOGAT (GLU2) may be responsible for the primary assimilation of ammonia in roots and also for the recapture of ammonia released during amino acid catabolism. Isoenzymes: AspAT: Aspartate aminotransferase, Fd-GOGAT: glutamate synthase (ferredoxin-dependent), GS1: cytosolic glutamine synthase, GS2: chloroplastic glutamine synthase. Correspondent genes: *ASP*, *GLU1,GLU2*, *GLT1*, *GLN2*, *GLN1*

a high concentration of oxygen relative to carbon dioxide. To simulate photorespiration for example, the user should constraint the ratio determines experimentally of carbon dioxide: oxygen (~3:1). The activity of metabolic flux that characterize the photorespiratory can be observed when phosphoglycerate and phosphoglycolate are produced (check reaction R03140_p); phosphoglycerate reenters the Calvin Cycle and is converted back to RuBP. When running the simulations it is important to check if the metabolic activity is consistent to each physiological scenario. Table 3 shows key flux values for the tested scenarios according to the use of constraints shown in Table 2. Note that the absolute flux values are subjected to changes according to the constraints (e.g., biomass composition, growth rate, etc.)

4.2 Modifying the AraGEM Biomass Equation

Two examples are provided to demonstrate modifications that can be performed on AraGEM in COBRA. The first is the addition of "Folate_c" to the biomass drain equation "BIO_L". Carry out the required steps as listed in Subheading 3.5. At Subheading 3.5,

Table 3
Key flux simulations for each scenario (FBA)

Rxn ID	Description	Case 1	Case 2	Case 3
Ex1	CO_2 uptake (+)/out(−)	2.315	2.315	−0.778
Ex17	Sucrose uptake	0	0	−0.256
Ex16	Photons uptake	23.29	35.217	0
R00024_p	RuBisCO; EC 4.1.1.39	2.49	2.88	0
R03140_p	Carboxylation:oxygenation ratio	1:0	3:1	0
R00021_p	Fd-GOGAT; EC 1.4.7.1	0.303	0.728	0
R00093_p	NADH-GOGAT; EC 1.4.1.14	0	0	0.284
Ex16	Optimization: minimize uptake	Photons	Photons	Sucrose
Bio_L	Biomass rate	Leaf	Leaf	Root[a]

CASE 1: photosynthesis; CASE 2: Photorespiration; CASE 3: respiration/nitrogen assimilation in non-photosynthetic cell. (−) flux limited: constraint to zero. (+) free flux; estimated through optimization. *RuBisCO* ribulose-bisphosphate carboxylase, *Fd-GOGAT* glutamate synthase (ferredoxin), *NADH-GOGAT* glutamate synthase (NADH). Flux unit: mmol/gbiomass day

[a]Note that for simplification we have used the biomass composition and growth rate of leaf tissue (0.11 g/g day) to capture only the respiratory metabolic adjustments without the interference of the compositional biomass effect

step 4, append "Folate_c" to the end of *hitMets* cell array, and add the folate's biomass composition value (e.g., a rate value with a small order of magnitude just to test the activity of this pathway) to the end of *hitStoic* vector. Continue with Subheading 3.5, **step 5**.

Note that modifications made to AraGEM in COBRA cannot be exported into an SBML file. This is because AraGEM used slightly different convention to tag metabolites for the various cellular compartments (e.g., "_p" is used for peroxisome and plastid in COBRA and AraGEM, respectively).

4.3 Adding Reactions to AraGEM

The second modification is the addition of an irreversible glycolate exchange reaction (meaning that glycolate can be secreted). Cytosolic glycolate is designated "Glycolate_c" in the model. Using a new reaction ID of "glycolate_transport", type "*model* = *addReaction*(*model*, 'glycolate_transport', 'Glycolate_c', [-1])" to add the new exchange reaction into the COBRA model. Note, that this glycolate exchange reaction is written in the direction of glycolate efflux because the stoichiometric coefficient assigned to "Glycolate_c" is −1. Observe that if the photorespiration scenario is tested after adding this reaction, the photorespiration cycle will no longer carry flux; instead the plant will secrete glycolate. While this example is not feasible in plants, it is indeed the strategy adopted by single-celled algae Chlorella [37, 38] and Chlamydomonas [39]; simulated by AlgaGEM [17].

5 Concluding Remarks

Significant initial successes have been achieved on plant genome-scale metabolic reconstruction and its application promises to be important in plant biotechnology. The principles of model formulation and its implementation were presented. COBRA Toolbox can greatly facilitate the use of plant reconstructions, if such are available in the proper format (SBML for COBRA). AraGEM SBML is readable into COBRA and is ready to be tested as demonstrated in this chapter.

6 Notes

1. libSBML is updated frequently, and the version currently compatible to COBRA Toolbox is 5.1.0-b0, which is an older version. Download libSBML that matches.
2. Gurobi Optimizer is an academic-free solver that fulfills all solver components required by COBRA Toolbox, namely Linear and Quadratic Programming. The only other complete alternative is TOMLAB, which is not freely available.
3. The SBML file read by COBRA contains more information than a standard SBML file. The additional requirements include correct unit definition, consistent compartment tags to the suffix of the metabolite IDs, "R_" and "M_" prefixes to the suffix of the metabolite IDs, "R_" and "M_" prefixes to reaction and metabolite IDs, respectively, Gene–Protein Reaction annotation in the "notes" element if available, and reaction parameters (flux lower and upper bound, optimized fluxes, and objective coefficient). The suffix "_b" is reserved for external metabolites that are excluded from the stoichiometric matrix. Metabolic models in SBML file format are made available publicly on BiGG Database and Model SEED. The SBML file specifications and requirements by COBRA Toolbox are expected to change over time as successive versions or updates are introduced. The current instructions issued by COBRA Toolbox should be followed according to the operating system being used (e.g., Windows, Mac OS X, Linux, 64-bit, 32-bit). libSBML with only MATLAB binding can be downloaded.
4. New user must register for an account with Gurobi. The activation of Gurobi's free academic license must be performed within the academic institution's network domain. The license is user- and node locked. A new license must be requested and activated in a separate computer by the same user. The license will last 12 months.
5. At least for flux simulation, the solvers that must pass the test are LP and MILP. QP is required if MOMA is to be performed.

NLP solver has less relevance because it is only used by 13C fitting. NLP solvers can be found in TOMLAB software package or as MATLAB's *fmincon* function.

6. In addition to generating the stoichiometric model from the SBML file, COBRA Toolbox will import all available annotation for reactions and metabolites. The (lower and upper) flux boundary values and the objective coefficient provided in the SBML file will be used as constraints and as objective function, respectively.
7. [*p1*,*p2*] are the input parameters to modify the approach of the FBA. While maintaining the value achieved by the original objective function, additional constraints/objectives can be imposed to reduce magnitude (or movement) of unconstrained internal fluxes by modifying the last two parameters. [*0*, *1*] configuration gives a standard FBA flux distribution. [*1*, *1*] configuration minimizes absolute fluxes by LP, and [*2*, *1*] by QP. [*p1*, *0*] introduces the Loopless constraints to eliminate thermodynamically infeasible metabolic loops [40].
8. More than one reaction's boundary values or the objective coefficients can be modified using functions *changeRxnBounds* or *changeObjective* once. Instead of providing individual reaction IDs, the list of reaction IDs are provided in a cell array, with the corresponding bound values and bound types in a vector and cell array, respectively. The reaction IDs must match the list found in *model.rxns*.
9. COBRA maximizes the objective function by default. To minimize the objective, either provide a negative objective coefficient or switch COBRA to minimization mode in *optimizeCbModel* function.
10. The function *changeObjective* clears away any existing objective coefficients. It is necessary to use a cell array input if more than one non-zero coefficients are required (*see* **Note 8**).
11. *metaboliteList* is a cell array containing new or existing metabolite IDs. For existing metabolites, inspect metabolite list *model.mets* to identify the correct metabolite IDs to be used. Any metabolite IDs in *metaboliteList* cell array different from ones listed in *model.mets* will lead to the creation of a new metabolite in the COBRA model. The correct compartment tag in the metabolite ID's suffice should be used (Table 1), e.g., glc-D[c], otherwise COBRA model cannot be exported properly.
12. The [*stoic coefficients*] is a row vector of stoichiometric coefficients for the reaction to be added or modified. The coefficients must be provided in the same order as the corresponding metabolites in *metaboliteList*. Reactants are denoted by negative coefficients, while products by positive coefficients.

13. Reaction reversibility, lower and upper flux boundary value, and the objective coefficient can be provided as optional parameters in *addReaction* function. The default values are used otherwise (reaction is reversible, null objective coefficient).
14. The *addReaction* function is built in such a way that the entire reaction equation is over-written when an existing reaction is modified. *addReaction* is not designed to modify a specific stoichiometry or metabolite in a reaction equation. Along with the required modifications, metabolite IDs (found in *model.mets*), and corresponding stoichiometric coefficients that are to be kept the same must also be provided. Nevertheless, the original flux parameters and annotations are preserved if the optional parameters are not provided in *addReaction* function.
15. An alternative to permanently removing a reaction is to set the lower and upper flux boundary value to zero.
16. An alternative to modify reactions is to directly change contents in the SBML file.
17. A modified metabolic model (in COBRA) can be exported into an SBML file. However, this require that any new metabolite introduce adhere to COBRA's existing suffix convention. All boundary (external) metabolites are lost during the export.

References

1. Durot M, Bourguignon PY, Schachter V (2009) Genome-scale models of bacterial metabolism: reconstruction and applications. FEMS Microbiol Rev 33(1):164–190
2. Feist AM, Herrgard MJ, Thiele I et al (2009) Reconstruction of biochemical networks in microorganisms. Nat Rev Microbiol 7 (2):129–143
3. Jang YS, Lee J, Malaviya A et al (2012) Butanol production from renewable biomass: rediscovery of metabolic pathways and metabolic engineering. Biotechnol J 7(2):186–198
4. Park JM, Kim TY, Lee SY (2011) Genome-scale reconstruction and in silico analysis of the Ralstonia eutropha H16 for polyhydroxyalkanoate synthesis, lithoautotrophic growth, and 2-methyl citric acid production. BMC Syst Biol 5:101
5. Edwards JS, Ibarra RU, Palsson BO (2001) In silico predictions of Escherichia coli metabolic capabilities are consistent with experimental data. Nat Biotechnol 19(2):125–130
6. Werner E (2007) Systems biology: properties of reconstructed networks. Nature 446 (7135):493–494
7. Edwards JS, Palsson BO (1999) Systems properties of the Haemophilus influenzae Rd metabolic genotype. J Biol Chem 274(25): 17410–17416
8. Covert MW, Schilling CH, Famili I et al (2001) Metabolic modeling of microbial strains in silico. Trends Biochem Sci 26(3):179–186
9. Fong SS, Marciniak JY, Palsson BO (2003) Description and interpretation of adaptive evolution of Escherichia coli K-12 MG1655 by using a genome-scale in silico metabolic model. J Bacteriol 185(21): 6400–6408
10. Felst AM, Henry CS, Reed JL et al (2007) A genome-scale metabolic reconstruction for Escherichia coli K-12 MG1655 that accounts for 1260 ORFs and thermodynamic information. Mol Syst Biol 3:121
11. Gonzalez O, Gronau S, Falb M et al (2008) Reconstruction, modeling & analysis of Halobacterium salinarum R-1 metabolism. Mol Biosyst 4(2):148–159
12. Famili I, Forster J, Nielson J et al (2003) Saccharomyces cerevisiae phenotypes can be predicted by using constraint-based analysis of a genome-scale reconstructed metabolic network. Proc Natl Acad Sci U S A 100 (23):13134–13139

13. Boyle NR, Morgan J (2009) Flux balance analysis of primary metabolism in Chlamydomonas reinhardtii. BMC Syst Biol 3:32
14. Huthmacher C, Hoppe A, Bulik S, Holzhütter H-G (2010) Antimalarial drug targets in Plasmodium falciparum predicted by stage-specific metabolic network analysis. BMC Syst Biol 4:120
15. Vanee N, Roberts SB, Fong SS et al (2010) A genome-scale metabolic model of cryptosporidium hominis. Chem Biodivers 7(5):1026–1039
16. Zhang P, Dreher K, Karthikeyan A et al (2010) Creation of a genome-wide metabolic pathway database for Populus trichocarpa using a new approach for reconstruction and curation of metabolic pathways for plants. Plant Physiol 153(4):1479–1491
17. de Oliveira Dal'Molin CGD, Quek L-E, Palfreyman RW, Nielsen LK (2011) AlgaGEM - a genome-scale metabolic reconstruction of algae based on the Chlamydomonas reinhardtii genome. BMC Genomics 12(Suppl 4):S5
18. Caspeta L, Shoaie S, Agren R, Nookaew I, Jens NJ (2012) Genome-scale metabolic reconstructions of Pichia stipitis and Pichia pastoris and in silico evaluation of their potentials. BMC Syst Biol 6:24
19. Sohn SB, Kim TY, Lee JH, Lee SY (2012) Genome-scale metabolic model of the fission yeast Schizosaccharomyces pombe and the reconciliation of in silico/in vivo mutant growth. BMC Syst Biol 6:49
20. Quek L, Nielsen LK (2008) On the reconstruction of the Mus musculus genome-scale metabolic network model. Genome Inform 21: 89–100
21. Duarte NC, Becker SA, Jamshidi N et al (2007) Global reconstruction of the human metabolic network based on genomic and bibliomic data. Proc Natl Acad Sci U S A 104 (6):1777–1782
22. de Oliveira Dal'Molin CG, Quek LE, Palfreyman RW et al (2010) AraGEM, a genome-scale reconstruction of the primary metabolic network in Arabidopsis. Plant Physiol 152(2):579–589
23. Mintz-Oron S, Meir S, Malitsky S et al (2012) Reconstruction of Arabidopsis metabolic network models accounting for subcellular compartmentalization and tissue-specificity. Proc Natl Acad Sci U S A 109(1):339–344
24. Poolman MG, Miguet L, Sweetlove LJ, Fell DA (2009) A genome-scale metabolic model of arabidopsis and some of its properties. Plant Physiol 151(3):1570–1581
25. Saha R, Suthers PF, Maranas CD (2011) Zea mays iRS1563: a comprehensive genome-scale metabolic reconstruction of maize metabolism. PLoS One 6(7):e21784
26. de Oliveira Dal'Molin CG, Quek LE, Palfreyman RW et al (2010) C4GEM, a genome-scale metabolic model to study C4 plant metabolism. Plant Physiol 154(4):1871–1885
27. Thiele I, Palsson BO (2010) A protocol for generating a high-quality genome-scale metabolic reconstruction. Nat Protoc 5(1):93–121
28. Henry CS, DeJongh M, Best AA et al (2010) High-throughput generation, optimization and analysis of genome-scale metabolic models. Nat Biotechnol 28(9):977–982
29. de Oliveira Dal'Molin CG, Nielsen LK (2012) Plant genome-scale metabolic reconstruction and modelling. Curr Opin Biotechnol 24: 271–277
30. Livingston DP, Henson CA (1998) Apoplastic sugars, fructans, fructan exohydrolase, and invertase in winter oat: respones to second-phase cold hardening. Plant Physiol 116(1): 403–408
31. Gowik U, Westhoff P (2011) The path from C3 to C4 photosynthesis. Plant Physiol 155 (1):56–63
32. Segre D, Vitkup D, Church GM (2002) Analysis of optimality in natural and perturbed metabolic networks. Proc Natl Acad Sci U S A 99 (23):15112–15117
33. Shlomi T, Berkman O, Ruppin E (2005) Regulatory on/off minimization of metabolic flux changes after genetic perturbations. Proc Natl Acad Sci U S A 102(21):7695–7700
34. Lonien J, Schwender J (2009) Analysis of metabolic flux phenotypes for two Arabidopsis mutants with severe impairment in seed storage lipid synthesis. Plant Physiol 151(3):1617–1634
35. Schellenberger J, Que R, Fleming RM et al (2011) Quantitative prediction of cellular metabolism with constraint-based models: the COBRA Toolbox v2.0. Nat Protoc 6(9): 1290–1307
36. Lam HM, Coschigano K, Schultz C et al (1995) Use of arabidopsis mutants and genes to study amide amino-acid biosynthesis. Plant Cell 7(7):887–898
37. Hess JL, Tolbert NE (1967) Glycolate pathway in algae. Plant Physiol 42(3):371–379
38. Miller RM, Meyer CM, Tanner HA (1963) Glycolate excretion & uptake by chlorella. Plant Physiol 38(2):184–188
39. Moroney JV, Wilson BJ, Tolbert NE (1986) Glycolate metabolism and excretion by chlamydomonas reinhardtii. Plant Physiol 82 (3):821–826
40. Schellenberger J, Lewis NE, Palsson BO (2011) Elimination of thermodynamically infeasible loops in steady-state metabolic models. Biophys J 100(3):544–553

Part IV

Flux Analysis at the Whole Plant Level

Chapter 20

^{34}S and ^{15}N Labelling to Model S and N Flux in Plants and Determine the Different Components of N and S Use Efficiency

Christophe Salon, Marie-Paule Bataillé, Karine Gallardo, Christian Jeudy, Anne-Lise Santoni, Jacques Trouverie, Anne-Sophie Voisin, and Jean-Christophe Avice

Abstract

In order to highlight our understanding on ecosystems functioning and resource sharing/competition, either in artificial environment or agrosystems, according to changes in the climatic conditions, it is necessary to measure accurately element fluxes within plants. Stable isotopes allow tracking safely and accurately on a short time frame the behavior of elements in plants. After a short review devoted to isotopic studies of elemental flux within plants, we explain how a direct multiple labelling study might be conducted in a plant, so as to measure over short time nitrogen and sulfur acquisition, and assimilates arising from a labelled source.

Key words Nutrient use efficiency, Flux, Isotope, Allocation, Remobilization

1 Introduction

Due to agroecological transition, the main agronomic challenge is now to reduce chemical preparation and supply target new plant varieties having enhanced mineral/nutrient use efficiency (NUE, Fig. 1) [1–3]. NUE, according to the considered element, is comprised of (1) uptake processes (NUpE, Nutrient Uptake Efficiency) mediated by various transporters [4], (2) assimilation, transfer and storage (NAE, Nutrient Assimilation Efficiency), and (3) remobilization (NRE, Nutrient Remobilization Efficiency) [5, 6] towards seeds as for an example [7, 8].

Among the mineral elements necessary for plant development and growth, nitrogen (N) and sulfur (S) have key roles. Even if N is usually the more limiting, S is becoming a more and more limiting element, due to reduction in S deposits arising from industrial activities. S shares similar pathways than N, for example for amino

Martine Dieuaide-Noubhani and Ana Paula Alonso (eds.), *Plant Metabolic Flux Analysis: Methods and Protocols*, Methods in Molecular Biology, vol. 1090, DOI 10.1007/978-1-62703-688-7_20,

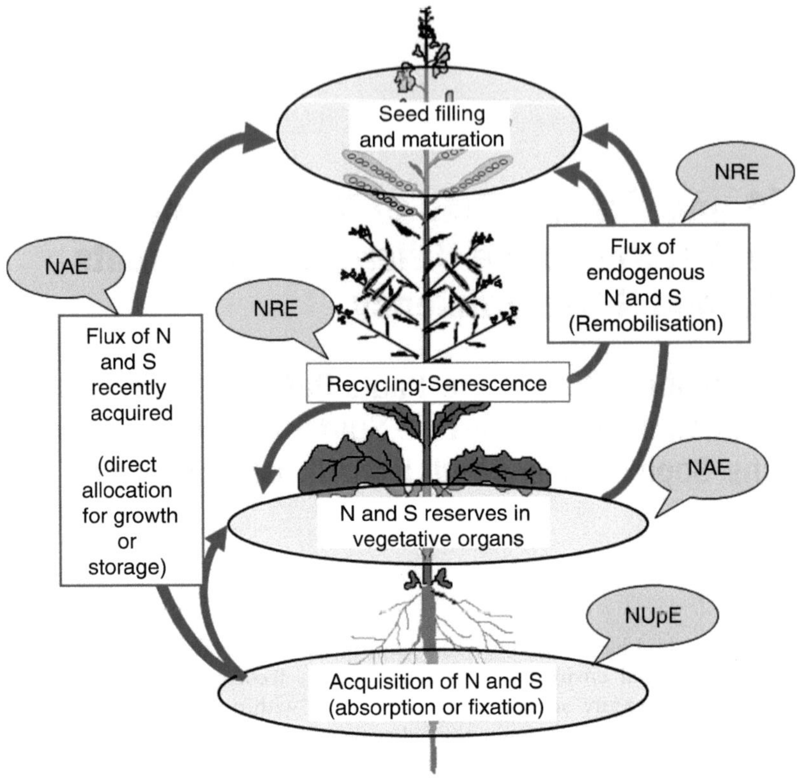

Fig. 1 Schematic representation of nitrogen (N) and sulfur (S) fluxes and associated components of the nutrient use efficiency (NUE) at whole plant level. *NUpE* nutrient uptake efficiency, *NAE* nutrient assimilation efficiency, *NRE* nutrient remobilization efficiency

acid biosynthesis, and influences N use efficiency [9–11], including N assimilation and fixation [12–14].

Elements such as N and S are either transported as mineral or in reduced-form within plants and can be exuded in their surrounding environment. The characterization of these elemental fluxes is of tremendous importance in the understanding of the interactions of ecosystems. Although a large amount of data is already available concerning the various components of NUE at the molecular, cellular, or organ level, up to now little is known at the plant level. As such, (a) the regulation of N and S flux partitioning within plants, its coupling with other plant partners, (b) interaction with soil microbial communities, (c) the molecular basis involved in such trophic regulations, still constitute a major black box of the understanding of ecosystems functioning. This results from the difficulty to track accurately the fate of elements in plant, on sometimes short time windows depending on the physiological mechanism of interest.

The apparent allocation or remobilization efficiency for a given element can be measured through the increment of this element in a compartment (i.e., a plant or its organs). However, if the period spanned by the measurements is small or if plant development and growth are heterogeneous, this method is far from precise. Moreover, such methods give only information on the net flux but in no case can provide an estimation of turnover within or between various compartments (i.e., pools, cellular compartments, plant organs, ecosystem component).

Stable isotopes allow tracking safely and accurately on a short time frame the behavior of elements in plants, from seconds [15] or weeks [16]. The way elements are tracked changes according to species. As for an example for plants relying exclusively on soil mineral resources, the more common N forms to be tracked (and so labelled) are nitrate, and less usually ammonium and ammonium nitrate. Symbiotic plants have the ability to establish symbiosis with microbes and acquire N through the additional way of N_2 fixation via nitrogenase. Here the substrate is a gas, which implies special equipment and methodology if it wished to introduce labelled element (^{15}N) at the site of assimilation (see below). Otherwise when impossible for either technical or research objectives reasons, N fluxes may be tracked indirectly through its dilution [17]. Book chapters, which record and compare all of these methods, are available elsewhere [18, 19].

Element flux in the whole plant might also be characterized throughout its phenology. At the beginning of the growth period, such study can provide information on how resources are allocated to the various organs, according to their exogenous availability/ form [20, 21]. Later on, storage and remobilization strategies might be studied to understand where, when, and how plants transiently mobilize the stored resources [5, 11], how these flux impact on organ morphogenesis and intimate the storage life of seeds [22].

Transfer to other biotic partners/competitors can also be studied using isotopic studies: this has tremendous importance in understanding trophic links mediated by rhizodeposits between telluric microorganisms and plants [23, 24] or even between complex microbial communities.

In this book chapter, we describe a method allowing measuring precisely S and N fluxes within plants and calculating NUE components. This method is applicable to various plant species and growth conditions.

2 Materials

2.1 Nutrient Solution for Plant Growth

1. Prepare all solutions using deionized water and use analytical grade reagents. Prepare and store all reagents of the nutrient solution at room temperature (unless indicated otherwise).

2. Concentrated solution of oligonutrients for preparation of nutrient solution: for 1 L weigh 1.731 g of H_3BO_3 (final concentration: 14 μM), 1.69 g of $MnSO_4 \cdot 2H_2O$ (final concentration: 5 μM), 1.725 g of $ZnSO_4 \cdot 7H_2O$ (final concentration: 3 μM), 1.729 g of $(NH_4)_6Mo_7O_{24}$ (final concentration: 0.7 μM), 0.35 g of $CuSO_4 \cdot 5H_2O$ (final concentration: 0.7 μM), and 0.0475 g of $CoCl_2$ (final concentration: 0.1 μM).
3. 25 % Hoagland nutrient solution: for 10 L of nutrient solution add 5 mL of 1.25 M $Ca(NO_3)_2 \cdot 4H_2O$ (final concentration: 1.25 mM), 25 mL of 0.5 M KNO_3 (final concentration: 1.25 mM), 5 mL of 1 M $MgSO_4$ (final concentration: 0.5 mM), 5 mL of 0.25 mM KH_2PO_4 (final concentration: 0.25 mM), 25 mL of 80 mM EDTA, $2NaFe \cdot 3H_2O$ (final concentration 0.2 mM), and 5 mL of concentrated solution of oligonutrients (see above). Note that for legumes, final concentrations of $Ca(NO_3)_2 \cdot 4H_2O$ and KNO_3 should be 4 mM each.

2.2 Preparation and Introduction of Labelled ^{15}N and ^{34}S Compounds

When mineral forms of the elements are used to measure flux, there is no special equipment need, however, when it is a gas (such as for N_2), special containers/methods are necessary (*see* **Note 1**).

2.2.1 Introduction of ^{15}N Compounds

1. When the label is $^{15}N_2$ (providers: Sigma Aldrich, St Quentin France), the gas is either available under pressure in a gas tank (Sigma Aldrich 364 584 SPEC) or at atmospheric pressure in a glass vial (Cambridge Isotope laboratories Andover, MA, USA, ref. NLM-363-1-BS).
   (a) In the first case, labelled gas injection in the system (*see* **Note 1**) is realized using a mass flow regulator and a gas injector.
   (b) In the second case, the glass vial is turned and the aperture is drowned in a water container. A syringe allows taking off through a cork the necessary amount of gas, which is replaced within the glass vial by ascending water. The labelled gas is then injected in the loop through a septum.
   (c) If the volume of labelled gas to be injected in the system is large as compared to the final volume, add to the system a supple plastic pouch. It will absorb all the pressure differentials, including those arising from temperature variations.
2. When $^{15}NO_3^-$ (or $^{15}NH_4^+$) is applied (providers: Euriso-top, Saint Aubain France; CortecNet, Voisins le Bretonneux France), for a short-term labelling, unlabelled nitrate (or ammonium) should be instantaneously and totally replaced by labelled nitrate (or ammonium) so as to induce constant enrichment of mineral N supply during the labelling period

and to allow calculations afterwards. For a long-term labelling, for example until flowering, the quantity of labelled nitrate is adjusted to obtain 2–5 % of ^{15}N enrichment. Pots should be rinsed using the labelled solution until the unlabelled solution is totally drained and replaced (*see* **Note 2**).

2.2.2 Production of ^{34}S-Sulfate

1. The labelled sulfate (^{34}S-sulfate) is obtained by oxidation of 100 mg ^{34}S-sulfur (Sigma Aldrich) with 10 mL of nitric acid (HNO_3 at 65 %) in a tube of mineralization at 200 °C for 2 h.
2. After 30 min at room temperature, the liquid containing sulfuric acid produced during oxidation is poured in a beaker and the sulfur residue is eventually subjected to a new step of oxidation (*see* **Note 3**).
3. The sulfuric acid solution is then neutralized by the addition of potassium carbonate (K_2CO_3) until the reaction of effervescence stops.
4. The labelled solution of potassium sulfate ($K_2{}^{34}SO_4$) obtained is then filtered on a membrane of 0.45 μm using a vacuum pump system. The filtered solution is adjusted to 100 mL with ultrapure water.
5. The concentration of ^{34}S-sulfate is determined using high performance liquid chromatography (ICS3000 Thermo Scientific Dionex, Voisins le Bretonneux, France) using an IonPac AS22 column and a buffer containing 4.05 mM carbonate and 1.26 mM bicarbonate as eluant. The ^{34}S-sulfate solution also contains nitrate that is necessary to take into account for the calculation of nitrate supply by the nutrient solution (*see* **Note 4**).

2.3 Label Detection

1. Detection of $^{15}N/^{14}N$ ratio and $^{34}S/^{32}S$ ratio is realized using an elemental analyzer and an isotopic mass spectrometer (in example ANCA-GSL_20-20, Sercon). In some cases, ^{15}N and ^{34}S can be analyzed simultaneously: for example, an elemental analyzer such as EA3000 (EuroVector) can be coupled to a mass spectrometer Horizon (Nu Instruments).
2. Inter laboratories trials with different standards for ^{15}N and ^{34}S should be realized to guarantee both fiability and precision of the results (*see* **Note 5**).

3 Methods

3.1 Calculations Prior Labelling

Prior starting the experiment, it is necessary to make some rather simple calculations to adjust isotope enrichment of the source. This ensures compounds of interest will be sufficiently labelled at the end of the labelling period. Here is an example on how processing for nitrogen.

1. At first, take into account the precision (P in atom% unit) of your mass spectrometer and assume you will be much less precise.
2. Then estimate the N content (Q_N) of your biological sample at the time you will have to measure its ^{15}N enrichment.
3. The minimum amount of ^{15}N enriched than can be detected by the mass spectrometer in the sample is $P \times Q_N$
4. Knowing the rate of N assimilation (V_N) by the plant during the period (T), the initial ^{15}N enrichment ($A\%^{15}N$) is calculated as:

$$A\%^{15}\text{N} = 100 \times P \times Q_N/(V_N \times T)$$

3.2 Experimental Design for Double Labelled Pulse-Chase Method

In this section, we describe an experiment that is designed in order to determine the N and S fluxes in a leguminous or not leguminous plant (Fig. 2) in a given condition of culture (optimal conditions) and between two stages of developments of interest (beginning of reproductive phase until the stage corresponding to mature grains). This long-term-pulse-chase double labelling is particularly reliable for the determination of endogenous S and N fluxes and could be applied under various conditions of culture (different climatic or nutrient conditions), in the presence of other plants or organisms (mycorhizal or bacteria inoculum ...) or at diverse stages of development.

1. After surface sterilization and germination of seeds, seedlings are transferred to individual pot containing mixed perlite/vermiculite, which mimics soil texture and is free of N or S mineral or organic forms.
2. During all the experiment, plants are grown under controlled conditions with a given thermoperiod (e.g., 22 °C during the day and 16 h, 18 °C during the 8 h of night) and are illuminated by natural light, supplied with Philips® Green Power lamps (400 μmol/m^2/s photosynthetically active radiation in the canopy). The 25 % Hoagland nutrient solution (with or without labelled compounds) is supplied automatically (2 or 3 times per week) in an increasing volume as a function of the growth stages.
3. After a first phase of pre-culture that is extended until obtaining the "3-4 leaves" stages, the double labelling ("pulse" period) is applied during a long period (from 1 day up to 3 weeks) in order to obtain plants with homogeneous ^{34}S and ^{15}N labelling. Thus, plants are supplied with $^{34}SO_4^{2-}$ (1 atom% excess) and $^{15}NO_3^-$ (2 atom% excess) for non-leguminous or with $^{34}SO_4^{2-}$ (1 atom% excess) and $^{15}N_2$ (2–5 atom% excess) for leguminous plants. Recommendations about the preparation of the double-labelled nutrient solution or $^{15}N_2$ gas are indicated in previous sections (*see* Subheadings 2.2.1 and 2.2.2).

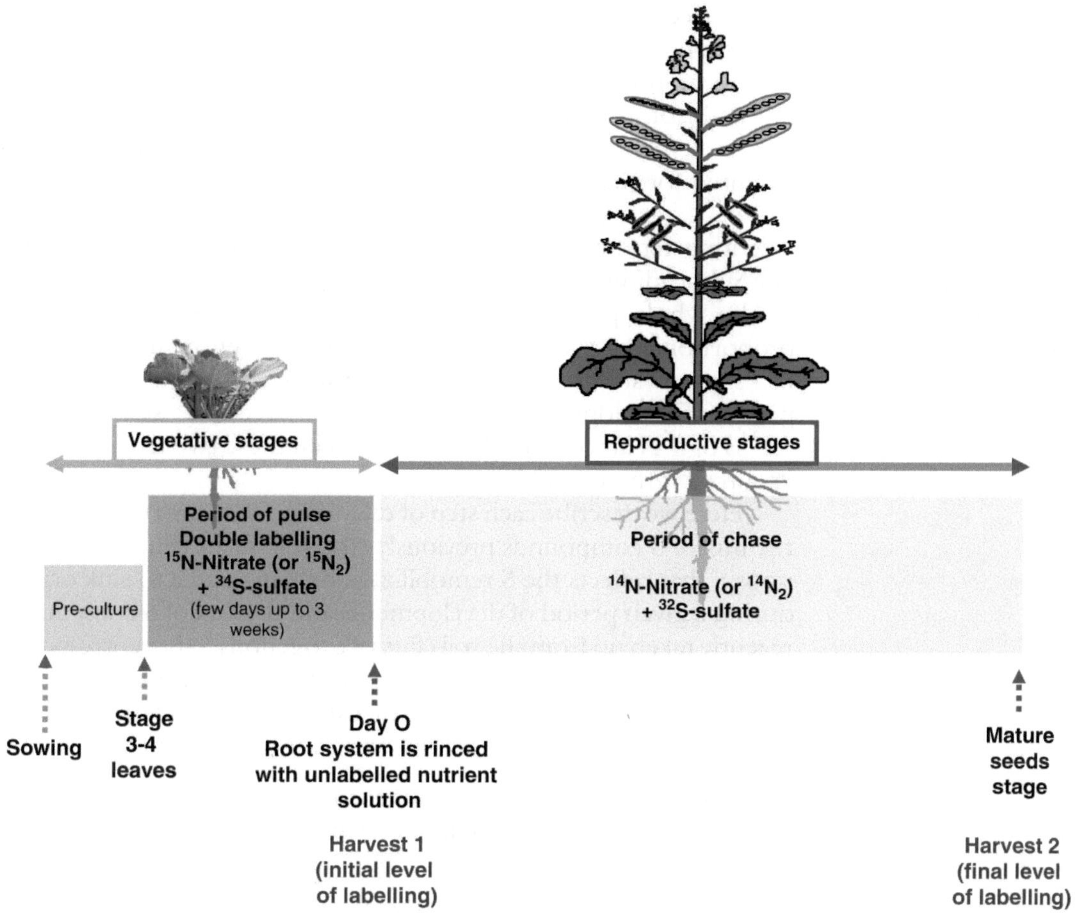

Fig. 2 Schematic diagram of experimental design of double labelling method (long-term pulse chase). For details *see* Subheadings 2 and 3

4. When the beginning of the reproductive phase occurs, the double labelling is stopped and plants are carefully rinsed with unlabelled 25 % Hoagland nutrient solution until labelled solution was totally drained and replaced. This date corresponds to the onset of the "chase" period. At this date, it is imperative to schedule a harvest in order to determine the initial level of ^{34}S and ^{15}N labelling in the different organs (including the root compartment). This point is crucial for the calculation of endogenous S and N flux (*see* Subheading 3.3).
5. At the end of the chase period, which corresponds to the final stage of the seed maturation in the present experiment, plants are harvested in the aim to determine the final level of ^{34}S and ^{15}N labelling in the different organs (including the root compartment).
6. At each date of harvest, at least four plants are harvested (biological replicates) in order to obtain reliable data and

proceed to statistical analyses. The different plant parts (roots, leaves, stem, flowers, pod walls, seeds, etc.) are weighed, freeze-dried and then ground to a fine powder for elemental and isotope analyses as described in Subheading 2.3. Variation of elemental and isotope amounts in the different organs are used for the calculations of S and N fluxes (see below).

3.3 Calculations of S and N Flux of Remobilization and Allocation

The long-term period of double isotope labelling described above (*see* Subheading 3.2) allows a homogenous incorporation of tracers in the whole plant. Therefore, ^{34}S or ^{15}N amount in excess is proportional to S or N amount and allows identifying the "sink" or "source" status of organs during the chase period. For a given period of development, an organ is considered as a "sink" if a gain of ^{34}S or ^{15}N amount is observed between these two dates, and as a "source" if there is a loss.

Here we describe each step of calculation that aims to determine the flux of S compounds previously stored in plant (flux of endogenous S that reflects the S remobilization from source to sink organs during a given period of development) and the flux of S compounds recently taken up from the soil (flux of exogenous S that corresponds to the S allocation from soil to sink organs during a given period of development). This method of calculation could be used for the determination of endogenous and exogenous *N* fluxes as well.

1. The value of isotope abundance (A%) given by the IRMS in sample or natural standard is calculated as:

$$A_{\%} = 100 \times \frac{^{34}\mathrm{S}}{\left(^{34}\mathrm{S} + ^{34}\mathrm{S}\right)}$$

2. The determination of *S* fluxes relies on the quantification of the ^{34}S amount in excess ($\mathrm{Q}^{34}\mathrm{S}_{\text{in excess}}$) in each organ:

$$\mathrm{Q}^{34}\mathrm{S}_{\text{in excess}}(\mathrm{mg}) = \frac{\left(A_{\%\ \text{sample}} - A_{\%\ \text{natural standard}}\right)}{100} \times \mathrm{QS}$$

where $A_{\%\ \text{natural standard}}$ corresponds to natural ^{34}S abundance, i.e., 4.21 %, QS is the amount of total S in a given organ in mg, and $A_{\%\ \text{sample}}$ corresponds to ^{34}S isotope abundance in a given organ.

3. For every organ at each date, ^{34}S amounts is normalized ($\mathrm{Q}^{34}\mathrm{S}_{\text{norm}}$) with the following calculation at every date (t):

$$\mathrm{Q}^{34}\mathrm{S}_{\text{norm}} = \frac{\mathrm{Q}^{34}\mathrm{S}_{\text{organ}}(t) \times \overline{\mathrm{Q}^{34}\mathrm{S}_{\text{whole plant}_{(\text{all date})}}}}{\mathrm{Q}^{34}\mathrm{S}_{\text{whole plant}}(t)}$$

Where $\mathrm{Q}^{34}\mathrm{S}_{\text{organ}}$ (t) indicates ^{34}S amount in excess in a given organ at the date t, $\mathrm{Q}^{34}\mathrm{S}_{\text{whole plant}_{(\text{all date})}}$ represents the average of

^{34}S amount in whole plant at every studying date and $Q^{34}S_{whole\ plant}\ (t)$ indicates ^{34}S amount in the whole plant at the date t.

4. When a loss of ^{34}S amount between two dates (t and $t + \Delta t$) is observed, the organ is considered as a source and the S amount remobilized (QSR_{source}) was defined as:

$$QSR_{source} = QS_{tot}(t) \times \frac{[Q^{34}S_{norm}\ (t) - Q^{34}S_{norm}(t + \Delta t)]}{Q^{34}S_{norm}\ (t)}$$

where $QS_{tot}\ (t)$ indicates total S amount in the source organ at the date t, $Q^{34}S(t)$ and $Q^{34}S(t + \Delta t)$ represent the ^{34}S amount in excess in the source organ at the date t and $t + \Delta t$, respectively.

5. When a gain of ^{34}S amount is observed, the organ is considered as a sink and the S amount remobilized (QSR_{sink}) corresponded to:

$$QSR_{sink} = \frac{[Q^{34}S_{norm}(t + \Delta t) - Q^{34}S_{norm}\ (t)]}{\left[\left(\sum Q^{34}S_{norm\ source}(t) - \sum Q^{34}S_{norm\ source}\left(t + \Delta t\right)\right] / \sum QSR_{source}}$$

where $\sum Q^{34}S_{norm\ source}(t)$ and $\sum Q^{34}S_{norm\ source}(t + \Delta t)$ indicate total ^{34}S amount in excess from all source organs at t and $t + \Delta t$, respectively. $\sum QSR_{source}$ indicates total S amount in excess remobilized from all source organs.

6. The inflow of S taken up is also calculated between t and $t + \Delta t$. For source organ, S amount taken up (QSI_{source}) was defined as:

$$QSI_{source} = [QS\ (t + \Delta t) - QS\ (t)] + QSR_{source}$$

For sink organ, S amount taken up (QSI_{sink}) corresponded to:

$$QSI_{sink} = [QS\ (t + \Delta t) - QS\ (t)] - QSR_{sink}$$

where QSR_{source} indicates the S amount from the studying source organ and QSR_{sink} indicates the S amount from the studying sink organ.

7. These different steps for the calculations of endogenous and exogenous flux can be easily performed using a commercial spreadsheet application. Nevertheless, in case of a study including numerous organs (up to 4) and/or several periods of development, calculation becomes tedious. Moreover, for statistical considerations determining the S content requires several plants which complex calculation (*see* **Note 6**). Then, a specific tool is recommended to facilitate the calculation of data fluxes (*see* **Note 7**).

3.4 Determining of the Components of N and S Use Efficiency

The determination of endogenous and exogenous N and S fluxes via the utilization of elemental and isotope data gives also the opportunity to calculate accurately various N or S harvest indexes (NHI, SHI) as well as the principle components of NUE [11]. Here, we present two examples among others:

1. Specific NHI or SHI could be determined as the percentage of total N or S amount observed in seeds at end of reproductive stage coming from the remobilization of N or S acquired during the vegetative phase.
2. N and S Remobilization Efficiency in leaves (NRE_{leaf} or SRE_{leaf}), which represent the main source organ for N and S, could be also calculated as the percentage of the total amount of N or S remobilized from leaves (between the end of vegetative phase and the end of reproductive stage) that is finally redistributed into the seeds.

4 Notes

1. In legumes, various equipments have been designed for that purpose, either to understand how N flux varies according to plant phenology [20] or how plants adapt their strategy towards maintaining a constant N status [25, 26].
2. A preliminary experiment is required to measure the volume of labelled solution needed until the enrichment of drainage solution becomes constants and reaches the enrichment of the labelled solution.
3. In order to increase the yield of the reaction of oxidation of ^{34}S-sulfur and finally improve the level of ^{34}S-sulfate production, the step of mineralization could be renewed two times more by add 10 mL of HNO_3 (65 %) on the ^{34}S-sulfur residue and incubate for additional mineralization (200 °C, 90 min). After each supplemental oxidation the liquid is pooled with the first products of oxidation.
4. Due to the chemical reactions used for the production ^{34}S-sulfate, the resulting solution also contains nitrate. In order to adjust the level of nitrate given by the nutrient solution, it is necessary to determine the concentration of nitrate into the ^{34}S-sulfate solution using high performance liquid chromatography (ICS3000 Thermo Scientific Dionex). This part of nitrate that accompanied the ^{34}S-sulfate is deduced to the 25 % Hoagland nutrient solution (*see* Subheading 2.1, **item 3**).
5. Each mass spectrometer has its own range of measures, both in isotopic enrichment and in mass enrichment.

6. For statistical consideration, elemental content is determined on several plants (at least four biological replicates). As IRMS analysis are destructive, determination at each developmental stages needs new plants. As a consequence, elemental fluxes between two dates need to be calculated for each organ for all possible plant combinations (i.e., $i \times n^2$ for i organs and n replicates).
7. An automated calculation tool named PEF for Plant Elemental Flux was recently developed in Visual Basis Applications (VBA) in order to improve integration of isotope data and facilitate the calculations of elemental fluxes in the framework of a pulse-chase labelling experiment with numerous plants, compartments, and/or several periods of development. For details please *see* https://www6.rennes.inra.fr/umreva/Equipes/INFlux/Amelioration-de-l-EUN

References

1. Good AG, Johnson SJ, De Pauw M et al (2007) Engineering nitrogen use efficiency with alanine aminotransferase. Can J Bot 85: 252–262
2. Masclaux-Daubresse C, Daniel-Vedele F, Dechorgnat J et al (2010) Nitrogen uptake, assimilation and remobilization in plants: challenges for sustainable and productive agriculture. Ann Bot 105:1141–1157
3. Kant S, Bi YM, Rothstein SJ (2011) Understanding plant response to nitrogen limitation for the improvement of crop nitrogen use efficiency. J Exp Bot 62:1499–1509
4. Salon C, Avice JC, Larmure A et al (2011) Plant N fluxes and modulation by nitrogen, heat and water stresses: a review based on comparison of legumes and non legume plants. In: INTECH (eds) Abiotic stress in plants: mechanisms and adaptations, http://www.intechweb.org/
5. Schiltz S, Munier-Jolain NG, Jeudy C et al (2005) Dynamic of exogenous nitrogen partitioning and nitrogen remobilisation from vegetative organs in pea revealed by ^{15}N in vivo labeling throughout seed filling. Plant Physiol 137:1463–1473
6. Desclos M, Etienne P, Coquet L et al (2009) A combined ^{15}N tracing/proteomics study in *Brassica napus* reveals the chronology of proteomics events associated with N remobilization during leaf senescence induced by nitrate limitation or starvation. Proteomics 9:3580–3608
7. Noquet C, Avice JC, Rossato L et al (2004) Effects of altered source-sink relationships on N allocation and vegetative storage protein accumulation in *Brassica napus* L. Plant Sci 166:1007–1018
8. Larmure A, Salon C, Munier-Jolain NG (2005) How does temperature affect C and N allocation to the seeds during the seed filling period in pea? Effect on seed nitrogen concentration. Funct Plant Biol 32:1009–1017
9. Schnug E, Haneklaus S, Murphy D (1993) Impact of sulphur fertilization on fertilizer nitrogen efficiency. Sulphur Agric 17:8–12
10. Fismes J, Vong PC, Guckert A, Frossard E (2000) Influence of sulfur on apparent N-use efficiency, yield and quality of oilseed rape (*Brassica napus* L.) grown on a calcareous soil. Eur J Agron 12:127–141
11. Dubousset L, Etienne P, Avice J-C (2010) Is the remobilisationof S- and N-reserves for seed filling of winter oilseed rape modulated by sulphate restrictions occurring at different growth stages? J Exp Bot 61:4313–4324
12. Kopriva S, Rennenberg H (2004) Control of sulphate assimilation and glutathione synthesis: interaction with N and C metabolism. J Exp Bot 55:1831–1842
13. Scherer HW, Pacyna S, Manthey N, Schulz M (2006) Sulphur supply to peas (*Pisum sativum* L.) influences symbiotic N_2 fixation. Plant Soil Environ 52:72–77
14. Varin S, Cliquet J-B, Personeni E, Avice J-C, Lemauviel-Lavenant S (2010) How does sulphur availability modify N acquisition of White Clover (*Trifolium repens* L.)? J Exp Bot 61: 225–234
15. Salon C, Raymond P, Pradet A (1988) Quantification of carbon fluxes through the

tricarboxylic acid cycle in early germinating lettuce embryos. J Biol Chem 263:12278–12287

16. Avice J-C, Ourry A, Lemaire G, Boucaud J (1996) Nitrogen and carbon flows estimated by ^{15}N and ^{13}C pulse chase labelling during regrowth of alfalfa. Plant Physiol 112(1): 281–290
17. Voisin AS, Salon C, Munier-Jolain NG et al (2002) Quantitative effect of soil nitrate, growth potential and phenology on symbiotic nitrogen fixation of pea (*Pisum sativum* L.). Plant and Soil 243:31–42
18. Moreau D, Bourion V, Fournier H et al (2006) Medicaco truncatula protocols: phenotyping. In: Malthesius U, Journet EP, Sumner LW (eds) The medicago truncatula hand book, http://www.noble.org/MedicagoHandbook/. ISBN: 0-9754303-1-9
19. Voisin AS, Salon C, Amarger N et al (2008) Theoretical evaluation of isotopic ^{15}N methods for measurement of symbiotic nitrogen fixation in the field. In: Couto GN (ed) Nitrogen fixation research progress. Nova Science Publishers Inc, New York, NY, pp 331–354
20. Voisin AS, Salon C, Jeudy C et al (2003) Symbiotic N_2 fixation in relation to C economy of *Pisum sativum* L. as a function of plant phenology. J Exp Bot 54:2733–2744
21. Malagoli P, Lainé P, Rossato L, Ourry A (2005) Dynamic of nitrogen uptake and mobilisation in field grown winter oilseed rape from stem extension to harvest. I. Global N flows between vegetative and reproductive tissues in relation with leaf fall and their residual N. Ann Bot 95:853–861
22. Munier-Jolain NG, Salon C (2003) Can sucrose content in phloem sap reaching field pea (*Pisum sativum* L.) seeds be a useful indicator of seed growth potential? J Exp Bot 54 (392):2457–2465
23. Lepinay C, Rigaud T, Salon C et al (2012) Interaction between *Medicago truncatula* and *Pseudomonas fluorescens* C7R12: evaluation of costs and benefits across an elevated atmospheric CO_2. PLoS ONE 7(9):e45740
24. Zancarini A, Mougel C, Voisin AS et al (2012) Soil Nitrogen Availability Modifies the Nutrition Strategies of the Interaction between Medicago truncatula and the Rhizosphere Bacterial Communities. PLoS ONE 7(10):e47096
25. Jeudy C, Ruffel S, Freixes S et al (2010) Plasticity of nodule development has a major role in the adaptation of Medicago truncatula to N-limitation. New Phytol 185:817–828
26. Ruffel S, Freixes S, Balzergue S et al (2008) Systemic signalling of the plant N status triggers specific transcriptome responses depending on the N source in *Medicago Truncatula*. Plant Physiol 146:2020–2035

Chapter 21

Ecophysiological Process-Based Model to Simulate Carbon Fluxes in Plants

Gilles Vercambre, Valentina Baldazzi, Nadia Bertin, Hélène Gautier, and Michel Génard

Abstract

Carbon fluxes in plants have been subject to many modeling studies. The conceptual framework of models of carbon acquisition, allocation, and metabolism in plants are first introduced, together with methods to calibrate and evaluate the validity of the resulting models. The possibility to combine different models within an integrated plant–organ system is illustrated. In the last part of the chapter, methods used to measure the carbon flows at the plant scale are discussed.

Key words Photosynthesis, Carbon allocation, Phloem transport, Unloading, Respiration, Sugar metabolism, Parameter estimation, Model selection, Sensitivity analysis

1 Introduction

Plants invest in competing activities for carbon such as growth, reproduction, or maintenance costs. Carbon acquisition, allocation, and metabolism within a plant depend on complex rules linking carbon source organs (mainly leaves) and carbon sink organs (mainly roots, apices, and fruits). The complexity comes essentially from regulatory mechanisms, interactions among specific functions and the spatial distribution of individual plant compartments within the plant architecture. In order to analyze this complexity, carbon models of plant growth have been developed during the last 30 years. The objective of this chapter is to address (1) the conceptual framework of models of carbon acquisition, allocation, and metabolism in plants, (2) the methods for modeling these processes, and (3) their experimental measurement.

Martine Dieuaide-Noubhani and Ana Paula Alonso (eds.), *Plant Metabolic Flux Analysis: Methods and Protocols*, Methods in Molecular Biology, vol. 1090, DOI 10.1007/978-1-62703-688-7_21,

2 Conceptual Basis of Ecophysiological Model

2.1 Carbon Acquisition

The three main steps of plant carbon acquisition modeling are (1) the leaf photosynthesis, allowing the prediction of carbohydrates from CO_2, (2) the leaf conductance, allowing the integration of the CO_2 transfer process from the air to the leaf cell, and (3) the up scaling from the leaf scale to the whole canopy.

1. Photosynthesis: Photosynthesis is recognized to be limited by the two processes of ribulose 1,5-bisphosphate (RuBP) carboxylation and RuBP regeneration. Initially designed at the cell scale with the pioneer work of Farquhar et al. [1], and widely used subsequently (*see* for example [2–4]), the photosynthesis Farquhar model was formalized by the minimum of the two limiting processes:

$$A = \min(A_c, A_j) - R_d$$

where A_c (mmol/m^2/s) is the rate of photosynthesis limited by Rubisco activity and A_j (mmol/m^2/s) is the rate of photosynthesis limited by RuBP regeneration by electron transport. R_d is the rate of respiration. The rate of photosynthesis limited by Rubisco activity could be expressed as:

$$A_c = Vc_{max} \times \frac{C_i - \Gamma^*}{C_i + K_c \cdot \left(1 + \frac{O_i}{K_o}\right)}$$

where Vc_{max} is the maximum activity of Rubisco; C_i and O_i are the intercellular concentrations of CO_2 and O_2, respectively, K_c and K_o are the Michaelis constants for CO_2 and O_2, respectively, and Γ^* is the CO_2 compensation point in the absence of mitochondrial respiration. When the rate of photosynthesis is limited by RuBP regeneration through electron transport, the rate of assimilation is given by:

$$A_j = \frac{J}{4} \times \frac{C_i - \Gamma^*}{C_i + 2\Gamma^*}$$

where J is the rate of electron transport for a given Photon Flux Density (PFD). A non-rectangular hyperbolic function is classically used to relate J to PFD:

$$J = \frac{\alpha\mathrm{PFD} + J_{max} - \sqrt{(\alpha\mathrm{PFD} + J_{max})^2 - 4\Theta J_{max}\alpha\mathrm{PFD}}}{2\Theta}$$

where J_{max} is the potential maximum rate of electron transport, Θ is the convexity of the response curve of J to light, and α is the quantum yield of electron transport under limiting light.

Vc_{max} and J_{max} are temperature dependent as well as K_c, K_o, and Γ^*. Variations of these parameters with temperature have been integrated through Arrhenius function [3, 5].

2. Stomatal conductance: In order to predict leaf photosynthesis rate from external (C_a) rather than internal (C_i) CO_2 concentration, the Farquhar model have been coupled with a conductance model, allowing integration of the CO_2 transfer process from the air to the cell. Internal CO_2 concentration is related to the total conductance to CO_2 transfer between the air and the inner part of the leaf, the CO_2 assimilation and to dilution by transpiration stream as:

$$C_i = \frac{\left(g_c - \frac{E}{2}\right) C_a - A \cdot P}{g_c + \frac{E}{2}}$$

with E, the leaf transpiration rate, P, the atmospheric pressure and g_c, the total conductance to CO_2. Stomatal conductance (g_s) was formerly considered as the lower conductance in g_c, and mesophyll and boundary conductance are often neglected. Various stomatal conductance model have been developed, one of the most used is the Ball–Berry model [6–9]. Although this model is satisfactory for describing the gas exchange of well-watered plants, it is inadequate for plants subjected to some degree of water stress. An additional reduction function is therefore added [10–12] in the case of limiting water conditions. Another way to account for changes in stomatal conductance is to use an empirical relationship, with a multiplicative algorithm that adjusts the reference value of stomatal conductance according to changes in environmental conditions [13]. The Jarvis model has been shown to perform well across many different species [14, 15], including drought-stressed ecosystems [16].

3. Up scaling from the leaf scale to the whole canopy: Up scaling is done following assumptions on plant representation and physics of light attenuation through the canopy. Big-leaf model, multilayer model, and sunlit-shaded model [17–19] have been developed to estimate the plant photosynthesis. Light attenuation is linked with canopy architecture and leaves properties. Therefore, canopy architecture parameters, such as ratio of horizontal versus vertical projected leaf area ratios, leaf area index, and randomness of leaf angle distribution, are classically integrated in model of canopy photosynthesis.

2.2 Allocation

Three main approaches have been used to model carbon allocation. These approaches are either teleonomic, driven by source–sink relationships, or based on transport concepts [20].

1. Teleonomic: Many models of plant carbon allocation are based on teleonomic approaches in which the plant is assumed pursue

an intrinsic goal, namely a specific relationship between foliage and conducting tissue as in the pipe-model [21] or a functional balance between shoot and root [22]. The pipe model theory states that the sapwood area (A) at a given height and the foliage C biomass above (W) are related through a constant ratio (k) [21]:

$$W = k \cdot A$$

The pipe model is one of the most commonly used models to distribute resources between foliage and woody structure in process-based models [23].

Another important teleonomic approach concerns the root: shoot functional balance theory, described by [22] as:

$$W_s / W_r = k(S_r / S_s)$$

where W_s and W_r are the shoot and root C biomasses, respectively, S_s and S_r are the shoot and root specific activities, respectively, and k is a constant. The specific activities are usually allowed to vary with soil N or water availability (for S_r) and with above-ground environmental resources such as light intensity and atmospheric CO_2 concentration (for S_s).

2. Source–sink: In models based on source–sink relationships, carbon allocation is assumed to depend on the respective ability of the different sinks to import available assimilates from the same sources [24]. According to Grossman and DeJong [25], this ability or "sink strength" is related to a genetically determined "potential net sink strength," defined as the maximum rate at which the organ can accumulate dry matter per unit time. The actual sink strength of an organ (also termed conditional net sink strength, CSS) can be decreased by suboptimal environmental conditions. The resulting rate of sink growth is calculated from the conditional net sink strength, considering the resources availability.

   Lescourret et al. [26] proposed an equation for the conditional net sink strength that emphasizes the role of organ "history" by means of the accumulated growth (M), both in terms of sink size and sink activity. It also emphasizes the role of time and temperature by means of accumulated degree-days (dd):

$$\mathrm{CSS} = \mathrm{RGR_{ini}} \times M \times \left(1 - \frac{M}{M_{\max}}\right) \times f(\mathrm{dd})$$

with $f(\mathrm{dd}) = 1$ if $\mathrm{dd} < \mathrm{dd_{min}}$

$$f(\mathrm{dd}) = \frac{\mathrm{dd_{max}} - \mathrm{dd}}{\mathrm{dd_{max}} - \mathrm{dd_{min}}} \text{ if } \mathrm{dd_{min}} < \mathrm{dd} < \mathrm{dd_{max}}$$

$$f(\mathrm{dd}) = 0 \quad \text{if } \mathrm{dd} > \mathrm{dd_{max}},$$

where $\mathrm{RGR_{ini}}$ is the initial relative growth rate, $M_{\max}$ refers to the maximum final mass potentially attainable by the sink, and

dd_{min} and dd_{max} are parameters that account for age-dependent decrease of growth rate.

Once the quantification of organ demands and carbon availability in source organs has been performed, the carbon has to be allocated from sources to sinks. When demand is less than supply, each sink gets its own demand and excess supply goes to reserves. In the case of supply shortage, two different approaches have been proposed. In the "proportional" approach, the carbon supply by sources is shared by the sink organs that all get the same proportion of their demand. Alternatively, in the "hierarchical-priority" approach, the sink with the highest priority is served first, and then the component with the next priority level is considered, and so on.

3. Transport: Models of the carbon transfer within the plant are based on the Münch's theory developed in the 1930s. Accordingly, active loading of photosynthate at the source lead to local accumulation of sugars in the phloem, whereas unloading at the sink decreases the sugars concentration. The loading/unloading processes are accompanied by a water flow into or from the phloem, leading to region of high and low turgor pressure (respectively, source and sink). The resulting turgor pressure gradient drives the circulation of phloem sap from source to sink.

   Carbohydrates translocated in the phloem sap can then be transported to the sinks by active transport, mass flow or diffusion. Fishman and Génard [27] proposed the following equation to represent the total sink uptake of carbohydrates (U):

$$U = Ua + (1 - \sigma) \times \left(\frac{Cp + Cf}{2} \right) \times Up + A \times ps \times (Cp - Cf)$$

   where Ua is the uptake rate due to active transport, Up is the phloem flow of liquid entering the organ, Cp and Cf are the sugar concentrations in the phloem and sink, respectively. σ is the reflexion coefficient of the cell membrane which is inversely proportional to the density of plasmodesmata. If $\sigma = 1$, the membranes are not permeable to solutes and there is no sugar uptake through mass flow. A is the membrane area through which solutes diffuse and ps is the permeability coefficient of the solute. As ps is usually small, the diffusion component can often be neglected. The uptake rate Ua obeys to a Michaelis-Menten equation:

$$Ua = vmCp/(K_m + Cp)$$

   where vm is the maximum uptake rate and K_m is the Michaelis-Menten constant, controlling the velocity of the transport inside the sink.

2.3 Metabolism

Respiration as well as the transformation of phloem sugars (sucrose, sorbitol, etc.) into sink soluble carbohydrates (sucrose, glucose, fructose, acids, etc.), starch, or cell walls are major metabolic processes driving organ composition and growth.

1. Respiration: The growth-maintenance paradigm [28] is used to calculate the respiration in terms of CO_2. The growth respiration is considered proportional to organ growth rate and the maintenance respiration to dry mass and temperature [29, 30]. The organ respiration is then calculated as:

$$\mathrm{Resp} = q_\mathrm{g}\frac{\mathrm{d}M_\mathrm{dry}}{\mathrm{d}t} + q_\mathrm{m}M_\mathrm{dry}Q_{10}^{\frac{(T-20)}{10}}$$

where M_dry is the organ dry mass, q_g is the growth respiration coefficient, q_m is the maintenance respiration coefficient at 20 °C, Q_{10} is a temperature coefficient (often close to 2), and T is temperature in °C.

2. Sugars metabolism : Génard et al. [31] designed a mechanistic SUGAR model to predict the changes in sugar composition during sink development. The unloaded sugars can be (1) directly stored in the tissues, (2) transformed in CO_2 through the respiratory process, (3) transformed into other sugars, or (4) used to synthesize other compounds (structural carbohydrates, etc.). The enzymatic reactions are described according to the "rate law" of chemical kinetics [32], which states that the reaction rate is proportional to the reactant concentration. Thus, the rates of change in the concentrations of a given sugar j (e.g., glucose, fructose, sucrose, etc.) depend on sugars already accumulated in the sink and on dilution due to the increase in organ volume (last term of the equation hereafter). The rate of change in sugars concentrations is thus described through a set of differential equations, each equation being of the following form:

$$\frac{\mathrm{d}[C_j]}{\mathrm{d}t} = \frac{U_j}{c_jM} + \frac{1}{c_j} \times \sum_{i\neq j} k_{ij}(x)c_i[C_i] - [C_j]\sum_{i\neq j} k_{ji}(x) - \frac{\mathrm{Resp}_j}{c_jM} - \frac{[C_j]}{M} \times \frac{\mathrm{d}M}{\mathrm{d}t}$$

where M is the organ fresh mass, $[C_j]$ the concentration of sugar j, and c_i the carbon content of sugar i. U_j and Resp_j represent the sugar uptake from the phloem and the carbon loss by respiration, respectively. Depending on the sugar, U_j and Resp_j can be zero. k_{ij} is a genotype-dependent function of the physiological state x of the sink describing the relative rate of sugar transformation of sugar i into sugar j (e.g., x can be the age of the sink, its relative growth rate, etc.).

2.4 Models Integration in a System

An important point concerns the integration of the submodels presented above in a plant or organ system. In order to unravel the complexity of the system, a hierarchical approach can be followed [33].

Large variation in carbohydrate availability within the plant can be computed, resulting from the carbon acquisition through photosynthesis and its following transport within the plant architecture. Sink strength or sugar concentration in the phloem then provides the boundary conditions for a fruit model, integrating sugar unloading mechanisms, respiration, and sugar metabolism. As shown in Fig. 1 in the case of a peach tree, fruits located at different positions can receive distinct resources, resulting in a marked variability in terms of both mass and dry content. Changes in fruit internal fluxes can be computed, too. Figure 1 (panel B) shows a comparison between two fruits located at the bottom

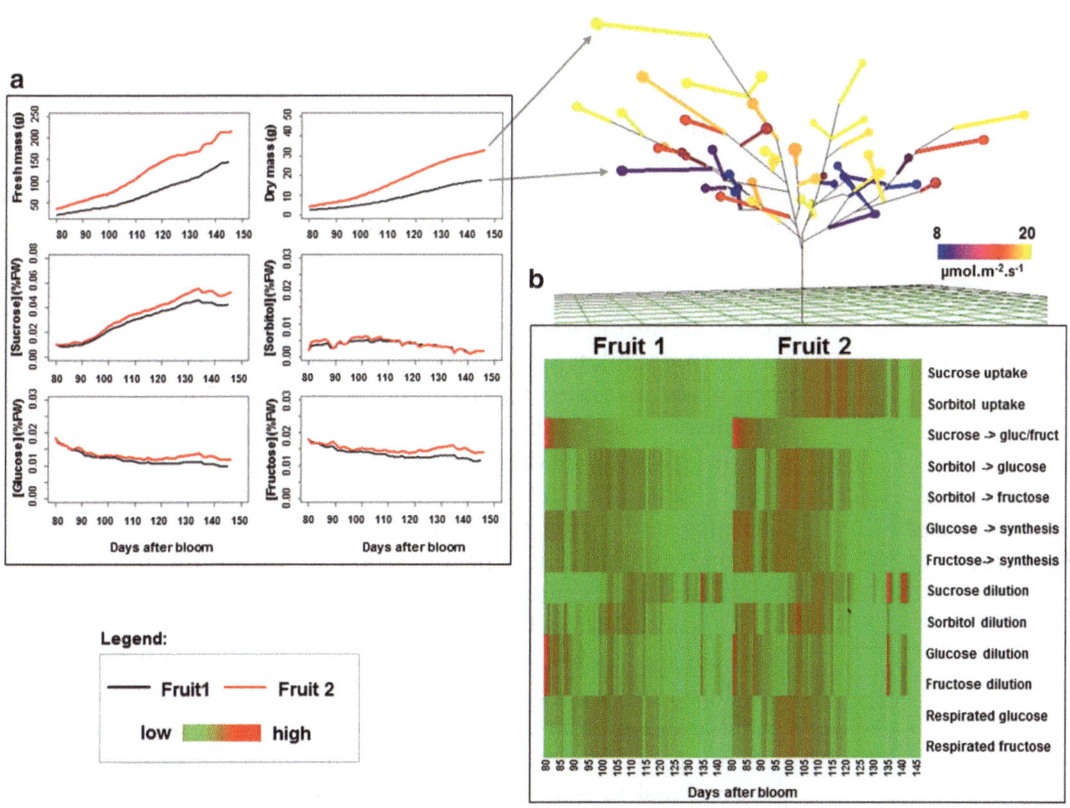

Fig. 1 Simulation results of an integrated plant-fruit model. Changes in photosynthesis rates within a plant architecture are computed by means of the peach tree model Qualitree [57, 58]. Differences in local resources affect organ growth, resulting in a large variability of fruits mass and composition (panel **a**). Changes in sugar content, in particular, result from an overall modification of fruit metabolism, computed according to Génard et al. [31]. A comparison of internal fluxes between two fruits, respectively, at the bottom and at the top of the tree, shows differences in both flux magnitude and dynamics depending on the availability of local resources (panel **b**)

(low photosynthesis rate) and top (high photosynthesis rate) part of the tree. Metabolic fluxes as well as respiration rates appear modified in both magnitude and temporal patterns. Sucrose uptake during the expansion phase, for instance, is greatly increased in fruits located at the top of the tree, whereas wall synthesis is fostered in the first part of the development, compared to fruits receiving a lower amount of resources. At maturity, a 30 % increases in sugar concentration is expected for fruits grown in high photosynthesis regions, an important aspect for fruit quality control.

3 Methods for Modeling

In the previous section, we introduced various models and their equations to describe the main carbon fluxes in plants. However, in order to use these equations, the value of the different parameters has to be fixed and initial conditions provided for all model variables.

While the latter can be usually inferred from experimental measurements or literature data, parameters values are more difficult to assess. Indeed, with the exception of some physical quantities (e.g. conductance), most of the above parameters are an integrated function of the genetic and metabolic properties of the system and a direct measurement of their value is difficult to achieve.

In these cases, computational methods for parameter estimation become an inevitable step to calibrate the unknown parameters values, according to overall behavior of the system. Once the model has been completely specified, approaches from sensitivity analysis then allow investigating the response of the model to changes in parameters values, helping to identifying key-point, in the perspective of experimental design or model reduction.

3.1 Parameter Estimation

The idea underlying the procedure of parameter estimation is to find that combination of parameters values that optimize the agreement between the model and the real data.

Classical approaches for parameter estimation (as opposed to Bayesian approaches, that we will not treat here [34]) are based on a likelihood principle according to which the estimated parameters values are those that maximize the probability (or likelihood, L) of observing real data. For practical convenience, a merit function, proportional to $-\ln(L)$, is usually defined and minimized instead of maximizing the likelihood function.

Different parameter estimation strategies exist, depending on the choice of the merit function and related hypothesis on data distribution. In the case of normal-distributed data, a common approach is the chi-square fitting (also known as weighted least-squares fitting), in which the merit function to be minimized is defined as

$$\chi^2 = \sum_{i=1}^{N} \frac{(y_i - y(x_i, a_1, a_2, , a_k))^2}{\sigma_i},$$

where y_i are the experimental data of variance σ_i, and $y(x_i, a_1, a_2, \ldots, a_k)$ is our mathematical model with parameters (a_1, a_2, …, a_k) [35]. In the special case of constant variance $\sigma_i = \sigma$, the chi-square fitting reduces the standard least-squares fit.

Minimization of the merit function is generally performed by algebra methods or local-search algorithms. In the case of complex nonlinear models, the merit function is often non-convex, with many local minima that make difficult to find a global optimum solution to the problem. In these cases, global optimization methods like evolutionary algorithm or simulated annealing, to name the most common ones, allow for a better exploration of the parameter space [36].

Once the fit have been performed, the quality of the adjustment (goodness-of-fit) between the selected model and data can be quantified. In the case of chi-square fitting, the χ^2 value can be used directly to estimate the resulting goodness-of-fit by means of chi-square statistics, with ($N - k$) degrees of freedom. In alternative, the root-mean-square error (RMSE) is a common measure of the (average) difference between data and model:

$$\text{RMSE} = \sqrt{E((y - y(x, a_1, a_2, , a_k)^2)} = \sqrt{\frac{1}{N}\sum_{i=1}^{N} (y_i - y(x_i, a_1, a_2, , a_k))^2}.$$

When comparing different datasets, it can be useful to use a normalized version of the RMSE with respect to the range of observed values x in order to remove differences due to sample dispersion:

$$\text{NRMSE} = \frac{\text{RMSE}}{x_{\max} - x_{\min}}$$

3.2 Model Selection and Parameters Uncertainty

In the above examples, the structure of the model $y(x, a_1, a_2, \ldots, a_k)$ was not called into question. Sometimes, however, the mathematical form of the function $y(x, a_1, a_2, \ldots, a_k)$ is not completely defined and it can be interesting to compare different model formulations in the perspective of model selection [37].

In this situation, the Akaike Information Criterion (AIC) or the Bayesian Information Criterion (BIC) allows to compare the goodness-of-fit of models having a different number of parameters:

$$\text{AIC} = 2k - 2\ln(L)$$

$$\text{BIC} = -2\ln(L) + k\ln(N)$$

where L is the likelihood function for the estimated model and N the number of experimental data points. Both criteria, in particular, introduce a penalty (stronger for the BIC criterion) for models having a large number k of parameters, the selected formulation being a compromise between accuracy of the fit and quality of parameter estimation.

In many biological problems, indeed, the estimated parameters values can have an importance on its own, as an indirect measure of specific biochemical properties of the system. In these cases, both the best estimated value and its statistical dispersion are of primary importance in order to assess the plausibility and biological significance of our parameters estimation procedure. The bootstrap method offers a simple and effective strategy to calculate confidence limits of estimated parameters by using a Monte Carlo resampling methods. In practice, the idea is to generate any number p of "synthetic" datasets by randomly draw N elements with replacement from the original dataset x and then use these synthetic data to get new estimated values for model parameters. The mean and variance of each parameter value can then be easily computed from the results of the p estimations, using the standard statistical formula.

3.3 Sensitivity Analysis

As illustrated in the previous section, parameter estimation is far from simple and estimated parameters values have to be taken with caution. For this reason, it is good practice to study the behavior of the model face to changes in the parameters values. The aim of sensitivity analysis (SA) is to provide a measure of the relative importance of the different parameters on the outcome of the model. Depending on the scientific aim of the study, results from SA can then be used to (1) prioritize parameters, identifying the ones which deserve additional measurements, (2) to fix non-influential parameters to their nominal value, in the perspective of model reduction, or (3) to reduce the output variance, by reducing the uncertainty of a few key parameters [38].

Several methods for sensitivity analysis exist and their choice depends on the size of the model, its degree of nonlinearity and the computational cost (CPU time) to run a simulation.

Classical local (also called one-at-time) approaches are based on the computation of partial derivatives and are easy to implement. Local approaches however are informative only for linear models and independent parameters. For most biological situation this is not the case and global approaches have to be used instead. Indeed, contrary to local approaches, global techniques test the sensitivity of the model against changes in the value of one parameter, but allowing the other parameters to vary as well. This allows to simultaneously account for individual effects and for correlations among parameters, or group of parameters, in a model-independent way, i.e., irrespective of its degree of nonlinearity.

For models having up to 20 parameters, variance-based methods are the best choice as they allow to decompose the output variance into a sum of individual contributions. The first-order sensitivity index for parameter i is simply defined as

$$S_i = \frac{V_i}{V(\Upsilon)}$$

where $V(\Upsilon)$ is the total variance of the output variable Υ, and V_i is the variance due to parameter i. Intuitively, the more S_i is close to one, the more the parameter i is influential, as it accounts for a large fraction of the output variability. In addition to first-order indices, higher-order and total sensitivity indices can be used to evaluate the degree of interaction among parameters, whenever the variance explained by the first-order terms is low [38, 39].

Different techniques have been developed for the computation of sensitivity indices, the most common being the Fourier Amplitude Sensitivity Test (FAST) and the Sobol' method [39]. The Sobol' approach, in particular, is easy to implement but both methods are available under free software packages, like SIMLAB.

When model complexity increases (large number of parameters or long CPU time), variance-based methods may prove too expensive and the method of Morris can become helpful.

The idea of the method is to provide a cheap strategy for preselection of important parameters by applying a one-at-time strategy at different points of the parameter space. The mean and the variance of the different local responses, indeed, provide information on the parameter's importance and can be used to select and rank those parameters to which more detailed approaches are worth to be applied [39].

4 Data Acquisition

We will focus on some methods used to measure the flows at the plant scale.

4.1 Photosynthesis Measurements

Today, a range of portable systems are available allowing to make real-time measurements of leaf photosynthetic CO_2 uptake (A), leaf conductance (g_s) and intercellular CO_2 mole fraction (C_i) and allow detailed inferences to be made about biochemical limitations [1]. Photosynthesis is classically estimated at various light intensity to evaluate the photosynthesis response curve to light at the ambient CO_2 concentration. Parameters of the Farquhar model (Subheading 2.1) can be estimated using nonlinear regression fitting, especially when CO_2 exchange measurements are combined with chlorophyll fluorescence measurements (used to evaluate changes in photosystem II photochemistry, linear electron flux, and CO_2 assimilation in vivo [40]). A series of A/C_i, responses are measured at different PFD and C_i. The majority of CO_2 exchange studies involve enclosure, i.e., the leaf, plant, or stand is sealed into a transparent exchanges chamber for measurement (e.g., [41, 42]). However, apart from being highly dependent on the instantaneous physical conditions of the environment (light, humidity), very large and careful sampling is needed to derive representative results at whole plant or canopy levels.

4.2 Phloem Transport Measurement

Phloem transport in plants is particularly difficult to measure, resulting from the slow flow velocities and the very small flowing volume. Several methods exist with specific interests and drawbacks. The subtractive method, developed by Lang and Thorpe [43], estimates the contribution of xylem and phloem by girdling a stem or the pedicel (fruit stalk) of a fruit. Girdling (physical removing of the phloem for trees or heat girdling) destroys the sieve tubes, stopping phloem influx, while the xylem is assumed to remain intact and functional. By comparing the growth of non-girdled fruits to that of girdled fruits, the phloem contribution can be estimated. Fishman et al. [44], however, showed that the cou pling between phloem and xylem influx could give rise to significant errors when using such subtractive technique.

Another technique is associated with the exudation of phloem sap from an incision. Exudation in most plants is generally stopped by the formation of a callus in the wounded part of the phloem. Treating the cut plant surfaces with EDTA, a chelating agent that binds divalent ions such as Ca^{2+}, inhibits callus formation [45]. This type of technique, both with free or enhanced exudation of phloem sap, is commonly used to investigate phloem sucrose flow [46]. In alternative, pure samples of phloem sap can be obtained using the aphid stylets method [47, 48]. Aphids suck plants by inserting their proboscis directly in a phloem sieve tube. Severing the stylet allows to collect the phloem sap in order to estimate phloem rate from individual sieve tube as well as phloem composition.

A number of radioisotope labeling techniques have been used to estimate flow velocities and mass flows in the phloem (*see* [49] for a review on the topic) by monitoring changes in the radioactivity levels along a path. This approach gives interesting result in term of carbohydrate translocation rate and allocation within the plant.

Over the last 15 years, NMR flow imaging has demonstrated to provide an excellent tool to measure xylem and phloem transport [50]. NMR flow imaging does not only give information about the average flow velocity, but gives access to the flow conducting area, the distribution of flow velocities, and the volume flow [51]. So far, studies have been conducted measuring flow in the stem of a variety of plants, ranging from castor bean seedlings [52] to fully developed tomato, castor bean, and tobacco (*Nicotiana tabacum*) plants, up to a small poplar tree [53]. The technique has been used to study the diurnal variation in long-distance transport [53, 54], the effects of cold girdling [55]. The technique has been used to measure long-distance transport to fruits as well [56].

References

1. Farquhar G, Caemmerer S, Berry J (1980) A biochemical model of photosynthetic CO2 assimilation in leaves of C3 species. Planta 90:78–90
2. Hikosaka K (1997) Modelling optimal temperature acclimation of the photosynthetic apparatus in C3 plants with respect to nitrogen use. Ann Bot 80:721–730
3. Medlyn BE, Dreyer E, Ellsworth D et al (2002) Temperature response of parameters of a biochemically based model of photosynthesis. II. A review of experimental data. Plant Cell Environ 25:1167–1179
4. Diaz-Pérez JC, Shackel KA, Sutter EG (1995) Relative water content and water potential of tissue-cultured apple shoots under water deficits. J Exp Bot 46:111–118
5. Hikosaka K, Ishikawa K, Borjigidai A et al (2006) Temperature acclimation of photosynthesis: mechanisms involved in the changes in temperature dependence of photosynthetic rate. J Exp Bot 57:291–302
6. Ball J, Woodrow I, Berry J (1987) A model predicting stomatal conductance and its contribution to the control of photosynthesis under different environmental conditions. In: Biggins J (ed) Progress in photosynthesis research. Martinus Nijhoff Publishers, Dordrecht, The Netherlands, pp 221–224
7. Collatz GJ, Ball JT, Grivet C et al (1991) Physiological and environmental regulation of stomatal conductance, photosynthesis and transpiration: a model that includes a laminar boundary layer. Agric Forest Meteorol 54:107–136
8. Sellers PJ, Berry JA, Collatz GJ et al (1992) Canopy reflectance, photosynthesis, and transpiration. III. A reanalysis using improved leaf models and a new canopy integration scheme. Rem Sens Environ 42:187–216
9. Leuning R, Kelliher FM, Pury DGG et al (1995) Leaf nitrogen, photosynthesis, conductance and transpiration: scaling from leaves to canopies. Plant Cell Environ 18:1183–1200
10. Wang Y-P, Leuning R (1998) A two-leaf model for canopy conductance, photosynthesis and partitioning of available energy I. Agric Forest Meteorol 91:89–111
11. Tuzet A, Perrier A, Leuning R (2003) A coupled model of stomatal conductance, photosynthesis and transpiration. Plant Cell Environ 26:1097–1116
12. Yin X, Struik PC (2009) C3 and C4 photosynthesis models: an overview from the perspective of crop modelling. NJAS - Wagn J Life Sci 57:27–38
13. Jarvis P (1976) The interpretation of the variations in leaf water potential and stomatal conductance found in canopies in the field. Philos Trans R Soc Lond B Biol Sci 273:593
14. Massman WJ, Kaufmann MR (1991) Stomatal response to certain environmental factors: a comparison of models for subalpine trees in the Rocky Mountains. Agric Forest Meteorol 54:155–167
15. Bosveld F, Bouten W (2001) Evaluation of transpiration models with observations over a Douglas-fir forest. Agric Forest Meteorol 108:247–264
16. Running SW, Coughlan JC (1988) A general model of forest ecosystem processes for regional applications I. Hydrologic balance, canopy gas exchange and primary production processes. Ecol Model 42:125–154
17. Pury DD, Farquhar G (1997) Simple scaling of photosynthesis from leaves to canopies without the errors of big-leaf models. Plant Cell Environ 20:537–557
18. Greer DH, Seleznyova AN, Green SR (2004) From controlled environments to field simulations: leaf area dynamics and photosynthesis of kiwifruit vines (Actinidia deliciosa). Func Plant Biol 31:169
19. Dai Y, Dickinson RE, Wang Y-P (2004) A two-big-leaf model for canopy temperature, photosynthesis, and stomatal conductance. J Climate 17:2281–2299
20. Génard M, Dauzat J, Franck N et al (2007) Carbon allocation in fruit trees: from theory to modelling. Trees 22:269–282
21. Shinozaki K, Yoda K, Hozumi K et al (1964) A quantitative analysis of plant form. The pipe model theory. II. further evidence of the theory and its application in forest ecology. Jpn J Ecol 14:133–139
22. Davidson RL (1969) Effect of root/leaf temperature differentials on root/shoot ratios in some pasture grasses and clover. Ann Bot 25:59–104

23. Sievanen R, Nikinmaa E, Nygren P et al (2000) Components of functional-structural tree models. Annu Forest Sci 57:399–412
24. Lacointe A (2000) Carbon allocation among tree organs: a review of basic processes and representation in functional-structural tree models. Annu Forest Sci 57:521–533
25. Grossman YL, Dejong TM (1994) Peach - a simulation-model of reproductive and vegetative growth in peach-trees. Tree Physiol 14:329–345
26. Lescourret F, Ben-Mimoun M, Génard M (1998) A simulation model of growth at the shoot-bearing fruit level. I. Description and parameterization for peach. Eur J Agron 9:173–188
27. Fishman S, Génard M (1998) A biophysical model of fruit growth: simulation of seasonal and diurnal dynamics of mass. Plant Cell Environ 21:739–752
28. Cannell MGR, Thornley JHM (2000) Modelling the components of plant respiration: representation and realism. Ann Bot 85:54–55
29. Thornley JHM, Johnson IR (1990) Plant and crop modelling. Oxford University Press, Oxford
30. Penning de Vries FWT, van Laar HH (1982) Simulation of plant growth and crop production. Pudoc, Wageningen, The Netherlands
31. Génard M, Lescourret F, Gomez L et al (2003) Changes in fruit sugar concentrations in response to assimilate supply, metabolism and dilution: a modeling approach applied to peach fruit (Prunus persica). Tree Physiol 23:373–385
32. Chang R (2000) Physical chemistry for the chemical and biological sciences. University Science, Sausalito, CA
33. Génard M, Bertin N, Borel C et al (2007) Towards a virtual fruit focusing on quality: modelling features and potential uses. J Exp Bot 58:917–928
34. Ellison AM (2004) Bayesian inference in ecology. Ecol Lett 7:509–520
35. Vetterling WT, Teukolsky SA, Flannery BP et al (1992) Numerical recipes in C. The art of scientific computing. Cambridge University Press, Cambridge
36. Moles CG, Mendes P, Banga JR (2003) Parameter estimation in biochemical pathways: a comparison of global optimization methods. Gen Res 13:2467–2474
37. Johnson JB, Omland KS (2004) Model selection in ecology and evolution. Trend Ecol Evol 19:101–108
38. Cariboni J, Gatelli D, Liska R et al (2007) The role of sensitivity analysis in ecological modelling. Ecol Model 203:167–182
39. Saltelli A, Ratto M, Tarantola S et al (2005) Sensitivity analysis for chemical models. Chem Rev 105:2811–2828
40. Baker R (2008) Chlorophyll fluorescence: a probe of photosynthesis in vivo. Annu Rev Plant Physiol Plant Mol Biol 59:89–113
41. Daudet F-A, Claustres J (1995) Continuous monitoring of gas exchange in open-top chambers in pollution experiments: methodological aspects. New Phytol 130:81–88
42. Poni S, Magnanini E, Rebucci B (1997) An automated chamber system for measurements of whole-vine gas exchange. Hort Sci 32:64–67
43. Lang A, Thorpe MR (1989) Xylem, phloem and transpiration flows in a grape: application of a technique for measuring the volume of attached fruits to high resolution using Archimedes' principle. J Exp Bot 40:1069–1078
44. Fishman S, Génard M, Huguet JG (2001) Theoretical analysis of systematic errors introduced by a pedicel-girdling technique used to estimate separately the xylem and phloem flows. J Theor Biol 213:435–446
45. King RW, Zeevaart JAD (1974) Enhancement of phloem exudation from cut petioles by chelating agents. Plant Physiol 53:96–103
46. Araki T, Kitano M, Eguchi H (1997) Evaluation of photoassimilate flux through a tomato pedicel. Biotronics 26:21–29
47. Girousse C, Faucher M, Kleinpeter C et al (2003) Dissection of the effects of the aphid Acyrthosiphon pisum feeding on assimilate partitioning in Medicago sativa. New Phytol 157:83–92
48. Lohaus G, Moellers C (2000) Phloem transport of amino acids in two Brassica napus L. genotypes and one B. carinata genotype in relation to their seed protein content. Planta 211:833–840
49. Minchin PEH, Troughton JH (1980) Quantitative interpretation of phloem translocation data. Annu Rev Plant Physiol Plant Mol Biol 31:191–215
50. Van As H (2007) Intact plant MRI for the study of cell water relations, membrane permeability, cell-to-cell and long distance water transport. J Exp Bot 58:743–756
51. Scheenen T (2000) Quantification of water transport in plants with NMR imaging. J Exp Bot 51:1751–1759
52. Köckenberger W, Pope JM, Xia Y et al (1997) A non-invasive measurement of phloem and xylem water flow in castor bean seedlings by nuclear magnetic resonance microimaging. Planta 201:53–63

53. Windt C, Vergeldt F (2006) MRI of long-distance water transport: a comparison of the phloem and xylem flow characteristics and dynamics in poplar, castor bean, tomato and tobacco. Plant Cell Environ 29:1715–1729
54. Peuke AD, Rokitta M, Zimmermann U et al (2001) Simultaneous measurement of water flow velocity and solute transport in xylem and phloem of adult plants of Ricinus communis over a daily time course by nuclear magnetic resonance spectrometry. Plant Cell Environ 24:491–503
55. Peuke AD, Windt C, Van As H (2006) Effects of cold-girdling on flows in the transport phloem in Ricinus communis: is mass flow inhibited? Plant Cell Environ 29:15–25
56. Windt CW, Gerkema E, Van As H (2009) Most water in the tomato truss is imported through the xylem, not the phloem: a nuclear magnetic resonance flow imaging study. Plant Physiol 151:830–842
57. Mirás-Avalos JM, Egea G, Nicolás E et al (2011) QualiTree, a virtual fruit tree to study the management of fruit quality. II. Parameterisation for peach, analysis of growth-related processes and agronomic scenarios. Trees 25:785–799
58. Lescourret F, Moitrier N, Valsesia P et al (2010) QualiTree, a virtual fruit tree to study the management of fruit quality. I. Model development. Trees 25:519–530

Index

Martine Dieuaide-Noubhani and Ana Paula Alonso (eds.), *Plant Metabolic Flux Analysis: Methods and Protocols*, Methods in Molecular Biology, vol. 1090, DOI 10.1007/978-1-62703-688-7, © Springer Science+Business Media New York 2014

F

G

H

I

K

L

M

N

O

P

Q

R

S

T

V

MIX
Papier aus verantwortungsvollen Quellen
Paper from responsible sources
FSC® C105338

If you have any concerns about our products,
you can contact us on
ProductSafety@springernature.com

In case Publisher is established outside the EU,
the EU authorized representative is:
Springer Nature Customer Service Center GmbH
Europaplatz 3, 69115 Heidelberg, Germany

Printed by Libri Plureos GmbH
in Hamburg, Germany